OBJECTIVE PLANT PHYSIOLOGY

MCQ in Plant Physiology

OBJECTIVE PLANT PHYSIOLOGY

MCQ in Plant Physiology

for

ARS, CSIR, TIFT/NCBS/IISc, GATE, IIT-JAM, JRF SRF, BIOLOGY OLYMPIADS AND OTHER COMPETITIVE EXAMINATIONS

4th Revised Edition

Prof. Padmanabh Dwivedi
Department of Plant Physiology
Institute of Agricultural Sciences
Varanasi - 221 005

Dr. Prasann Kumar
Assistant Professor
School of Agricultural
Lovely Professional University
Punjab - 144 411

Published by:

Jodhpur
5 A, New Pali Road, P.O. Box 91
Jodhpur 342 001 (India)

New Delhi
4806/24, Ansari Road, Daryaganj,
New Delhi - 110 002 (India)

E-mail: info@scientificpub.com
Website: www.scientificpubonline.com

First Edition : 2013
Second Edition : 2018
Third Edition : 2019
Fourth Edition : 2023

ISBN : **978-93-88172-00-4** [Paperbound]
eISBN : **978-93-88449-03-8**

Printed in India

Dedicated in loving memory of the founder of the Department of Plant Physiology, Institute of Agricultural Sciences, Banaras Hindu University, Varanasi

Late Professor Bhola Nath Singh, D.Sc.
(September 20, 1901 - June 6, 1984)

About the Authors

Prof. Padmanabh Dwivedi, Professor & Head in the Department of Plant Physiology, Institute of Agricultural Sciences, Banaras Hindu University, Varanasi, specializes in stress physiology, plant tissue culture, physiology of plant microbe/pathogen interaction, nitric oxide signaling and biodiversity management. A doctorate from University of Cambridge (U.K.), he has teaching and research experience of 24 and 29 years, respectively. He has to his credit around 300 publications including research papers in national and international journals (like Ind Crops Products, Fron Pl Sci, BMC Pl Biol, JPGR, SAJB, ACPP, Fron Microb, 3 Biotech, JSSPN, Phytotherapty Res, J Ethnopharmacol, AMB, etc), book chapters, abstracts, etc. He also has authored 11 books published by national and international publishers, which find place in national as well as various Overseas Institutions' libraries. He has guided 15 Ph.D. and 43 M.Sc. students for their thesis. Dr. Dwivedi has carried out research work in 04 Overseas Research laboratories (England, Wales, Czech Republic, Germany). He has handled/completed several research projects funded by CST(UP), UGC, State Horticulture Mission, DBT, NMPB, DST. He is a recipient of several scholarships/fellowship/awards such as Gold Medal (BHU), University Medal (BHU), Malviya Medal (BHU), Young Scientist award (CST, Govt. of U.P.), Cambridge Nehru Scholarship award, Overseas Research Studentship award of U.K., Downing Muller Scholarship award of University of Cambridge (U.K.), UNESCO-Czech Academy of Sciences Fellowship, INSA-DFG Fellowship, R D Asana Medal (Indian Society for Plant Physiology), Prof. Hira Lal Chakravarty Medal (Indian Science Congress Association), Prof. Y S Murty Medal (Indian Botanical Society), Commonwealth Academic Staff Fellowship (U.K.), International Achievement award, Best Teacher award (BHU-ICAR), Award of Excellence in Research, DST International Travel Grant award to deliver lecture in Spain, ISCA Platinum Jubilee lecture award, etc. He is a Life Fellow of Cambridge Commonwealth Society (UK), Cambridge Philosophical Society (UK), Indian Society for Plant Physiology, Indian Botanical Society, Purvanchal Environmental Association. He served as Vice President of Indian Society for Plant Physiology for the year 2013, and Sectional member, Plant Science Section of ISCA. He has organized several International as well as National Workshop/Conference/ Seminars/ICAR-Winter & Summer School as Convener, Organizing Secretary and Course Co-ordinator/Director, and is a reviewer of more than 30 International and National research journals having high SCI impact factor. He has delivered more than 50 lectures as Invited speaker in various National and International conferences, both in India and abroad. He has presented research papers in more than 50 national and international conferences/seminars including ten abroad. He has been chairing/Co-chairing technical sessions in conferences. Currently he is Technical Editor of IJAEB and acting as Review Editor, Fron Pl Sci, besides guest editing SI of reputed journals like JPGR and Agriculture.

He has wide educational administrative experience including being a member of Academic Council of BHU and Chairman of Departmental Policy & Planning Committee. Currently he is heading the Department of Plant Physiology at BHU for the second term.

Dr. Prasann Kumar is the Assistant Professor in the Department of Agronomy, School of Agricultural, Lovely Professional University, Punjab, doing his research work on heavy metals toxicity. He has completed International Agricultural Rural Development Programe from Cornell University, USA. Dr. Kumar is the recipient of ITS (United Kingdom), TWAS-BIOVISION.NXT.FELLOW 2014 (FRANCE), TWAS-BIBLIOTHECA ALEXANDRIA (EGYPT, 2014), YOUNG SCIENTIST AWARD 2014 (BRC, ALLAHABAD), YOUNG SCIENTIST (EDU.EXPO), OIRA (NEPAL), BFA (EDU.EXPO.), SAMMAN PATRA(BHU), JOS(SAFNA, LPU) . He has qualified GATE, ICAR-NET, ICAR-SRF, UGC-JRF, UGC-SRF, INSA Fellowship (New Delhi) and NTS (MSJE, New Delhi) besides being a UGC-Research Fellowship of BHU. He already has more than hundred publications including research paper, review articles, book chapters, popular articles, abstracts and books of International and National Repute. Dr. Kumar has published 18 books of National and International repute on the aspect of heavy metal stress and others.

Dr. Kumar has visited the foreign countries France, United Kingdom and Nepal. He acts as a Potential Reviewer in Elsevier Editorial Systemfor the Journal Ecotoxicology and Environmental Safety(Impact factor 2.69), Medicinal Plant-International Journal of Phytomedicines and Related Industries [NAAS Rating: 4.25], Journal of Crop Research, Journal of Dynamics in Agricultural Research, SKY journal of soil science and environment, Lournal of Agriculture Science, G-Journal of Environmental Sciences and Technology(GLEST) [Published by Grace and Peace Welfare Society], and acting as the Managing Editor in the Journal of the Kalash Sciences (International Journal) (The Official Publication of the Kalash Research and Welfare Society) (ISSN 2321-7634). Dr. Kumar is the Life member/Annual member of several scientific societies like Indian Society for Plant Physiology (Life member), Indian Society of Cell Biology (Life member), Kalash Welfare Society (Life member), Indian Botanical Society (Annual member), Indian Science Congress Association (Annual member), Hind Agri-Horticultural Society (Annual member) and Member of Times Higher Education World University Ranking selected by Elsevier, England, 2015-16. He has completed six projects/ training programes of two to four months duration that were funded by INSA, IAS, NASI, ILRI, BHU and Fruit and Vegetable Preservation and Processing at BHU. He has completed successfully Edusat based Distance Learning Programe organized by IIRS, Dehradun and ISRO, Ahmadabad. Next to his course credit load, he has also completed level II Diploma in English and Yoga from BHU. Dr. Kumar has been awarded 24th and 25th Youth Fest National Award by Association of Indian Universities, New Delhi. He has attended and presented his research papers more than twenty one in conference/seminar/ symposium at National and International repute. Dr. Kumar has been awarded up to twenty six prizes for standing first at National level, University level and Institute level in extra-curriculum activities.

PREFACE (4th Edition)

The revised edition of this book contains a new chapter, Chapter 31 (Cell Biology), that covers questions from various aspects of cell physiology in a comprehensive manner. These kinds of questions/ patterns of questions can be found in many competitive examinations, such as the ones conducted by ARS, CSIR, GATE, IIT-JAM, ICAR-JRF, ICAR-SRF, etc. In this edition of the book, there are more than 2500 questions covering a wide range of topics related to the subject. The present edition should satisfy students in their quest to solve moderately difficult to very difficult questions. We would greatly appreciate to receive any criticism and suggestion.

It is our pleasure to extend thanks to M/s Scientific Publishers (India), Jodhpur for printing this revised edition of the book.

Padmanabh Dwivedi
Prasann Kumar

Banaras Hindu University, Varanasi

13th February, 2023

PREFACE (3rd Edition)

We are delighted to present this revised edition of the book where we have added a new chapter, Chapter 30 (Seed Physiology of Crops) which covers questions from various aspects of plant physiology. Such sort/pattern of questions is regular feature in competitive examinations including that of ARS, CSIR, GATE, IIT-JAM, ICAR-JRF, JRF, SRF, etc. There are more than 2400 questions in this edition of book covering different aspects of the subject. We hope the present edition would satisfy the students. We welcome criticism and comments from readers.

Our thanks are due to M/s Scientific Publishers (India), Jodhpur for printing this revised edition of the book.

Padmanabh Dwivedi
Prasann Kumar

Banaras Hindu University, Varanasi
5th September, 2019

CONTENTS

1 THE BIOLOGICAL CLOCK: RHYTHMS OF LIFE

1. The rhythmicities or periodicities whose periods match to those of lunation are called

[A] Circaannual rhythms [B] Circatidal rhythms
[C] Circalunar rhythms [D] None of these

2. The rhythmicities or periodicities whose periods match to those of tide are called

[A] Circaannual rhythms [B] Circatidal rhythms
[C] Circalunar rhythms [D] None of these

3. The rhythmicities or periodicities whose periods match to those of season are called

[A] Circaannual rhythms [B] Circatidal rhythms
[C] Circalunar rhythms [D] None of these

4. The rhythmicities or periodicities whose periods match to those of time between successive spring low water are called

[A] Circaannual rhythms [B] Circatidal rhythms
[C] Circalunar rhythms [D] None of these

5. Which of the following is the example of earliest observation of circadian rhythms in plant group?

[A] Algae [B] Fungi
[C] Pteridophyta [D] Angiosperms

6. Which of the following is the example of earliest observation of circadian rhythms in plant group?

[A] *Gonyaulax polyedra* [B] *Pilobolus sphaerosporous*
[C] *Selaginella serpens* [D] *Phaseolus multiflorus*

7. Circadian rhythms are

[A] Exogenous in nature [B] Endogenous in nature
[C] Both [D] None of these

8. Classical experiments in understanding the nature of circadian rhythms in plants were carried out first by

[A] Bunning [B] Tazawa
[C] De Mairan [D] None

9. **The findings of De Mairan marked the beginning of one of the most fascinating and mysterious areas of study (i.e., periodicities of rhythmicities in organisms) for plant physiologists and other biologists which in recent years has been termed as,**

[A] Biological chronometry [B] Biochronometry

[C] Rhythmicities [D] More than One

10. **Which of the following term is commonly used in Biochronometry?**

[A] Oscillation, period, phase and phase shift

[B] Amplitude, entrainment and free running period

[C] Zeitgerber, transient and damping

[D] All of above

11. **Five rules to establish endogenous nature of rhythm were first proposed by**

[A] De Mairan [B] Pittendrigh

[C] Bunning and Tazawa [D] None of these

12. **The classical book on "The Power of Movement in Plants" [1880] was authored by**

[A] De Mairan [B] Charles Darwin

[C] F.W. Went [D] None of these

13. **Which chemical caused a considerable increase in the phototaxis period of *Euglena* and also in the bean-leaf movement, and it showed the measurement of time in flowering is controlled by photoperiod.**

[A] Urethane [B] Colchicines

[C] Deuterium oxide [D] All of these

14. **Which ion lengthens the period of rhythm and affect the rhythm of several plant and animals (reviewed by Engelmann & Schrempf, 1980), suggesting the involvement of membranes.**

[A] Na [B] Mg

[C] Li [D] All of these

15. **Valinomycin, a K^+ inophore, caused phase shifts in .... movement, but some other Ca^{2+} and K^+ inophores were ineffective in ....**

[A] Gonyaulax, bean leaf [B] Bean leaf, Gonyaulax

[C] Acetabularia, Chlorella [D] None of these

16. **Eclosion rhythm of *Drosophila* pupae was discovered by**

[A] Bunning [B] Pittendrigh

[C] H.Kalmus [D] None

17. **............. of ultradian rhythms are sensitive to temperature?**

[A] Most [B] All

[C] Few [D] None

18. **The term phase is used as any point on a cycle that is recognizable by its relationship to the rest of the cycle. If the cycle is described by a sinusoidal function (approximately fits a sine of cosine curve), then the maximum phase on the cycle is called..........?**

[A] Acrophase [B] Metaphase
[C] Geophase [D] None of these

19. When plants or animals are exposed to an environment that fluctuates according to some periods and the rhythms exhibit the same period, they are said to be .... to the environment, as opposed to free - running.

[A] Entrained [B] Free energy
[C] Entropy [D] None of these

20. The processes such as Glycolysis (1.8 min in *Saccharomyces carlsbergensis*), Protoplasmic streaming (2-2.5 min in *Physarum polycephalum*), Leaf move-ment (3 min in *Desmodium gyrans*), Sap flux (20 min in *Gossypium areysianum*), Auxin transport (25 min in *Zea mays*), Transpiration (30 min in *Avena sativa*), Leaf movement (36 min in *Gossypium hirsutum*), Enzyme activity (1-5h in *Pisum sativum*), Leaf cell viscosity (2-3h in *Helodea densa*), Enzyme activity (12-15h in *Chenopodium rubrum*), are the examples of which type of rhythms in nature?

[A] Circadian [B] Infradian
[C] Tidal [D] Ultradian

21. The infraannual rhythms are observed in which plant?

[A] Various Bamboos [B] Various zooplanktons
[C] Various Diatoms [D] None of these

22. *Many rhythmic phenomena besides leaf movement have been observed in higher plants (Koukkari and Warde, 1985). These include petal movement, growth rates of various organs, concentration of pigments and hormones, stomatal opening and closing, discharge of fragrance from flowers, times of cell division, metabolic activities (for example, photosynthesis and respiration) and even the volume of the nucleus. Many plant species exhibit a rhythm in sensitivity to such environmental factors as light and temperature. Some species flower or grow well only when temperature during the part of cycle that normally comes at night called .... are lower than temperature during day called .... ?*

[A] Subjective night and subjective day
[B] Subjective day and subjective night
[C] Both
[D] None of these

23. The period of an endogenous oscillator or rhythm under uniform environmental conditions i.e. in the absence of a zeitgeber e.g. at constant temperature, and in continuous darkness or continuous light, is called

[A] Free running period [B] Entrainment
[C] Phase shift [D] None of these

24. Circadian rhythms have been found in some members of almost all major groups of plants except?

[A] Bryophyta [B] Gymnosperms
[C] Both [D] None

25. Zeitgeber is the .... which entrains a biological oscillation e.g. the environmental cycle of light and temperature, tide, moonlight and season. In other words, it is a rhythmically fluctuating environmental parameter.

[A] Synchronizer [B] Forcing oscillation
[C] Both [D] None

26. Circadian plant rhythm. The below picture showed the position of primary leaves of .... during .... and ....?

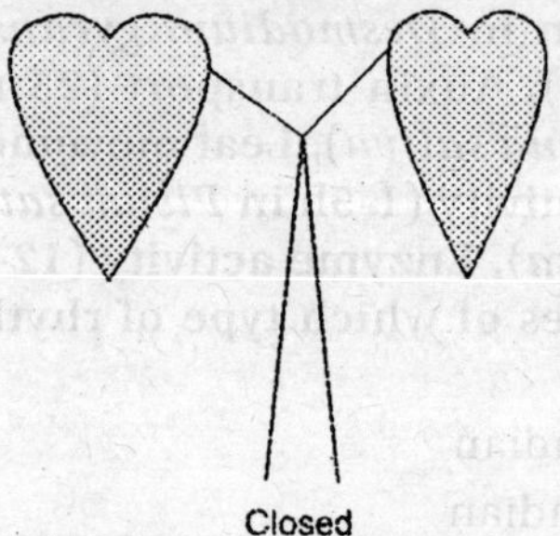

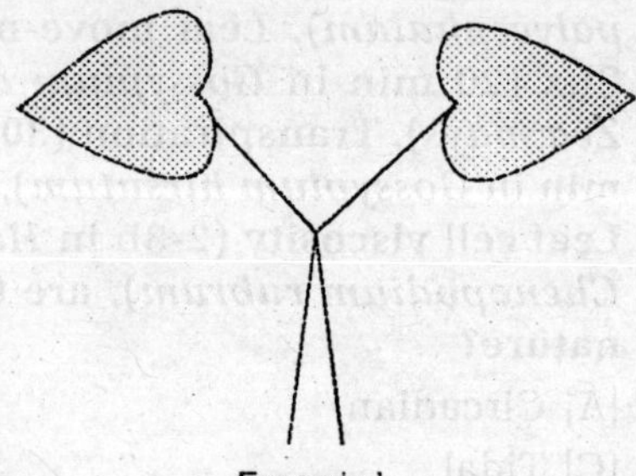

[A] *Phaseolus multiflorus*, Day, Night
[B] *Phaseolus multiflorus*, Night, Day
[C] *Gossypium* spp., Day, Night
[D] *Gossypium* spp., Night, Day

27. In primary leaves of *Phaseolus multiflorus*, the circadian rhythm was first demonstrated by

[A] Bunning and Tazawa [B] De Mairan
[C] Darwin [D] None

28. Which of the following is an example of circadian rhythm in plants?

[A] Sleep movement of leaves in *Phaseolus multiflorus*.
[B] Bioluminescence in *Gonyaulax polyedra*
[C] Spore discharge in *Pilobolus sphaerosporus*
[D] All of above

29. In organisms showing circadian rhythms, the time measurement system in them is called as

[A] Oscillator [B] Pendulum
[C] Biological Clock [D] None of the above

30. Which of the following statement is correct?

[A] Circadian rhythms are not dependent on environmental parameters but may be synchronized with them
[B] Circadian rhythms are dependent on the environmental parameters
[C] Circadian rhythms are neither dependent on environmental parameters, nor they are synchronized with them
[D] None of the above

2 PHOTOMORPHOGENESIS

1. **Fill in the blanks (A, B, C and D) in the following statement with a proper combination of m, n, o and p.**
Wherein, m represents-longer, n represents-shorter, o represents-prevents, p represents-induces
Short day (SD) plants flower when night lengths are .... than a critical dark period. Interruption of the dark period by a brief light treatment .... flowering in SD plants. Long day (LD) plants flower when night length is ...... than a critical period. Shortening of the night with a brief light treatment .... flowering in LD plants.

	1	2	3	4
[A]	m	o	n	p
[B]	n	p	m	o
[C]	n	o	m	p
[D]	m	p	n	o

2. **Atmospheric CO_2 contains the naturally occurring stable carbon isotopes ^{12}C and ^{13}C in the proportion of 98.9% and 1.1%, respectively. Which of the followings are the statements regarding CO_2 assimilation:**
[A] Both C_3 and C_4 plants assimilate $^{13}CO_2$ than $^{12}CO_2$.
[B] Both C_3 and C_4 plants assimilate less $^{12}CO_2$ than $^{13}CO_2$.
[C] C_3 plants assimilate lesser $^{13}CO_2$ than $^{12}CO_2$ ascompared to C_4 plants
[D] C_4 plants assimilate lesser $^{13}CO_2$ than $^{12}CO_2$ as compared to C_3 plants

3. **Which one of the following wavelength is most effective for the photomorphogenetic response?**
[A] 730 nm [B] 680 nm
[C] 660 nm [D] 560 nm

4. **When a short day plant (SD) and a long day plant (LD), both kept under favourable conditions for flowering, are exposed to a short flash of red light in the middle of the dark period:**
[A] Both SD and LD plants will flower
[B] Both SD and LD plants will remain vegetative
[C] SD plant will remain vegetative and LD plant will flower
[D] SD plant will flower and LD plant will remain vegetative

5. **Phototropic response of coleoptiles is mediated through**
   [A] Anthocyanin [B] Chlorophyll a
   [C] Cryptochrome [D] Phytochrome

6. **Charles Darwin and his son Francis Darwin experimented with phototropism of grass seedlings by placing a metal for blindfold over different parts of the seedlings coleoptiles. Which of the following statements best explains their results?**
   [A] Both the seedlings' root and coleoptiles perceive and respond to light in the same manner
   [B] The light signal is perceived below the tip, and these cells cause the coleoptiles to grow toward the light.
   [C] A chemical messenger must travel from the base of the coleoptiles to the tip
   [D] The light signal is perceived at the tip of the coleoptiles, but the growth response occurs a few millimeters below the tip.

7. **Which of the following is incorrect about Phytochrome?**
   [A] Phytochrome pigment present in two photo reversible forms Pfr and Pr
   [B] Phytochrome in all plants is synthesized entirely as Pr
   [C] Chromophore present in Phytochrome is an open chain tetra pyrole ring
   [D] Phytochrome is involved in light energy transfer

8. **Type I Phytochrome**
   [A] Is destroyed in the light
   [B] Is synthesized in the far-red absorbing form (Pfr)
   [C] Reverts to type II Phytochrome in the dark
   [D] Senses red light, while type II senses far-red light

9. **Which of the following is true for photomorphogenesis?**
   [A] It is non-directional developmental process
   [B] It involves Phytochrome pigment
   [C] It does not depend on direction of light stimulus
   [D] All of the above

10. **The role of light in regulating morphogenesis is known as**
    [A] De-etiolation [B] Morphogenesis
    [C] Photomorphogenesis [D] None of these

11. **The development of seedling in darkness is called**
    [A] Photomorphogenesis [B] Skotomorphogenesis
    [C] Pattern specification [D] None of these

12. **According to ...., there are two important stages of photomorphoge-nesis (i) ...., in which cells and tissue develop specific ability or competence to respond to light during certain developmental stage, and (ii) ...., during which time the photoresponse occurs.**

[A] Hans Mohr, pattern specification, pattern realization
[B] McAlister, pattern realization, pattern specification
[C] Flint, pattern specification, pattern realization
[D] None of these

13. The ratio of the P_{FR} concentration and the total phytochrome concentration (P_{total}) at a given wavelength is known as

[A] Photostationary equilibrium [B] Photochemical equilibrium
[C] Photoreversible equilibrium [D] None of these

14. Photostationary equilibrium can be measured by

[A] Spectrophotometer
[B] Difference spectroscopy
[C] Atomic absorption spectrophotometer
[D] None of these

15. Number of photons absorbed per unit surface area is called

[A] Far-red treatment [B] White light treatment
[C] Fluence [D] None of these

16. Phytochrome in two different forms was isolated in plant extract for the first time by

[A] Borthwick *et al.* [B] Butler *et al.*
[C] Both of them [D] None of them

17. Based on the amount of light required or the fluence, the phytochrome mediated photoresponse can be grouped into how many categories?

[A] One [B] Two
[C] Three [D] Four

18. Phytochrome was initially isolated from .... of etiolated turnip seedlings.

[A] Leaves [B] Cotyledons
[C] Shoots [D] None of these

19. .... scientist (s) responsible for a highly purified extract that led to further purifications and analysis of the phytochrome structure.

[A] Borthwick [B] Hendricks
[C] Siegelman and Firer [D] None of these

20. Phytochrome is a protein with a .... prosthetic group that resembles, in basic structure, the open-chain tetrapyrrole chromophore of the algal pigment named ....

[A] Chromophore, 6-phycocyanin
[B] Chromophore, 9-phycocynin
[C] Chromophore, 2-phycocynin
[D] Chromophore, 4-phycocynin

21. There is probably .... chromophore for each phytochrome molecule. The chromophore is linked to the protein at ring .... Apparently, the photoconversion of the Pr and Pfr forms

involved electronic changes in ring .... with either the addition or loss of a proton.

[A] One, III, I [B] Two, IV, III

[C] Three, I, IV [D] None of these

22. **Phytochrome term was coined by**

[A] Garner and Allard [B] Borthwick and Hendricks

[C] Both [D] None of these

23. **In addition to having isolated from such higher plants as tobacco, oat, corn and bean, phytochrome has been isolated from alga .... and liverwort ....**

[A] *Mestaenium, Sphaerocarpos* [B] *Sphaerocarpos, Mestaenium*

[C] *Mestaenium* only [D] None of these

24. **Apart from phytochrome mediated photoresponse, large number of photoresponse in plants is known which are controlled by .... and are believed to be mediated through a group of yet unidentified pigment called ....**

[A] Blue light, Cryptochrome [B] Cryptochrome, Blue light

[C] Both [D] None

25. **The action spectra of many blue-light responses in higher plants such as phototropism, stomatal movement, inhibition of hypocotyl elongation etc are similar in characteristic. They show .... in blue region (400-500 nm) of the spectrum of visible light. This three peaked action spectrum is also known as .... action spectrum and is typical of most blue light responses.**

[A] Three peaks, three fingers [B] Three fingers, three peaks

[C] Two peaks, two fingers [D] None of these

26. **The carotenoid .... is blue-light photoreceptor involved in stomatal opening.**

[A] Riboflavin [B] Zeaxanthin

[C] FAD [D] None of these

27. **Cryptochrome, the first protein with characteristics of blue-light receptor, was isolated in 1993 from ....**

[A] *Mougeotia* [B] *Onoclea*

[C] *Arabidopsis* [D] None

28. **It was found that .... mutant of *Arabidopsis* had lost the capacity to respond specifically to blue-light in that it showed an elongated hypocotyls even on irradiation with blue-light (In the wild type, blue – light causes inhibition of hypocotyls elongation).**

[A] hy4 mutant [B] hy3 mutant

[C] hy2 mutant [D] hy1 mutant

29. **hy4 gene related to photomorphogenesis was later named as ...... gene.**

[A] cry9 [B] cry3

[C] cry1 [D] cry6

30. Isolation of the hy4 gene (later named as cry1) showed that it encoded a 75kDa protein called............... with remarkable sequence similarity to .... in having two (1) .... (2)............... .

[A] cryptochrome, DNA photolyase, FAD, pterin

[B] DNA photolyase, cryptochrome, pterin, FAD

[C] Pterin, FAD, cryptochrome, DNA photolyase

[D] None of these

31. Diagrammatic representation of the chromophore binding domains of microbial photolyase

Pterin FAD

A=? B=?

Find the A and B in the question.

[A] A=COOH, B= NH_2 [B] A=NH_2, B=COOH

[C] A= COOH, B=COOH [D] A=NH_2, B= NH_2

32. .... is blue light receptor that mediate chloroplast movement in plants.

[A] Zeaxanthin [B] Phototropin

[C] Chromophore [D] None of these

33. Phototropin is also a flavoprotein with .... flavin mononucleotide (FMN) chromophores.

[A] One [B] Two

[C] Three [D] None

34. The carotenoid Zeaxanthin has been shown to be blue-light receptor in .... that plays a central role in blue light stimulated ....

[A] Guard cells, stomatal opening.

[B] Stomatal opening, guard cells.

[C] Both

[D] None

35. Which is a potent inhibitor of the enzyme that converts violaxanthin to zeaxanthin?

[A] Dithiothreitol (DTT) [B] Thiourea

[C] TIBA [D] None

36. In facultative CAM plant species such as .... there is a shift from C3 to CAM mode of carbon metabolism in response to accumulation of salts.

[A] *Mesembryanthemum crystallinum*

[B] *Phaseolus lunatus*

[C] *Lotus tenuis*

[D] None

37. The blue-light stimulated stomatal opening can be reversed by ....

[A] Red light [B] Green light

[C] Yellow light [D] None

38. Photomorphogenetic responses in plants are

[A] Dependent on photosynthesis

[B] Independent of photosynthesis

[C] Found only in higher plants

[D] Found in only lower plants

39. Which of the following is the most effective in inducing photomorphogenetic responses in plants?

[A] Blue and yellow light [B] Blue and red light

[C] Orange and red light [D] Red and far-red light

40. Phytochrome is a

[A] Flavoprotein [B] Glycoprotein

[C] Lipoprotein [D] Chromoprotein

41. Which of the following is not a phytochrome mediated photoresponse?

[A] Germination of lettuce seeds

[B] De-etiolation of seedlings

[C] Phototropism

[D] Rotation of chloroplast in alga *Mougeotia*

42. Which of the following category of phytochrome mediated photoresponses in plants show reversible photoresponses?

[A] Very Low Fluence Responses (VLFRs)

[B] Low Fluence Responses (LFRs)

[C] High Irradiance Responses (HIRs)

[D] None

43. Which one of the following is not an example of blue-light response in plants?

[A] Phototropism [B] Stomatal opening

[C] Geotropism [D] Phototaxis

44. LOV domains are characteristic of

[A] Photolyase [B] Cryptochrome

[C] Carotenoid [D] Phototropin

45. Three peaked or three fingers action spectrum is typical of

[A] All blue light responses

[B] Most blue light responses

[C] All phytochrome mediated photoresponses

[D] None of the above

46. Cryptochrome absorbs light rays mostly in .... region of spectrum.

[A] Violet-blue region [B] Red-blue region

[C] Violet-green region [D] None of these

47. Phototropism, stomatal opening, inhibition of hypocotyls elongation, sun tracking by leaves, phototaxis, movement of chloroplasts within the cells and stimulation of synthesis of carotenoid and chlorophylls etc. are the typical examples and most commonly known as .... responses in plants.

[A] Red-light response [B] Blue light response

[C] Green light response [D] None

48. When excised barley root tips are swirled gently in a glass beaker with IAA, ATP, ascorbic acid and certain mineral salts, a brief red-light treatment causes them to begin within seconds to stick to the walls of the beaker. A brief far-red light treatment rapidly decreases the positive charge of the root tips and causes their release from the beaker walls. This described phenomenon is called ....

[A] Tanada effect [B] Kurosawa effect

[C] Tindal effect [D] None

49. Production of flavonoids requires .... as a source of the .... and .... that provide carbon atoms needed for the flavonoid B ring and as a source of acetate units needed for the flavonoids A ring.

[A] Sugares, phosphoenolpyruvate, erythrose-4-phosphate

[B] Phosphoenolpyruvate, erythrose-4-phosphate, sugars

[C] Erythrose-4-phosphate, sugars, phosphoenolpyruvate

[D] None

50. Tiller (branch) formation is retarded in closely spaced cereals-grain crops or in thick pastures because

[A] The light transmitted to the stem bases is rich in far-red wavelength, causing low value .

[B] The light transmitted to the stem bases is rich in red-wavelength, causing low value

[C] The light transmitted to the stem base is rich in green wavelength, causing low value

[D] None of these

51. Light effects on chlorophyll formation and chloroplast development result first from a triggering action of P_{fr} that causes production of .... from ....

[A] Delta aminolevulinic acid (ALA), glutamic acid

[B] Glutamic acid, delta aminolevulinic acid

[C] Fatty acid, lipid

[D] None

52. Plant responses that require relatively high irradiance levels and that have action spectra that are atypical of commonly observed phytochrome responses came to be known as

[A] High irradiance reactions (HIR) [B] High energy reaction(HER)

[C] High photon reaction(HPR) [D] None

53. Chemically, phytochrome is a homodimer of two identical polypeptides, each with a molecular weight of about .... KDa.

[A] 120 [B] 110
[C] 111 [D] None

54. Type I phytochrome predominates in

[A] Etiolated seedlings [B] Green plants
[C] Both [D] None

55. Type II phytochrome predominates in

[A] Etiolated seedlings [B] Green plants
[C] Both [D] None

56. A pigment that absorbs red and blue light and becomes reduced to chlorophyll a is known as

[A] Protochlorophyllide a [B] UV-B photoreceptor
[C] Cryptochrome [D] None

57. One or more unidentified compounds (technically not pigments) that absorb ultraviolet radiation between about 280 and 320 nm are known as

[A] Protochlorophyllide a [B] UV-B photoreceptor
[C] Cryptochrome [D] None

58. Each polypeptide has a prosthetic group called a chromophore that is attached via a .... atom in a .... residue of the polypeptide

[A] Sulfur atom, cysteine [B] Cysteine, sulfur
[C] Hydrogen, proline [D] None

59. Studies with phytochrome and cryptochrome-deficient mutants have confirmed that de-etiolation responses in *Arabidopsis* seedlings involve complex interactions between three different photoreceptors, and these receptors are

[A] phyA, phyB, and cry1 [B] phyC, phyP, and cry 2
[C] Both [D] None of the above

60. Which of the following photoreceptor involves complex interaction within *Arabidopsis* indicate that .... is the principle photoreceptor for control of hypocotyls elongation

[A] phyB [B] phyA
[C] phyC [D] None of the above

61. Sun is the one degree source of light and plants need light for photosynthesis. Those plants which grow in the shade or canopy can adjust to the reduced availability of light in two ways. They can (1) adjust their light –harvesting capability by increasing the specific leaf area and the amount of chlorophyll a/b light –harvest complex, or they can (2) adjust their morphology in order to position their leaves out of the shade. Shading leads to early flowering and seed set in an effort to

"escape" shading by shortening generation time. These and other effects are collectively called

[A] Shade avoidance syndrome [B] Shade adjustment syndrome

[C] Shade logging syndrome [D] None of the above

62. The effect of canopy shading can be described in terms of the ratio of

[A] Red to far-red [B] Far-red to red

[C] Both [D] None of the above

63. Shading can be mimicked in the laboratory of growth chamber by supplementing .... with various amounts of far-red light through the entire photoperiod.

[A] White fluorescent [B] Green fluorescent

[C] Yellow fluorescent [D] None of the above

64. The approximate value of R/FR for wheat canopy filtered light is

[A] 0.5 [B] 0.20

[C] 0.12 [D] None of the above

65. The approximate value of R/FR for maize canopy filtered light is

[A] 0.5 [B] 0.20

[C] 0.12 [D] None of the above

66. The approximate value of R/FR for oak woodland canopy filtered light is

[A] 0.5 [B] 0.20

[C] 0.12 [D] None of the above

67. The approximate value of R/FR for maple woodland canopy filtered light is

[A] 0.5 [B] 0.20

[C] 0.14-0.28 [D] None of the above

68. The approximate value of R/FR for spruce forest woodland canopy filtered light is

[A] 0.5 [B] 0.20

[C] 0.15-0.33 [D] None of the above

3 THE POWER OF MOVEMENT IN PLANTS

1. *"We believe that there is no structure in plants more wonderful, as far as its functions are concerned, than the tip of the radical. If the tip be lightly pressed or burnt or cut, it transmits an influence to the upper adjoining part, causing it to be bend away from the affected side; .... if the tip perceives the air to be moister on one side than on the other, it likewise transmits an influence to the upper adjoining part, which bend toward the source of moisture .... when the tip is excited by gravitation the adjoining part bends towards the centre of gravity. In almost every case we can clearly perceive the final purpose or advantage of the several movements".*

   This statement was written by which famous scientist(s) in the famous book "The Power of movement in plant" ?

   [A] Elbert Einstein
   [B] Charles and Francis Darwin
   [C] J.C. Bose
   [D] None

2. **Which of the following pair of processes and terms denoting the processes are correctly matched?**

   1. **Movement of growing tips of twining plants .... Nutation**
   2. **Curvature caused by injury to the tissues of organs .... Traumatotropism**
   3. **Movement of the plant organs due to contact with surface ....Thigmotropism**
   4. **Movements caused by change in turgor pressure ....Seismonasty**

   Select the correct answer using the codes given below:

   [A] 2 and 3 [B] 1 and 3
   [C] 1 and 2 [D] 1, 2, 3 and 4

3. **The movement of antherozoids towards the archegonium (egg) due to the secretion of some chemical substances is known as**

   [A] Chemoreception [B] Chemotropism
   [C] Chemotaxis [D] None

4. Chemical nature of statolith which is associated with gravitropism in plants

[A] Lipid [B] Amyloplast
[C] Protein [D] Starch

5. Movement of hairs in *Drosera* is

[A] Seismonastic [B] Thermonastic
[C] Photonastic [D] None

6. Induced movement is also termed as

[A] Spontaneous movement [B] Paratonic movement
[C] Both [D] None

7. Cell of stamina hairs of .... plant exhibited .... movement.

[A] *Tradeschantia*, rotation [B] *Tradeschantia*, circulation
[C] *Hydrilla*, circulation [D] None

8. In living cells of many plants the cytoplasm including various cell-organelles moves around the .... This movement of the cytoplasm is called ....

[A] Mitochondria, Amoeboid movement
[B] Chloroplast, Ciliary movement
[C] Vacuoles, Cyclosis
[D] None

9. Amoeboid movements are exhibited by .... where the naked plasmodium moves by producing pseudopodia like an *Amoeba*.

[A] Vallisneria [B] Myxomycetes
[C] *Chara* [D] None

10. Cyclosis movement is of two types-rotation and circulation. In rotation, which is exhibited by plants like *Chara*, *Hydrilla*, *Vallisneria*, *Elodea* etc., the cytoplasm moves either .... or .... around a .... While in circulation, which is exhibited by the cell of stamina hairs of plants like *Tradeschantia*, the cytoplasm moves in both .... and .... direction around many ....

[A] Clockwise, anticlockwise, large central vacuole, clockwise, anticlockwise, smaller vacuoles.
[B] Anticlockwise, clockwise, smaller vacuole, anticlockwise, clock-wise, larger vacuole
[C] Linear, triangular, cell wall, triangular, linear, nucleus
[D] None of these

11. Which of the following is not vital movement?

[A] Movement of locomotion [B] Movement of curvature
[C] Hygroscopic movement [D] All of above

12. Movements of locomotion which occur in response to external stimulus are called

[A] Tactic movements [B] Tropic movement
[C] Amoeboid movements [D] None of these

13. In plants, the movement of curvature which occurs in response to external stimulus are called as

[A] Paratonic [B] Autonomous

[C] Both [D] None of these

14. Peristome and Elaters teeth in bryophyta show

[A] Movement of curvature [B] Movement of locomotion

[C] Hygroscopic movements [D] None

15. Both nastic and tropistic movement are often the result of differential (irreversible) growth, but they can also be caused by differential and reversible uptake of water into special cells, called ...., that collectively form a ....

[A] Pulvinus, motor cells [B] Motor cells, pulvinus

[C] Both [D] None

16. Haptonastic movements are found in the leaves of

[A] *Dionaea and Drosera* [B] *Oxalis*

[C] *Hydrilla* [D] None

17. *Mimosa pudica* exhibits

[A] Thigmonastic movement [B] Seismonastic movement

[C] Photonastic movement [D] None

18. Stimuli always act on some machinery that is part of the plant, and the part that receives (perceives) the stimulus is the .... once the stimulus has been perceived, it is changed (transduced) into some other form, often called a ...., that then leads to some .... that is the actual cause of plant movement.

[A] Receptor, signal, motor response

[B] Signal, receptor, motor response

[C] Motor response, receptor, signal

[D] None

19. Pollen tubes and fungal hyphae exhibit

[A] Phototropic movement [B] Phototactic movement

[C] Chemotactic movement [D] Chemotropic movement

20. Thermotatic movement is exhibited by

[A] *Chlamydomonas* [B] *Desmodium gyrans*

[C] *Crocus* [D] None

21. In higher plants which are fixed, the movements are restricted only to the bending or curvature of some of their parts. Such movements are called curvature movements and it is of two types- .... having, the curvature or the bending of the plant part is temporary and ... having, the curvature or the bending of the plant part is of permanent nature.

[A] Variation movement, growth movement

[B] Growth movement, variation movement

[C] Thermotactic movement, chemotactic movement

[D] None

22. .... is an excellent example of autonomic movement of variation.

[A] *Chlamydomonas* [B] *Crocus*

[C] *Desmodium gyrans* [D] None

23. Sometimes the growth of the stem apices occurs in a zig-zag manner. It is because the two sides of the stem apex alternatively grow more. Such growth movements are called as

[A] Epinastic movement [B] Hyponastic movement

[C] Nutational movement [D] None

24. The primary roots grow down into the soil and are .... The secondary roots growing at the right angles to the force of gravity are called as .... on the other hand, the primary stems are ....

[A] +ve geotropic, diageotropic, -ve geotropic

[B] Diageotropic, +ve geotropic, -ve geotropic

[C] –ve geotropic, +ve geotropic, diageotropic

[D] None

25. The geotropic curvature results due to unilateral gravity stimulus and this can be demonstrated by using a

[A] Clinostat [B] Photometer

[C] Pyrometer [D] None

26. .... is believed to be the hormone that control nastic movement.

[A] Turgorin [B] Auxin

[C] Florigen [D] None

27. It is now well established that almost any part of *Mimosa* plant can perceive the stimulus and transmits it to the pulvinus as electric pulses through .... at velocities up to ....

[A] Phloem sieve tubes, 2 cm s^{-1} [B] Xylem parenchyma, 5cm s^{-1}

[C] Phloem sieve tubes, 0.2 cm s^{-1} [D] None

28. Turgorins are chemically identified as

[A] -D-glucosides of gallic acid [B] -D-glucosides of gallic acid

[C] Both [D] None

29. Turgorins were previously known as

[A] Periodic leaf movement factors (PLMFs)

[B] Leaf movement factors (LMFs)

[C] Movement factors (MFs)

[D] None

30. Generally speaking, nastic movement is ...., whether they are controlled by pulvini or by changing relative growth rates on top and bottom of an organ.

[A] Reversible [B] Irreversible

[C] Both [D] None

31. Nyctinastic movements are rhythmic processes controlled by interaction between the .... and the ....

[A] Environment, biological clock [B] Pathogens, environment
[C] Fertilizers, rainfall [D] None

32. To study the sleep movements in plant, it is sometimes convenient to study species with .... leaves, in which each leaf bears several .... and each pinna bears several pairs of .... attached to a single rachilla.

[A] Double compound, pinnae, opposite leaflets
[B] Single, leaves, leaflets
[C] Both
[D] None

33. The danger of photoinhibition is minimized by

[A] Nyctonasty [B] Hydronasty
[C] Both [D] None

34. The bulliform cells lose water by transpiration faster than epidermal cells. This statement is

[A] True [B] False
[C] Not known [D] Under investigation

35. The bulliform cells have little or no cuticle, so they lose water by transpiration faster than epidermal cells. This statement is

[A] True [B] False
[C] Not known [D] Under investigation

36. The electrical evidence of transmission of signal in *Mimosa* has been investigated first time by

[A] J.C. Bose [B] A.L. Houwink
[C] Umrath [D] None

37. The chemical evidence of transmission of signal in *Mimosa* has been reported first time by

[A] Ubaldo Ricca [B] Simons
[C] Both [D] None

38. 'Ricca factor' is chemically

[A] Glucose [B] Enzyme
[C] Hormone [D] None

39. 'Ricca factor' is chemically

[A] Turgorin [B] Auxin
[C] Cytokinin [D] None

40. Reciprocity is

[A] The response proportional to the duration of the exposure
[B] The total dose response
[C] Both
[D] None

41. Few experiment suggests that light really has two effects in phototropism. First, it acts as a trigger for the bending response; second, it decreases the sensitivity of the organ to subsequent light. This effect is non directional and is referred as a

[A] Tonic effect [B] Tindal effect
[C] Tanada effect [D] None

42. Leaves respond phototropically to light in such a way that number of leaves would presumably move them from shade into the light whenever this is physically possible. The results are that leaves often barely overlap, instead, they form patterns called ....

[A] Leaf mosaics [B] Leaf building
[C] Both [D] None

43. Flat blade of the leaf remains nearly at the right angles to the sun throughout the day, maximizing the light harvested by the leaf, this is known as

[A] Solar tracking [B] Leaf mosaics
[C] Both [D] None

44. Solar tracking is a true tropism because the orientation of the leaves is determined by the direction of the sun's rays, but it is neither positive nor negative, as is a stem phototropism; it is .... in which organ orient at right angle to the light rays.

[A] Diaphototropism [B] Diaheliotropism
[C] Both [D] None

45. A movement in plant stimulated by oxygen is called

[A] Thermotaxis [B] Sensitivity
[C] Nutation [D] Aerotaxis

46. The minimum quantity of a stimulus required for bringing about a visible responses are called

[A] Threshold value [B] Sensitivity
[C] Perceptibility [D] Summation

47. The effects of stimulus subsides after a time called

[A] Reaction time [B] Relaxation time
[C] Presentation time [D] Conduction time

48. Time period from perception of stimulus and reaction is called

[A] Restitution time [B] Presentation time
[C] Relaxation time [D] Reaction times

49. Tropic movements are basically due to

[A] Cell elongation [B] Cell division
[C] Sudden water gain [D]Sudden water loss

50. The closure of lid of pitcher in pitcher plant is

[A] A Paratonic movement [B] A tropic movement
[C] A turgor movement [D] None

51. Opening and closing of flowers represents a kind of

[A] Autonomic movement [B] Nutation

[C] Tropic movement [D] Nastic movement

52. Auxin gradient do not always account for gravitropic bending, it is clear that bending does not occur when insufficient auxin is present. This was shown by .... and .... They removed the tips of sunflower hypocotyls and then, after four days, they gravistimulated the hypocotyls by turning them to the horizontal position. Bending did not occurs, but if hypocotyls were returned to the vertical and the cut stamp supplied with an auxin (IAA) solution, the hypocotyls bent in the predicted direction. This phenomenon is known as ....

[A] Loe Brauner, Achim Hager, gravitropic memory

[B] MacDonald, Hart, gravitropic memory

[C] Wright, Harrison, gravitropic memory

[D] None

53. .... suggested the name clinostat.

[A] Sachs [B] Seward

[C] Donald [D] None

54. .... is the increased xylem produced on either the upper or lower side of a branch by more rapid division of the vascular cambium on that side.

[A] Compression wood [B] Reaction wood

[C] Tension wood [D] None

55. In conifer limbs, reaction wood forms on the lower side and by expansion pushes such limbs more upright, maintaining a more constant angle. Tracheid walls become abnormally thick and contain more lignin and less cellulose than usual. Such wood is called ....

[A] Compression wood [B] Reaction wood

[C] Tension wood [D] None

56. In angiosperm trees, reaction wood forms on top and contracts to pull the branch toward the trunk by tension, called ....

[A] Compression wood [B] Reaction wood

[C] Tension wood [D] None

57. In 1900, G. Haberlandth and E. Nemec independently adapted the statolith theory to account for plant responses to gravity. Based on careful cytological studies, they proposed the .... hypothesis.

[A] Starch-statolith hypothesis

[B] Lipid-statolith hypothesis

[C] Protein- statolith hypothesis

[D] None of the above

58. Tissue that contains statocytes are known as

[A] Statenchyma [B] Statolith cell

[C] Statolith parenchyma [D] None of the above

59. The root and shoot of the primary plant axis do align themselves parallel with the direction of gravitational pull. Such an alignment is said to be

[A] Orthogravitropic. [B] Plagiogravitropic
[C] Diagravitropic [D] None of the above

60. Gravitational stimulation (stimulus quantity or dose) is the .... of the intensity of the stimulus and the time over which the stimulus is applied.

[A] Product [B] Addition
[C] Division [D] None of the above

61. Organs such as stolons, rhizomes and some lateral branches, which grow at right angles to the pull of gravity, are said to be

[A] Diagravitropic [B] Plagiogravitropic
[C] Agravitropic [D] None of the above

62. Organs oriented at some intermediate angle (between 0° to 90° to the vertical) are said to be ....

[A] Diagravitropic [B] Plagiogravitropic
[C] Agravitropic [D] None of the above

63. .... photoreceptor has properties of both phytochrome and phototropin.

[A] Neochrome [B] phy3
[C] Both [D] None of the above

64. A statocyte (columella cell) contains .... statoliths (amyloplasts)

[A] One [B] Two
[C] Three [D] Four

65. The sedimentations of statoliths create physiological ....

[A] Asymmetry [B] Symmetry
[C] Both [D] None of the above

66. Russian plant physiologist M. Chailakhyan was the first to suggest in 1936 that the floral stimulus might be a hormone, which he proposed be called ....

[A] Florigen [B] Floragonr
[C] Florisen [D] All of the above

4 PLANT GROWTH REGULATORS

1. **Kogal & Haagen-Smith first isolated auxin from human**
   [A] Blood [B] Urine
   [C] Sperm [D] None
2. **Concentration of auxin is more in urine of person suffering from ……………… due to deficiency of ………………**
   [A] Pellegra, riboflavin [B] Cancer, phylloquinones
   [C] Blindness, Tocopherols [D] Pellegra, Niacin
3. **Dicamba is chemically known as**
   [A] 2-methoxy, 3, 6 dichlorobenzoic acid
   [B] 2, 3, 6 trichlorobenzoic acid
   [C] 2, 2 dichloropionic acid
   [D] None
4. **The activity of each meristem influences the activity of the other meristems, especially those near to it, giving rise to what is called as**
   [A] Grand period of growth [B] Growth correlations
   [C] Sigmoid curve [D] None
5. **Who proposed that "organ forming substances" were produced in the leaves of plants and translocated downward?**
   [A] Darwin [B] Sachs
   [C] Starling [D] None
6. **Who demonstrated that the effects of light and gravity on the bending of both roots and shoots are mediated by the tip, and that this influence can be transmitted to other parts of the plant?**
   [A] Sachs [B] Starling
   [C] Darwin [D] [None]
7. **Auxin-A is chemically**
   [A] Auxentriolic acid [B] Auxenolonic acid
   [C] Both [D] None
8. **Auxin-B is chemically**
   [A] Auxentriolic acid [B] Auxenolonic acid
   [C] Both [D] None

9. A plant hormone is an .... compound synthesized in .... part of plant and .... to another part, where in very .... concentrations it causes a physiological responses.

[A] Organic, one, translocated, low [B] Inorganic, many, stored, high
[C] Both [D] None

10. Sucrose is not considered a hormone, even though it is synthesized and translocated by plants, because it causes growth only at

[A] High concentration [B] Medium concentration
[C] Low concentration [D] None

11. In the early 1980s .... repeatedly emphasized the concept that differential sensitivity is much more important in determining the effects of a hormone than is the concentration of that hormone within the plant cells.

[A] J. Trewavas [B] Went
[C] Darwin [D] None

12. The term auxin was first used by

[A] F. W. Went [B] Darwin
[C] Both [D] None of the above

13. The .... was the first hormone to be discovered in plants and at one time considered to be the only naturally occurring plant growth hormone

[A] Gibberellin [B] Auxin
[C] Cytokinin [D] Ethylene

14. Which one of the followings is not associated with discovery of auxin in plants?

[A] Darwin [B] Kurosawa
[C] F.W. Went [D] Boysen-Jensen

15. Lateral branching of plant is controlled by the relative proportion of

[A] Cytokinins and Auxins [B] ABA and Auxins
[C] Auxins and Ethylene [D] None

16. Which one of the followings is considered as true natural auxin in higher plants?

[A] Indole-3-acetic acid [B] Indole-3-acetaldehyde
[C] Indole-3-pyruvic acid [D] Indole-3-acetonitrile

17. Which of the following plant hormone is incorrectly paired with its function?

[A] Auxin-responsible for apical dominance
[B] ABA- regulates the rate of transpiration
[C] Cytokinin- delays senescence (aging and decay)
[D] GA- promotes bud and seed dormancy

18. In monocots e.g. grasses, oats, maize, etc the plumule in the seed remains covered by a protective cap-like structure called ....

[A] Coleoptiles [B] Mycorrhiza
[C] Root hairs [D] None

19. The primary physiological effect of auxin in plants is

[A] Elongation of internodes [B] Cell division
[C] Elongation of cells in shoot [D] None of the above

20. Which of the following statement is false about auxin

[A] Accumulation in root tips and branch root initials
[B] Transport away from leaf primordial causes cells to differentiated into vascular tissue
[C] In high concentration in tissue culture media will promote root formation
[D] Synthesis by shoot apex results in the phenomenon of apical dominance

21. For the determination of course of differentiation in plant cells, which combination of hormone is best suited?

[A] Auxin+ Cytokinin [B] Auxin + Ethylene
[C] Auxin + ABA [D] None

22. Which one of the followings is a physiological effect of auxin in higher plants?

[A] Apical dominance [B] Cell elongation
[C] Root initiation [D] All of these

23. .... was the first to use the term phytohormones for growth regulating chemical of plant.

[A] Sachs [B] Thimann
[C] Boysen & Jensen [D] None of these

24. Phytohormones are chemically

[A] Amino [B] Purine
[C] Steroidal [D] Gaseous [E] All of these

25. Thimann isolated auxin from

[A] Rhizopus [B] Mucor
[C] Penicillium [D] None of these

26. The main site of auxin synthesis is

[A] Roots [B] Meristem region of plant
[C] Shoots [D] None of these

27. An agar block containing 5μg of IAA was placed on a decapitated coleoptiles stump, as shown in the given figure. After sometime an angle of curvature, α was measured. If this agar block was replaced by another one containing 8 μg of IAA, then

α

[A] Angle α will become narrower
[B] Angle α will widen α
[C] Angle α will remains the same as before
[D] The coleoptiles will shrink

28. The induction of cell elongation by IAA is explained on the basis of

[A] Auxin ethylene balance hypothesis
[B] Sodium-Potassium pump hypothesis
[C] Osmotic pressure hypothesis
[D] None of the above

29. The most commonly used growth regulator to induce somatic embryogenesis is

[A] BAP [B] IAA
[C] GA_3 [D] 2, 4-D

30. .... first pointed out that the apical dominance might be under the control of auxin produced at the terminal bud and which is transported downward through the stem to the lateral buds and hinders their growth.

[A] Skoog and Thimann (1934)
[B] Skoog and Miller (1934)
[C] F.W. Went
[D] None

31. Conjugated or bound auxin formed complex with

[A] Carbohydrate
[B] Protein or Amino acid
[C] Inositol
[D] All of the above

32. Which of the following amino acid is considered as precursor of auxin?

[A] Tryptophan [B] Tyrosine
[C] Proline [D] None of these

33. Transport of auxin in plant is predominantly

[A] Polar [B] Nonpolar
[C] Lateral [D] None of the above

34. Transport of auxin in plant is predominantly

[A] Physiological Apex to Physiological Base
[B] Physiological Base to Physiological Apex
[C] Physiological lateral to Physiological End
[D] None of the above

35. Which of the following is an antiauxin?

[A] 2, 4-D [B] TIBA
[C] 2, 4-T [D] All of the above

36. Destruction of auxin in plants is mainly by

[A] Conversion into bound auxin
[B] Irradiance with x-rays and y-rays
[C] Oxidation of Auxin
[D] None

37. The movement of auxin is

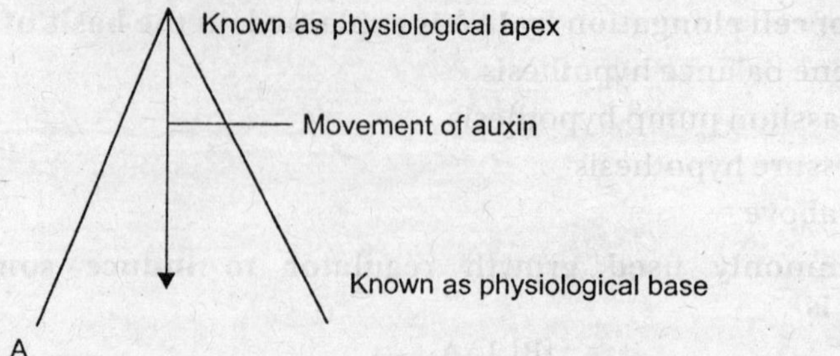

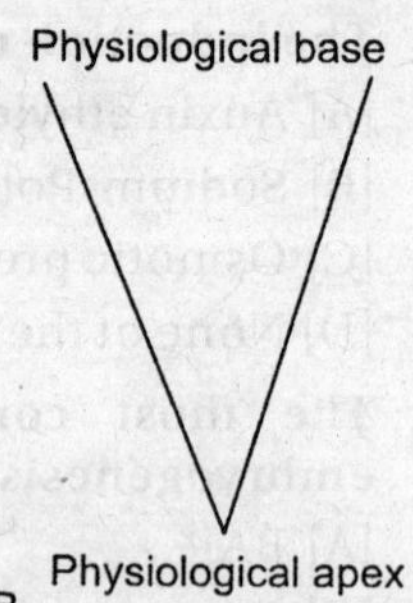

In figure: B, what will be the direction of movement of auxin?

[A] Toward Physiological apex
[B] Toward Physiological Base
[C] No clue
[D] None of these

38. Auxin increases the .... of cell walls

[A] Plasticity [B] Thickness
[C] Porosity [D] Rigidity

39. If an inhibitor of metabolism is applied to a tissue transporting auxin, which of the followings would happen

[A] It would have no effect
[B] The direction of transport would reverse
[C] Transport would be inhibited
[D] The rate of transport would increase

40. Environmental signals influence the distribution of an auxin in a plant by

[A] Decreasing the cell's sensitivity to the auxin
[B] Causing auxin to migrate to the lighted portion
[C] Destroying the auxin
[D] Causing auxin to migrate into the shaded portion

41. Which of the following statement(s) is/are correct?

[A] Acid growth hypothesis states that auxin causes receptive cells to secrete protons into their cell wall
[B] Unidirectional blue light causes a redistribution of auxin in the stem apex
[C] ABA synthesis occurs in chloroplast by the mevalonic acid pathway
[D] All of the above statements

42. Auxin induced gene fall under the category of

[A] One group [early genes]
[B] Two groups (early genes and late genes)
[C] Three groups (early genes, midway genes, late genes)
[D] None of the above

43. Minimum lag time required for auxin induced growth is

[A] 10 minutes [B] 15 minutes
[C] 20 minutes [D] 30 minutes

44. Which fungal phytotoxin mimics some of the physiological effects of auxin such as proton extrusion and growth?

[A] Fusicoccin [B] Tetracyclic diterpene acids
[C] Both [D] None

45. Lanolin is a soft fat prepared from wool and is a good solvent for

[A] Auxin [B] Cytokinin
[C] Ethylene [D] None

46. According to ...., the auxin may increase the rate of respiration indirectly through increased supply of ADP by rapidly utilizing the ATP in the expanding cells.

[A] French and Beevers [B] Gocal *et al.*
[C] Both [D] None of the above

47. The free forms of auxin are biologically .... forms.

[A] Active [B] Inactive
[C] Both [D] None of the above

48. In monocot seedling, the highest concentration of auxin is found in the ....

[A] Coleoptile tip [B] Seed
[C] Root [D] None

49. The indole-3-acetic acid can be formed from tryptophan by four different pathways, and these pathways are

[A] Bacterial pathway [B] IAN-pathway
[C] TAM-Pathway [D] IPA-pathway
[E] All of the above

50. IAN (Indole-3-acetonitrile) pathway occurs in some plants, especially belonging to families

[A] Brassicaceae [B] Poaceae
[C] Musaceae [D] All of the above

51. Bacterial pathway occurs in some pathogenic bacteria such as

[A] *Agrobacterium tumefaciens* [B] *Pseudomonas savastanoi*
[C] Both [D] None of the above

52. .... found polar transport of auxin in *Coleus* stem section to be both basipetal and acropetal in the ratio of ....

[A] Jacobs (1961), 3:1 [B] Audus (1959), 3:1
[C] Both [D] None of the above

53. Polar auxin transport is .... sensitive and requires metabolic energy.

[A] Temperature [B] Humidity
[C] Rainfall [D] None of the above

54. ………………. and ………………. are naturally occurring auxin transport inhibitor, found within the plant body.

[A] Genistein, quercetin [B] TIBA, NPA

[C] Both [D] None of the above

55. IAA may exists in two forms, one is ………………. form which is highly lipophilic and can cross the plasma membrane easily. The other form is ………………………… that does not cross plasmamembrane unaided.

[A] Protonated (IAAH), dissociated(IAA-)

[B] Symporter, antiporter

[C] Both [D] None

56. ………………. is the major end product of oxidation of auxin.

[A] 3-Methylene-oxindole [B] 2-Methylene-oxindole

[C] 1-Methylene-oxindole [D] None

57. A possible auxin-binding receptor protein identified in plants is called as

[A] ABP1 [B] ABP2

[C] ABP3 [D] ABP4

58. Auxin-binding protein 1(ABP1) appears to be a dimer made of two polypeptides of about ………………. KD each.

[A] 22 [B] 21

[C] 20 [D] 19

59. ………………. and ………………. prolong the growth response of auxin by providing osmotically active solutes.

[A] Glucose, NaCl [B] Sucrose, KCl

[C] Maltose, HCl [D] None

60. ………………. and ………………. in U.S.A. and ………………………… and ……………………… in Germany first proposed independently that H^+ may be involved in auxin-induced cell wall loosening.

[A] Rayle, Cleland, Hager, Cross

[B] Abel, Theologis, Hershko, Ciechanover

[C] Both

[D] None

61. The fungal phytotoxin fusicocein (obtained from the fungus *Fusicoccum amygdale*) is known to stimulate proton extrusion and growth only by activating the pre-existing H^+-ATPases in plasmamembrane. Fusicocein is sometimes called as

[A] Super auxin [B] Super chemical

[C] Both [D] None of the above

62. ………………………… and ………………………… are believed to be signaling intermediates in auxin induced cell division.

[A] Cyclic AMP, MAP (Mitogen Activated Protein)

[B] Ca^{++}, pH

[C] Both

[D] None

63. **Auxin has .... role in the stimulation of respiration rather than the .... role.**

[A] Indirect, direct [B] Direct, indirect

[C] No, direct [D] None of the above

64. **...................., a substance that inhibits oxidative phosphorylation increases the rate of respiration while inhibiting the growth.**

[A] Dinitrophenol [B] 2, 4-D

[C] Dicamba [D] None of the above

65. **The most important factor that controls abscission in leaf is**

[A] Auxin gradient across the abscission zone

[B] Leaf age

[C] Endogenous auxin concentration

[D] All of these

66.

[A] In Leaf A, rate of abscission increases

[B] In Leaf B, Rate of abscission decreases

[C] Both

[D] None

67. **.................... demonstrated that parthenocarpic development of fruit could be induced by application of IAA in lanolin paste to the stigma of the flower.**

[A] Muir [B] Gustafson

[C] Massart [D] None

68. **Statolith containing cells are termed as**

[A] Statocysts [B] Statocytes

[C] Both [D] None

69. **.. isolated an enzyme involved in the oxidation of IAA. This enzyme system is called IAA oxidase and is the major cause for the disappearance of the phytochrome *in vivo*.**

[A] Tang and Bonner

[B] Leopold and Hall

[C] Scott

[D] None

70.

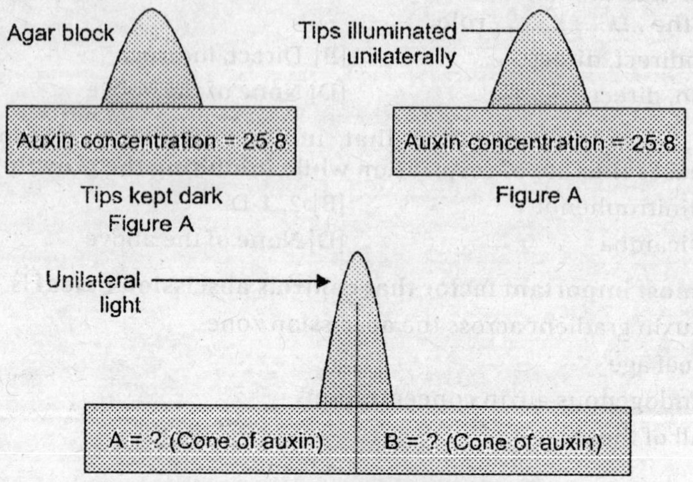

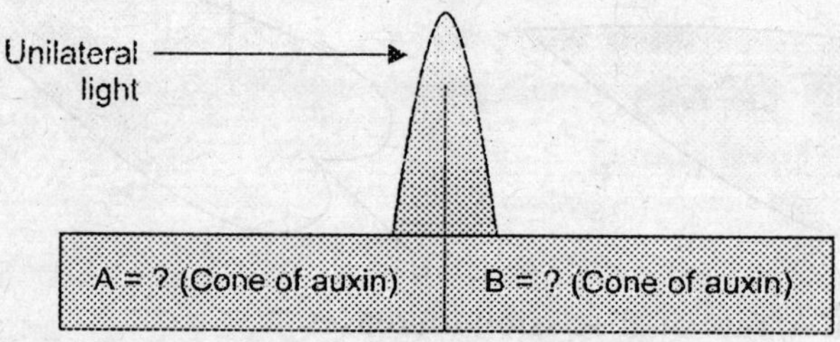

[A] In figure: C, A>B, in figure D; A>B
[B] In Figure: C, B>A, in figure D, A=B
[C] In figure: C, A=B, in figure D, A=B
[D] None

Note: The above experiment is popularly known as Cholodny-Went model which suggests that light from one side in some way causes a transport of auxin towards the shaded side, accounting for the basic transduction mechanism in phototropism.

71. Which of the following pair is correctly matched?

1. Auxin-Dwarf rice
2. Gibberellin-Elongation
3. Cytokinin- Coconut milk
4. ABA- Early leaf fall

Select the correct answer using the codes given below:

[A] 2, 3 and 4 [B] 1, 2 and 3
[C] 1, 2 and 4 [D] 1, 3 and 4

72. Which of the following statements are correct?

[1] Gibberellins can substitute light treatment

[2] Gibberellins can convert biennials into annuals

[3] Gibberellins favour flowering in long day plants.

[4] Gibberellins induce flowering in short day plants.

Select the correct answer using the codes given below

[A] 1, 2, 3 and 4 [B] 2 and 4

[C] 1, 3 and 4 [D] 1, 2 and 3

73. Higher concentration of auxins is inhibitory for ...., while stimulatory for ....

[A] Root, shoot [B] Shoot, root

[C] Both [D] None of the above

74. At supraoptimal concentrations, auxins interfere with the normal metabolism of plant, and may lead to their death. A mixture of 2, 4-D and 2, 4, 5-T was used in Vietnam war by United State to defoliate the citrus orchard under name ....

[A] Agent Orange [B] Toxins

[C] Little Boy [D] Fat men

75. .... and .... showed that flowering in *Nicotiana* can be accelerated by auxins.

[A] Hitchcock, Zimmerman [B] Kogl, Kostermans

[C] Confused [D] Quit

76. .... (1918) removed the tips of the coleoptile, replaced them .... on the stump. He found that the coleoptile bent away from the side with the tip, even in dark. It meant that there was some material substance coming down from the tip which could stimulate the growth of the cells below the tip.

[A] Paal, asymmetrically [B] Paal, symmetrically

[C] Paal, inside the cell [D] Quit

77. .... (1917) removed the tips of the coleoptile, replaced .... with dispersed coleoptile sap on the stump. He found that the curvature of coleoptile in opposite direction of agar block, even in dark.

[A] Stark, agar block [B] Fitting, gel block

[C] Seubert, agar block [D] Quit

78. The characteristics of a useful bioassay must include

[A] Specificity [B] Sensitivity

[C] Ease in measuring a detectable and relatively quick response,

[D] Relative ease in setting up and controlling

[E] Absence in the plant material, of the substance or related chemicals being tested.

[F] All of the above

79. 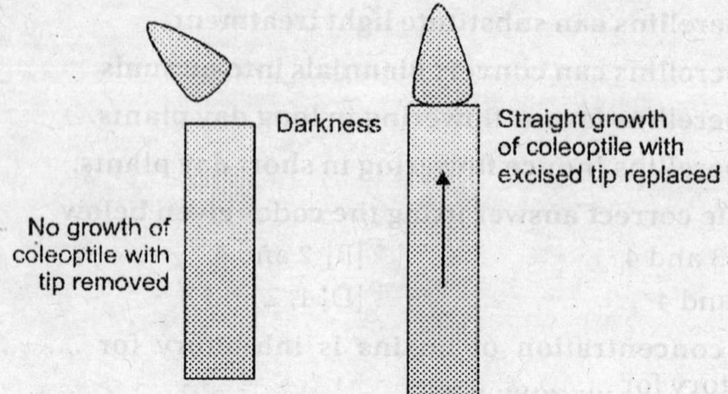

The above experiment was demonstrated by

[A] Darwin, 1880 [B] Fitting, 1907
[C] Seubert, 1925 [D] Quit

80. .... are much more sensitive to auxin than are ....

[A] Shoots, Roots [B] Roots, Shoots
[C] Leaves, Tendrils [D] None of the above

81. Which of the following bioassay is best for quantitative determination of auxin?

[A] *Avena* coleoptile curvature test
[B] Cress root inhibition test
[C] Split pea stem curvature test
[D] None

82. Cress root inhibition test is more .... than the *Avena* coleoptile curvature test since it can detect .... of IAA.

[A] Sensitivity, very low concentration
[B] Sensitivity, high concentration
[C] Both
[D] None

83. Gibberellin was named first by

[A] Kurosawa [B] Yabuta
[C] Mitchell [D] Stodola

84. The fungus *Gibberella fujikuroi* causes backanae disease in

[A] Seedlings [B] Leaves
[C] Flowers [D] Fruits

85. The fungus *Gibberella fujikuroi* causes bakanae disease in

[A] Oat [B] Wheat
[C] Rice [D] None

86. Gibberellins are derivatives of

[A] Monoterpens [B] Sesquiterpenes
[C] Diterpens [D] Triterpenes

87. Sprouting of Potato is prevented by use of ....

[A] Gibberellins [B] Auxin
[C] Cytokinin [D] None of the above

88. Auxin induces parthenocarpy in ....

[A] Tomato [B] Banana
[C] Both [D] None

89. In Banana natural parthenocarpy occurs due to more concentration of ....

[A] Auxin [B] Cytokinin
[C] Both [D] None

90. Auxin induces number of female flower in ....

[A] Cereals [B] Cucurbit
[C] Pulses [D] None

91. .... promotes flowering and fruiting in litchi and pineapple.

[A] Auxin [B] Cytokinin
[C] Both [D] None

92. Bakanae disease is also known as

[A] Foolish seedling disease [B] Farmers seedling disease
[C] Both [D] None

93. *Gibberella fugikuroi* is the perfect stage of ....

[A] *Fusarium moniliforme* [B] *Rhizopus*
[C] Both [D] None

94. *Fusarium moniliforme* is .... stage of Gibberella fugikuroi

[A] Asexual [B] Sexual
[C] Both [D] None

95. .... isolated GA from *Gibberella fugikuroi.*

[A] Yabuta & Sumiki [B] Kurosawa
[C] Both [D] None of the above

96. Transport of gibberellic acid in plant is

[A] Polar [B] Non polar
[C] Both [D] None

97. $C_{19}H_{24}O_6$ is

[A] GA_1 [B] GA_2
[C] GA_3 [D] None

98. $C_{19}H_{25}O_6$ is

[A] GA_1 [B] GA_2
[C] GA_3 [D] None

99. $C_{19}H_{22}O_6$ is

[A] GA_1 [B] GA_2
[C] GA_3 [D] None

100. Out of the numerous different gibberellins, which one of the following is called as Gibberellic acid that is widely used by scientist to show growth promoting properties of gibberellins?

[A] GA_3 [B] GA_4

[C] GA_5 [D] GA_6

101. The production of α-amylase in cereals grains

[A] Results in the degradation of sugar into starch

[B] Is induced by Gibberellin from the germinating embryo

[C] Will promote flowering

[D] Occurs throughout seed development

102. Seeds of the cucurbit .... contains at least .... gibberellins.

[A] *Sechium edule*, 20 [B] *Phaseolus vulgaris*, 16

[C] *Cucumis sativa*, 20 [D] None

103. Seed of common bean contains at least .... gibberellins

[A] 16 [B] 17

[C] 18 [D] 19

104. All gibberellins have .... carboxyl group attached to carbon ...., and some have an additional carboxyl attached to carbon ...., so all could be called Gibberellic acids.

[A] One, 7, 4 [B] Two, 6, 3

[C] Three, 5, 4 [D] None

105. GA_{17}, GA_{38}, GA_{44}, GA_9, GA_{20}, GA_{29} and GA_{51}-have been isolated from

[A] *Pisum sativum* [B] *Phaseolus vulgaris*

[C] Both [D] None

106. The gibberellins are easily interconvertible in the organism by change in the ....substitutions.

[A] Hydroxyl group [B] Carboxyl group

[C] Methyl group [D] None

107. Most of the information relating to gibberellins biosynthesis in plants comes from studies on ....

[A] Matured seeds [B] Immature seeds

[C] Matured bud [D] None of the above

108. The first few steps in gibberellins biosynthesis involve the formation of .... acetyl CoA molecules and their eventual condensation to form mevalonic acid.

[A] 2 [B] 3

[C] 4 [D] None

109. Isomerization of IPP to form .... constitutes the first step towards the synthesis of higher terpenoids

[A] Dimethylallyl pyrophosphate [B] Mevalonic acid pyrophosphate

[C] Isopentenyl pyrophosphate [D] None

110. .... showed that abscisic acid, a sesquiterpenoid, is synthesized from mevalonate and follows the same initial pathway as the gibberellins.

[A] Milborow [B] Sawada

[C] Bonow [D] None

111. .... postulated that the bakanae disease was caused by something secreted by the fungus *Gibberella fujikuroi.* This postulation was given experimental support by ...., who demonstrated that sterile filtrate of the fungus caused symptoms of bakanae disease in otherwise normal seedlings of rice

[A] Sawada, Kurosawa [B] Kurosawa, Sawada

[C] Yabuta, Hayashi [D] None

112. Which of the following is specific physiological effect of gibberellins?

[A] Internode elongation [B] Bolting

[C] Synthesis of -amylase [D] All of the above

113. In higher plants, the richest source of gibberellins is

[A] Immature seeds [B] Leaf

[C] Stem [D] Root

114. The primary precursor for the formation of isoprenoid unit and synthesis of gibberellin is

[A] Pyruvate [B] Acetate

[C] Oxaloacetate [D] None of the above

115. Antigibberllins such as cycocel (CCC) finds extensive use in

[A] Food Industry [B] Breweries

[C] Floriculture industry [D] None of the above

116. Which of the following is not an antigibberllins or growth retardant?

[A] Phosfon D [B] Cycocel

[C] Amo-1618 [D] NPA

117. Production of large sized seedless grapes and malting of barley can be enhanced by treatment with

[A] Auxins [B] Gibberellins

[C] Ethylene [D] Morphactins

118. Morphactins are synthetic derivatives of

[A] Fluorene- 9-carboxylic acid [B] Kinetin

[C] ABA [D] None of the above

119. Morphactins counteract the .....; hence, the treatment with Morphactins leads to growth in the lateral buds so that plants give a ..... appearance.

[A] Apical dominance, bushy [B] Lateral dominance, bushy

[C] Both [D] None

120. .... inhibits the growth of almost all the part of a plant.

[A] Morphactin [B] CCC

[C] TIBA [D] All of the above

121. There is a non-indole phenyolacetic acid compound, which plays a role similar to indole acetic acid. Which plant has been shown to be rich in it?

[A] Avena [B] Tomato
[C] Papaya [D] Potato

122. Gibberellins are more effective when applied to

[A] Intact plant [B] Coleoptiles
[C] Internodes [D] Roots

123. During seed germination stored food in seed is mobilized by

[A] Ethylene [B] Cytokinin
[C] ABA [D] Gibberellin

124. There is evidence that bound gibberellins exist in plant tissues as ....

[A] Gibberellin glycosides [B] Gibberellin amide
[C] Gibberellin steroid [D] None of these

125. Bound gibberellins are believed to exist in the

[A] Bleeding sap, of maple and elm
[B] Developing beans seeds
[C] Developing and germinating pea
[D] All of the above

126. GA's transport occurs in the .... according to flow pattern similar to .... and other organic substance.

[A] Phloem, carbohydrate [B] Phloem, protein
[C] Phloem, lipid [D] None

127. Antigibberellin compounds are growth ....

[A] Retardant [B] Promoter
[C] Inhibitor [D] None

128. Alar is chemically known as

[A] N-dimethylamminosuccinamic acid
[B] Tributyl-2, 4-dichlorobenzyphosphonium chloride
[C] Both
[D] None of the above

129. Phosfon-D inhibits the conversion of

[A] Copalyl pyrophosphate to Kaurene
[B] Kaurene to Copalyl pyrophosphate
[C] Geranylgeraniol pyrophosphate to Kaurene
[D] None of the above

130. .... breaks the genetic dwarfism of plant.

[A] GA [B] Auxin
[C] Cytokinin [D] None

131. .... corrects the light inhibited growth of stem.

[A] GA [B] Auxin

[C] Cytokinin [D] None

132. .... is used to increase the number and size of the grape in the bunch.

[A] GA [B] Auxin

[C] Cytokinin [D] None

133. .... acts as substitute of Red light for germination of positive photoblastic seeds.

[A] GA [B] Auxin

[C] Cytokinin [D] None

134. .... is more potent parthenocarpic than ....

[A] GA, Auxin [B] Auxin, GA

[C] GA, Cytokinin [D] None

135. GA induces parthenocarpy in .... and ....

[A] Pome, drupe [B] Pear, mango

[C] Tomato, apple [D] All of the above

136. GA treatment increases sugar yield of sugarcane by promoting ....

[A] Internodal growth [B] Leaf growth

[C] Flower growth [D] None of the above

137. A preparation called promalin is a mixture of

[A] Cytokinin (6-benzyladenine) + GA_4 + GA_7

[B] Auxin + GA_4 + GA_7

[C] GA_3 + GA_4 + GA_7

[D] None of the above

138. Currently .... is used to increase the number of grapes in cluster.

[A] GA_3 [B] GA_1

[C] GA_2 [D] None of the above

139. Space filling model describes highly precise interaction of .... with ....

[A] Gibberellins, DNA [B] Auxin, DNA

[C] Cytokinin, RNA [D] None

140. A mixture of GA_4 and GA_7 is now used to enhance seed production in seed orchards of

[A] Pinaceae [B] Legume

[C] Cereals [D] None

141. GA has no effect on ....

[A] Seed germination [B] Root elongation

[C] Internode elongation [D] None

142. Bioassay test of GA is ....

[A] It breaks the genetic dwarfism of plant

[B] Internode elongation

[C] Seed germination [D] None of the above

143. GA induces the formation of ....

[A] m-RNA [B] hn-RNA

[C] S-RNA [D] Both B & C

144. The pineapple which under natural conditions is difficult to blossom has been made to produce fruits throughout the year by the application of

[A] IAA, IBA [B] NAA, 2, 4-D

[C] Phenyl acetic acid [D] Cytokinin

145. .... are sprayed on the cotton fields to defoliate and hence facilitate machine picking of bolls.

[A] Antiauxins [B] Sodium salts of NAA

[C] Naphthalene acetic acid & related auxins

[D] Herbicides

146. Number of gibberellins present in a single plant organ can be

[A] One [B] More than one

[C] Five in each cell [D] Ten in each organ

147. The nature of all gibberellins is

[A] Alkaline [B] Neutral

[C] Acidic [D] Buffer

148. Phenolic inhibitors occur in plants as

[A] Glycosides [B] Terpenoids

[C] Indoles [D] Phospholipids

149. Sometimes phenolic inhibitors are excreted or washed down with rain, as a result, these inhibit the growth of other plants around the plant producing them. The phenomena is known as

[A] Antagonism [B] Antibiosis

[C] Allelopathy [D] Chemobiosis

150. *Gibberellins often increases cell-wall plasticity*. An excellent example of this occurs in internodes of ...., in which growth promotion of young cells derived from the intercalary meristems is unusually dramatic.

[A] Rice [B] Oat

[C] Wheat [D] None of the above

151. In grass seeds (including barley), gibberellins are probably synthesized in the .... and perhaps also in other parts of the ....

[A] Scutellum (cotyledon), embryo [B] Stem, root

[C] Leaves, fruit [D] None

152. Most of .... and some of .... respond by growing faster when treated with gibberellins.

[A] Dicots, monocots [B] Monocots, dicots
[C] Both [D] None

153. The credit for discovery of kinetin goes to

[A] Miller *et al.* [B] Letham *et al.*
[C] Zachau *et al.* [D] None

154. Chemical nature of kinetin is

[A] 6-furfuryl amino purine [B] 6-furfurylamino pyrimidine
[C] 5-furfurylamino purine [D] None

155. Which of the following does not show kinetin like activity in plants?

[A] 6-furfuryl amino purine [B] Zeatin
[C] Both [D] None of the above

156. .... is the most widely accepted term for those compounds that exhibit kinetin-like plant growth-regulating properties

[A] Cytokinin [B] Kinetin
[C] Zeatin [D] None of the above

157. Glucose is found to be conjugated to the 7th carbon position (N-atom) of Zeatin forming a glucosylzeatin named ....

[A] Raphanatin [B] Glucosylzeatin
[C] Both [D] None

158. During polar auxin transport in plant, PIN protein acts as

[A] Auxin influx carriers [B] Auxin efflux carriers
[C] Both [D] None

159. According to chemo-osmotic model of polar auxin transport, AUX1 carrier proteins are located on the plasma membrane

[A] Basally [B] Apically
[C] All around [D] None of the above

160. Which of the following is most active natural cytokinin in plants?

[A] *Trans-* Zeatin [B] *Cis-* Zeatin
[C] Kinetin [D] All of the above

161. Which of the following is not a natural cytokinin?

[A] Dihydrozeatin [B] Benzyl adenine
[C] Isopentenyl adenine [D] None of these

162. Biologically inactive bound form of cytokinin is known as

[A] Ribosides or ribotides [B] Glucosides
[C] Both [D] None of the above

163. If you want to convert etioplast into chloroplast, then its conversion can be increased by prior treatment with

[A] Cytokinin [B] Ethylene
[C] Auxin [D] ABA

164. [9R-5'-P] iP is precursor for synthesis of

[A] Zeatin [B] iP

[C] All natural cytokinin [D] None of the above

165. Which of the following is the cytokinin receptor in the plants?

[A] CRE1 [B] AHK2

[C] AHK 3 [D] All of these

166. Which of the following is not involved in the cytokinin signaling?

[A] AHP Protein [B] Type A ARR

[C] Type B ARR [D] None of the above

167. A number of bioassays have been devised for cytokinins which are based on their specific physiological activities. They are

[A] Cell division test [B] Chlorophyll preservation tests,

[C] Cell enlargement tests, [D] Germination tests

[E] Differentiation test [F] All of the above

168. A number of bioassays are known for gibberellins such as

[A] Pea test [B] Dwarf corn test

[C] Lettuce hypocotyls test [D] Barley endosperm test

[E] All of the above

169. Barley Endosperm Test is very specific for .... and it will detect a minimum of .... nanogram of gibberellins.

[A] GA_3, 0.2 [B] GA_1, 0.9

[C] GA_7, 0.8 [D] None of the above

170. A mixture of the .... (the preparation is known as promalin) is very effective in stimulating apple fruits to increase in size, especially in red delicious-type apples.

[A] Cytokinin benzyladenine and GA_4+GA_7

[B] Cytokinin benzyladenine and GA_9+GA_7

[C] Cytokinin benzyladenine and GA_1+GA_7

[D] None of the above

171. A mixture of .... is used to enhance seed production in conifers.

[A] GA_4+GA_7 [B] GA_1+GA_7

[C] GA_2+GA_7 [D] GA_3+GA_7

172. The inhibitory effect of anti-gibberellins can be counteracted by application of ....

[A] GA promoter [B] GA

[C] Both [D] None of the above

173. Cycocel (CCC) finds extensive use in floriculture industry to retard the growth of ornamental plants such as .... and ...., thereby increasing their cut flower quality

[A] *Chrysanthemum*, poinsettias [B] Marigold, *Chinodopodium album*

[C] Rose, carnation [D] None of the above

174. The conversion of ethephon into ethylene is ....

[A] Enzymatic process [B] Non enzymatic process

[C] Both [D] None

175. Breakdown of ethephon into ethylene is a very .... process.

[A] Slow [B] Fast

[C] Very fast [D] None of the above

176. Most commonly used hormone to hasten the fruit ripening in tomato and apple and degreening of citrus fruits is

[A] Auxin [B] Ethylene

[C] Cytokinin [D] Gibberellins

177. .... is an active or more active than .... in inhibiting virus increases and local lesions in various plant materials

[A] 6-benzyladenin, kinetin [B] Kinetin, 6-benzyladenin

[C] Both [D] None of the above

178. Green islands resulting from the clusters of the chlorophyll and starch-rich cells in chlorotic and necrotic leaf region are characteristic of plant disease induced by certain fungi and viruses. These green islands are maintained by .... that are either synthesized by the host cells and invading tissues

[A] Cytokinin [B] Auxin

[C] GA [D] ABA

179. Which of the following scientist introduced the concept of threshold concentration of ethylene which must accumulate in the tissue in order to initiate the ripening response?

[A] Kidd & West [B] Medawar

[C] Leopold [D] None of the above

180. Among the choices given below, which of the following is a cytokinin receptor in plants?

[A] CRE1 [B] AHK2

[C] AHK3 [D] All of the above

181. Which of the following is involved in cytokinin signaling?

[A] AHP protein [B] Type A ARR

[C] Type B ARR [D] All of the above

182. For degradation of unwanted protein in cells, Ubiquitination pathway operates in

[A] Prokaryotic cells [B] Eukaryotic cells

[C] Plants [D] Animals

183. The term cytokinin was coined by

[A] Letham [B] Miller

[C] Skoog [D] All of the above

184. The first naturally occurring cytokinin to be chemically identified was from .... in 1963, and was called ....

[A] Young Maize grain, zeatin [B] Young wheat grain, florigen
[C] Young oat grain, auxin [D] All of the above

185. .... delayed the senescence of the plant organs by controlling protein synthesis and metabolism of resources

[A] Auxin [B] Cytokinin
[C] Gibberellins [D] None of the above

186. Cytokinin interacts .... with nucleic acid

[A] Direct [B] Indirect
[C] Very less [D] None

187. Apical dominance appears to be controlled by a balance between endogenous concentration of .... and ....

[A] Cytokinin, auxin [B] Ethylene, auxin
[C] Gibberellins, cytokinin [D] None

188. Which of the following carotenoid in citrus peel is responsible for orange and red colour of oranges?

[A] -citraurin [B] - carotene
[C] - lycopene [D] None of the above

189. Traumatic acid is also known as

[A] Wound hormone [B] Stress hormone
[C] Anti senescence hormone [D] None of the above

190. Wound hormone was postulated by

[A] Haberlandt [B] Went
[C] Both [D] None of the above

191. 1-Methyl cyclopropene (MCP), a volatile synthetic compound, popularly used in post harvest agricultural practices is

[A] Antiauxin [B] Antigibberellins
[C] Antiethylene [D] None of the above

192. AVG, a potent .... of ethylene biosynthesis can be used to .... the .... and ....

[A] Inhibitor, retard, fruit ripening, flower fading
[B] Promoter, enhanced, fruit ripening, flower fading
[C] Both
[D] None of the above

193. Zeatin is considered as more active than any other cytokinins probably because of the presence of a highly reactive

[A] Allylic-OH group in its side chain
[B] Methyl group in its side chain
[C] Carboxylic group in its side chain
[D] None of the above

194. Because of the presence of the .... bound in its side chain, zeatin can exist either in trans or cis form

[A] Double bond [B] Single bond

[C] No [D] None of the above

195. In zeatin and other natural cytokinins, ribose sugar is attached to N at .... position of the .... ring.

[A] 9th, purine [B] 7th, pyrimidine

[C] 5th, purine [D] None of the above

196. In conjugated zeatin or dihydrozeatine, the glucose molecule is attached to N at ...., ... or .... position of the purine ring.

[A] 3, 7, 9 [B] 2, 5, 9

[C] 5,7, 9 [D] None of the above

197. 3-Methyl-7-(3-methylbutylamino) pyrazolo [4, 3-D] pyrimidine is an example of

[A] Cytokinin antagonists [B] Gibberellins antagonists

[C] Auxin antagonists [D] None of the above

198. Thidizuron that is used commercially as defoliant and a herbicide is

[A] Synthetic cytokinin [B] Natural cytokinin

[C] Mixed cytokinin [D] None of the above

199. A synthetic chemical compound which shows cytokinin activity but has not been isolated from plants is known as ....

[A] Benzyl adenine [B] Zeatin

[C] Both [D] None of the above

200. Gibberellins increases cell wall extensibility ....

[A] Without acidification [B] With acidification

[C] Both [D] None of the above

201. The minimum lag time for GA- induced growth is far .... auxins.

[A] Lower than [B] Greater than

[C] Both [D] None of the above

202. Which of the following hormone promotes cell expansion in cotyledons?

[A] Auxin [B] Gibberellins

[C] Cytokinins [D] None of the above

203. In pea stem there is induction of interfascicular cambium by the application of .... and this was discovered by ....

[A] Auxin, Sorokin [B] Cytokinin, Sorokin

[C] Gibberellin, Sorokin [D] None of the above

204. Seeds of *Striga asiatica* which requires the presence of their host for germination can also be induced to germinate by treating them with .... even in the absence of their host

[A] Auxin [B] Gibberellin

[C] Kinetin [D] None of the above

205. Effects of kinetin on delaying senescence is called as ....

[A] Richmond-lang effect [B] Mothers effects
[C] Osborne effects [D] None of the above

206. Reduction of the double bond in isopentenyl side chain of zeatin would give rise to ....

[A] Dihydrozeatine [B] Ribosyl-zeatin
[C] Free zeatin [D] None of the above

207. The amino acid ....can be conjugated with cytokinin through N at 9th position of purine ring forming ...., this conjugation is very stable and generally irreversible

[A] Alanine, lupinic acid [B] Glutamic acid, lupinic acid
[C] Both [D] None of the above

208. .... of cytokinin through side chain –OH group is also very common method of inactivation of cytokinin in plants

[A] Glucosylation [B] Gluconeogenesis
[D] Both [D] None of the above

209. Which of the following enzyme cleaves the isopentenyl side chain of cytokinins?

[A] Cytokinin oxidase [B] Cytokinin peroxidase
[C] Both [D] None of the above

210. The cytokinins are transported in their bound form as .... in plant.

[A] Ribonucleotides [B] Ribosides
[C] Both [D] None of the above

211. Which of the following is the cytokinin receptor?

[A] CRE1 [B] AHK2
[C] AHK3 [D] All of the above

212. Cytokinin promotes flowering in some short-day plants like .... and ....

[A] *Lemna, Wolffia* [B] Rice, wheat
[C] Maize, barley [D] None of the above

213. Isopentenyl adenine is found as constituents of ....

[A] t-RNA [B] m-RNA
[C] r-RNA [D] None of the above

214. Cytokinin synthesis occurs in ....

[A] Roots [B] Endosperm of seeds
[C] Both [D] None of the above

215. Isopentenyl adenine differs from zeatin in that a .... group is present at the third carbon position in place of the hydroxyl.

[A] Methyl [B] Ethyl
[C] Methoxy [D] None of the above

216. The most definitive information about cytokinin synthesis concerns the formation of ....

[A] Adenine [B] Thiamine

[C] Both [D] None of the above

217. The ring skeleton of adenine is derived from several small molecules. The nitrogen at the first carbon position is provided from ...., the second carbon position from ...., the nitrogen at the third position from ...., carbon at the fourth and fifth position and the nitrogen at the seventh position are formed from an intact .... molecule. Te sixth carbon position is derived from .... The eight carbon position is derived from ...., and the ninth nitrogen position comes from .... The amino group, at the sixth carbon position is derived from the ....

[A] Amino N of aspartate, formyl H_4 folate, glutamate amide N, glycine, CO_2, methylidyne H_4 folate, glutamine amide N, amino N of aspartate

[B] Formyl H_4 folate, glutamate amide N, glycine, CO_2, methylidyne H_4 folate, glutamine amide N, amino N of aspartate, amino N of aspartate

[C] Glutamate amide N, glycine, CO_2, methylidyne H_4 folate, glutamine amide N, amino N of aspartate, amino N of aspartate, formyl H_4 folate

[D] None of the above

218. The .... inhibition of succinate oxidation by the mitochondria in the presence of antimycin A is similar to that caused by salicylhydroxamic acid.

[A] Auxin [B] Cytokinin

[C] Both [D] None

219. Environmentally limited rosette plants such as spinach (*Spinacea oleraceae*) and cabbage (*Brassica* sp.) generally do not flower in the .... form

[A] Elongated form [B] Rosette form

[C] Both [D] None of the above

220. Bolting is normally triggered by an environmental signal, either photoperiod as in .... or a combination of low temperature and photoperiod as in ....

[A] Spinach, cabbage [B] Cabbage, spinach

[C] Both [D] None of the above

221. .... contains six gibberellins, including GA_{19} and GA_{20}. GA_{20} will cause bolting in short day condition while GA_{19} are biologically inactive.

[A] Spinach [B] Cabbage

[C] Both [D] None of the above

222. The relationship between dwarfing or internode-length genes and gibberellins was pioneered by the work of .... on maize and .... on the garden pea.

[A] B.O. Phinney, P.W. Brian [B] P.W. Brian, B.O. Phinney

[C] Both of the above [D] None of the above

223. In maize, more than .... mutants that influence plant height have been described.

[A] 10 [B] 20

[C] 30 [D] None of the above

224. There are certain structural requirements for the GA activity. A carboxyl group at carbon- .... is a feature of all GAs and is required for biological activity.

[A] 8 [B] 7

[C] 9 [D] 10

225. .... are a functionally and chemically diverse group of molecule. They are probably the most diverse and largest class of organic compounds in the plants.

[A] Terpenes [B] GA_{20} groups

[C] Alkaloids [D] None of the above

226. ...., states the product of a photochemical reaction is determined by the total amount of energy presented, regardless of the fluence rate or presentation time.

[A] Bunsen-Roscoe reciprocity law [B] First positive curvature law

[C] Second negative curvature law [D] None of the above

227. There are three principle sites for gibberellins synthesis

[A] Developing seeds and fruits

[B] The young leaves of developing apical buds and elongating shoots, and

[C] The apical regions of roots [D] All of the above

228. The first higher-plant gibberellin to be characterized was isolated from immature seeds of .... and found to be identical with gibberellin A_1.

[A] Renner bean (*Phaseolus coccineus*) [B] *Pisum sativum*

[C] *Marabmacrocarpus* [D] All of the above

229. IPP is a product of two different biosynthetic pathways, one located in the .... and another located in ....

[A] Mitochondria, nucleus [B] Cytoplasm, chloroplast

[C] Ribosome. Chloroplast [D] None of the above

230. In general, sesquiterpenes, triterpenes and polyterpenes such as latex and other large molecules are produced in the .... while isoprene, monoterpenes, diterpens and tetraterpens originate in ....

[A] Cytosolic compartment, plastids

[B] Plastids, cytosolic compartment

[C] Both

[D] None of the above

231. The main effect of growth retardants is to reduce

[A] Stem elongation [B] Plants are shorter & more compact

[C] Darker green foliage [D] All of the above

232. The first naturally occurring cytokinins were isolated from milky endosperm of .... and developing fruits.

[A] Coconut [B] Maize

[C] Plum fruits [D] More than one

233. Witch's broom on white pine (*Pinus strobes*) is the result of a fungal infection that stimulates an over production of ....

[A] Gibberellin [B] Auxin

[C] Cytokinin [D] None of the above

234. The *Agrobacterium* TMR gene encodes for the enzyme ...., which catalyzes the rate limiting step in cytokinin biosynthesis.

[A] Isopentenyl transferase [B] Isopentenyl epimerase

[C] Both [D] None of the above

235.

The above drawn figure is the schematic representation of which of the cytokinin receptors?

[A] Monomer of CRE1 [B] Dimer of CRE2

[C] Monomer of CRE3 [D] Dimer of CRE4

236. Abscisic acid is degraded to phaseic acid by ....

[A] Oxidation [B] Reduction

[C] Both [D] None of the above

237. Which of the following is the inhibitor of the ABA synthesis?

[A] Fluridone [B] Chlordone

[C] Bromidone [D] None

238. Precocious germination will occur when the ABA concentration is reduced to .... per g fresh weight of seed, a level that is not normally reached until the late stages of seed maturation.

[A] 3 to 4 μg [B] 9 to 10 μg

[C] 1 to 2 μg [D] None of the above

239. .... is the powerful inhibitor of stomatal opening.

[A] ABA [B] Cytokinin

[C] Gibberellins [D] None of the above

240. Two tomato mutants, known as .... and ...., fail to accumulate normal levels of ABA and both wilt very readily.

[A] Flacca, sitiens [B] Commelina, communis

[C] Both [D] None of the above

241. ABA biosynthesis begins in ....

[A] Cytoplasm [B] Chloroplast

[C] Mitochondria [D] All of the above

242. The critical enzyme NCED (nine-cis-expoycarotenoid dioxygenase), is known to be targeted in the ....

[A] Chloroplast [B] Mitochondria

[C] Nucleus [D] All of the above

243. At low pH, ABA exists in the .... which freely permeates most cell membranes.

[A] Protonated from [ABAH] [B] Dissociated form [ABA-]

[C] Both [D] None of the above

244. ABA is highly mobile and moves quickly out of the leaves to other parts of the plant, especially sink tissues. For example, radioactively labeled ABA applied to .... leaves can be detected in the roots within .... minutes.

[A] Soyabean, 15 [B] Mustered, 20

[C] Onion, 25 [D] None of the above

245. Vivipary can be chemically induced in maize by treatment of the developing ear at the appropriate time with ....

[A] Fluridone [B] Chlordane

[C] Bromodone [D] None of the above

246. ABA perception and signaling appears to be very complex. A more recent approach has made use of the antigen-antibody reactions with what is called ....

[A] Anti-idiotypic antibodies [B] Anti-idiotypic antigens

[C] Both [D] None of the above

247. Anti-idiotypic antibodies approach led to the identification of ABAP1, a protein that is located in the plasma membrane of the barley aleurone cells and that specifically and reversibly binds ABA *in vitro*.

[A] ABAP1 [B] ABAP2

[C] ABAP3 [D] ABAP4 [E] None of the above

248. ABA is a .... acid

[A] Strong [B] Week

[C] Both [D] None of the above

249. The principle receptor for brassinosteroides requires the interaction of two proteins that form a plasma membrane-associated heterodimer. The first is a serine/threonine kinase known as .... The second is the protein ....

[A] BRASSINOSTEROID INSENSITIVE 1(BRI1), BRI1-ASSOCIATED RECEPTOR KINASE(BAK1).

[B] BZR1, BZR2

[C] BIN2, BIN1

[D] MORE THAN ONE

250. Brassinosteroids are .... plant sterols.

[A] Monohydroxylated [B] Polyhydroxylated

[C] Dihydroxylated [D] None of the above

251. The precursor of the brassinolide is

[A] Campesterol [C28] [B] Campesterol [C29]

[C] Campesterol [C30] [D] Campesterol [C31]

252. ACC (1-aminocyclopropane-1-carboxylic acid) is a

[A] Non protein amino acid [B] Protein amino acid

[C] Both [D] None

253. In order to sustain normal rates of ethylene production, the sulfur released during ethylene formation must be recycled back to methionine. This is accomplished by what is commonly referred to as the methionine cycle. This cycle is also known as the

[A] Yang cycle [B] Cousins cycle

[C] Neljubow cycle [D] None of the above

254. Those whose business involves the shipping and storing of the fruit have long aware that ripe and rotting fruit could accelerate the ripening of the other fruit storéd nearby. One of the earliest reports that these effects were due to a volatile substance given off by plant tissue was published in 1910 by .... in an annual report of the Jamaican Department of Agriculture.

[A] H.H. Cousins [B] D.N. Neljubow

[C] R. Gane [D] None of the above

255. Who found that abnormal growth of the dark-grown pea seedlings could be traced to ethylene emanating from illuminating gas?

[A] H.H. Cousins [B] D.N. Neljubow

[C] R. Gane [D] None of the above

256. ABA occurs in plants predominantly in

[A] Stems [B] Mature green leaves

[C] Roots [D] Flowers

257. ABA is inactivated in plant cells by converting into

[A] Phaseic acid [B] Dihydrophaseic acid
[C] ABA-glucosyl ester [D] All of the these

258. ABA is a

[A] Sesquiterpene [B] Diterpene
[C] Triterpene [D] Tetraterpene

259. Ethylene acceptors are located on

[A] Plasmamembrane [B] Tonoplast
[C] Endoplasmic reticulum [D] All of the above

260. A potent inhibitor of ethylene action is

[A] Silver ions [B] Cobalt ions
[C] Manganese ions [D] All of the above

261. Which of the following is an inhibitor of ethylene biosynthesis?

[A] AOA [B] AVG
[C] CO^{2+} [D] All of the above

262. Immediate precursor of ethylene biosynthesis in plant is

[A] SAM [B] ACC
[C] Methionine [D] All of the above

263. Among the followings, which technique is used for detection and estimation of ethylene?

[A] Gas Chromatography technique
[B] Triple response bioassay method
[C] Photoacustic detector
[D] All of the above

264. The popular absorbant of ethylene is

[A] Formalin [B] Potassium permanganate
[C] Potassium hydroxide [D] All of the above

265. In plant cells, ethylene can easily diffuse through plasma membrane because this gaseous hormone has

[A] Hydrophobocity [B] Hydrophilic structure
[C] Amphoteric nature [D] None of the above

266. Jasmonic acid (JA) is structurally similar to

[A] Prostaglandins [B] Sucrose
[C] Prolactine [D] All of the above

267. Both Jasmonic acid and prostaglandins are derived form

[A] Fatty acid [B] Protein
[C] Carbohydrate [D] None of the above

268. .... and structurally related compound, *cis*-jasmone, are well known in the perfume industry as fragrant components of the essential oils of jasmine.

[A] Methyl jasmonate [B] Ethyl jasmonate
[C] Both [D] None of the above

269. Salicylic acid (SA) is far better known for its .... properties.

[A] Medicinal properties [B] Root initiation
[C] Shoot initiation [D] None of the above

270. The *Pseudomonas putida* .... gene product catalyzes the metabolism of SA to catechol and thereby preventing accumulation of SA.

[A] Nah G [B] Nah H
[C] Nah U [D] Nah I

271. The rate limiting step in the SA biosynthesis is the conversion of .... to ....

[A] Trans-cinnamic acid to benzoic acid
[B] Benzoic acid to trans-cinnamic acid
[C] Both
[D] None of the above

272. Phenylalanine is converted into trans-Cinnamic acid in the presence of enzyme

[A] Phenylalanine ammonia lyase
[B] Benzoic acid-2-hydroxylase
[C] Both
[D] None of the above

273. Benzoic acid is converted into Salicylic acid in the presence of enzyme

[A] Phenylalanine ammonia lyase
[B] Benzoic acid-2-hydroxylase
[C] Both
[D] None of the above

274. Which of the following is the important inhibitors of the conversion of S-adenosylmethionine (SAM) to 1-aminocyclopropane-1-carboxylic acid (ACC) in ethylene biosynthesis pathways?

[A] Aminoethoxyvinylglycine (AVG) [B] Aminooxyacetic acid (AOA)
[C] Both [D] None of the above

275. Jasmonic acid is formed biosynthetically from free ....

[A] Linolenic acid [B] Batasin
[C] Violaxanthin [D] None of the above

276. Batasins are compounds in yam plants that seem to cause dormancy in .... vegetative reproductive structure that arises from swelling of the aerial lateral buds.

[A] Bulbils [B] Gemmae
[C] Both [D] None of the above

277. The importance of ABA as a stress hormone was first suggested by .... and ....

[A] S.T.C. Wright & R.W.P. Hiron [B] A.B. Bleecker & M.A. Estelle
[C] Both [D] None of the above

278. ABA is often detected by .... because it is probably one of the most optically active natural products known to humans

[A] Optical rotatory dispersion techniques

[B] Gas chromatographic analysis

[C] Both

[D] None of the above

279. The external coincidence model explains day length measurement in .... plants.

[A] Long-day [B] Short-day

[C] Long-short day [D] Short-long day

5

PHOTOPERIODISM

1. **The phenomenon of photoperiodism was first discovered by**
   [A] Garner and Allard [B] M. Chailakhyan
   [C] Both [D] None of the above

2. **Photoperiodism term was coined by**
   [A] Garner and Allard [B] M. Chailakhyan
   [C] H. Klebs [D] None of the above

3. **W.W. Garner and H.A. Allard were scientists with the U.S. Department of Agriculture near Washington, D.C. The initial focus of their work was a mutant cultivar of**
   [A] *Nicotiana tabacum* [B] *Glycine max*
   [C] Both [D] None of the above

4. **.................... plants are those that flower only, or flower earlier, in response to day lengths that are shorter than a certain value within a 24-hour cycle.**
   [A] Short-day [B] Long-day
   [C] Day-neutral [D] None of the above

5. **Plants that have an absolute requirement for a particular photoperiod before they flower are considered .... photoperiodic types.**
   [A] Obligate [B] Facultative
   [C] Both [D] None of the above

6. **Plants that have no absolute requirement for a particular photoperiod before they flower are considered .... photoperiodic types.**
   [A] Obligate [B] Facultative
   [C] Both [D] None of the above

7. ***Xanthium strumarium* is an .... short-day plants.**
   [A] Obligate [B] Facultative
   [C] Both [D] None of the above

8. ***Bryophyllum* is an example of**
   [A] Long-short-day plants [B] Short-long-day plants
   [C] Both [D] None of the above

9. *Trifolium repens* (White clover) is an example of

[A] Long-short-day plants [B] Short-long-day plants

[C] Both [D] None of the above

10. *Madia elegans* (tarweed) is an example of

[A] Intermediate-daylength plants

[B] Amphophotoperiodism

[C] Long-short-day plants

[D] Short-long-day plants

11. *Cucumis sativus, Gomphrena globosa, Helianthus annuus, Phaseolus vulgarus, Pisum sativum, Zea mays* are the examples of

[A] Day-Neutral Plants

[B] Long-short-day plants

[C] Short-long-day plants

[D] None of the above

12. *Chenopodium rubrum, Chrysanthemum sp., Cosmos sulphureus, Euphorbia pulcherrima, Glycine max, Nicotiana tabacum, Perilla crispa, Pharbitis nil, Xanthium strumarioum* are the examples of ....

[A] Short-Day Plants [B] Long-Day Plants

[C] Both [D] None of the above

13. *Anethum graveolens, Beta vulgaris, Hyoscyamus niger, Lolium* sp., *Raphanus sativus, Secale cereal, Sinapis alba, Spinacea oleracea, Triticum aestivum are* the best examples of

[A] Long-Day plants [B] Short-Day plants

[C] Both [D] None of the above

14. In which of the followings, flowering is delayed under intermediate daylength (12 to 14 hours) but occurs rapidly under day length of 8 hours or 18 hours?

[A] Amphophotoperiodism [B] Intermediate- daylength plants

[C] Both [D] None of the above

15. Dark period is critical for

[A] Long-day plants [B] Short day plants

[C] Amphophotoperiodism [D] Non of the above

16. The inhibitory effect of red light on flowering during critical dark period in short day plants can be overcome by

[A] Far-red light [B] Blue light

[C] Infra-red rays [D] Ultraviolet rays

17. In which case, flowering will be promoted in long day plants?

[A] Increasing the number of inductive cycles

[B] Continues inductive cycles

[C] Prolongation of light period

[D] All of the above

18. In plants, photoperiodic induction is perceived by

[A] Stems [B] Leaves
[C] Roots [D] None of the above

19. The hormone produced as a result of photoperiodic induction is known as

[A] Cytokinin [B] Vernalin
[C] Polyamines [D] None of the above

20. Which of the followings is supposed to be the precursor of Florigen?

[A] Auxin [B] Gibberellins
[C] Cytokinin [D] None of the above

21. Phytochrome is a soluble protein which is a

[A] Heterodimer [B] Homodimer
[C] Both [D] None of the above

22. In plants, richest source of phytochrome are

[A] Seedling grown in light [B] Etiolated seedlings
[C] Both [D] None of the above

23. Which of the following phytochrome predominates in etiolated seedlings?

[A] Type II [B] Type I
[C] Both [D] None of the above

24. "Florigen" is translocated from the leaf to the apical tips (or meristems) via .... in photoinduced plants

[A] Xylem [B] Phloem
[C] Both [D] None of the above

25. Successful purification of intact native phytochrome (from etiolated Oat seedlings) was first reported by

[A] Vierstra and Quail [B] Brian
[C] Both [D] None of the above

26. *Arabidopsis* is a .... plant

[A] Long day plant [B] Short day plant
[C] Long-short day plant [D] None of the above

27. Which of the following gene has the prominent role in flowering?

[A] CONSTANS (CO) [B] COSNTANS (CS)
[C] Both [D] None of the above

28. A likely candidate for the mobile floral signal is ...., a small protein (ca. 20 KDa) that is known to promote flowering in a dose dependent manner.

[A] FLOWERING LOCUS T (FT)
[B] FLOWERING LOCUS F (FT)
[C] FLOWERING LOCUS G (GT)
[D] NONE OF THE ABOVE

29. Only a few plants respond to photoperiod when they are small seedlings. .... responds to SDs in the cotyledonary stage, and some species of .... respond and flower as minute seedlings.

[A] *Pharbities, Chenopodium* spp.
[B] *Anagallis arvensis, Corchorus olitorius*
[C] Both
[D] None of the above

30. The leaves of the *Anagallis arvensis* are most sensitive to .... when the plant is a seedling; sensitive actually declines in leaves that are most recently produced, so the plant becomes more difficult to induce as it gets older.

[A] LDs [B] SDs
[C] Both [D] None of the above

31. One of the famous scientists, Klebs called the condition a plant must achieve before it will flower in response to the environmental bluhreife, which translates as ....

[A] Ripeness to flower [B] Ripeness to respond
[C] Both [D] None of the above

32. The meristematic change is called

[A] Evocation [B] Competent
[C] Both [D] None of the above

33. If evocation occurs, the meristems are said to be

[A] Evocation [B] Competent
[C] Both [D] None of the above

34. The plant produces .... from seedling to earliest flower under the most ideal conditions for flowering.

[A] Minimum number of leaves
[B] Minimum number of formation
[C] Minimum number of vegetative stage
[D] None of the above

35. *Arabidopsis* has .... phytochrome genes.

[A] One [B] Two
[C] Three [D] Four
[E] Five

36. *Cattleya trianae, Chrysanthemum morifolium, Cosmos sulphureus, Glycine max, Kalanchoe biossfeldiana, Perilla crispa, Zea mays* are the best examples of

[A] Qualitative short day plant [B] Qualitative long day plant
[C] Quantitative short day plant [D] None of the above

37. Which of the following plant is the best examples of SDP at high temperature and day-neutral at low temperature?

[A] *Pharbtis nil* [B] *Nicotiana tabacum*
[C] Both [D] None of the above

38. Which of the following plant is the best examples of SDP at low temperature and day-neutral at low temperature?

[A] *Cosmos sulphureus* [B] *Nicotiana tabacum*

[C] Both [D] None of the above

39. Which of the following plant is the best examples of SDP at low temperature and LDP at low temperature?

[A] *Euphorbia pulcherrima* [B] *Ipomea purpurea*

[C] Both [D] None of the above

40. Which of the following plant is the best examples of LDP at low temperature and day-neutral at low temperature?

[A] *Delphinium cultorum* [B] *Rudbeckia bicolor*

[C] Both [D] None of the above

41. Which of the following plant is the best examples of LDP at high temperature and day-neutral at low temperature?

[A] *Cichorium intybus* [B] *Delphinium cultorum*

[C] *Rudbeckia bicolor* [D] None of the above

42. Which of the following plant is the best examples of LDP at low temperature and quantitative LDP at high temperature?

[A] *Beta vulgans* [B] *Cichorium intybus*

[C] *Rudbeckia bicolor* [D] None of the above

43.

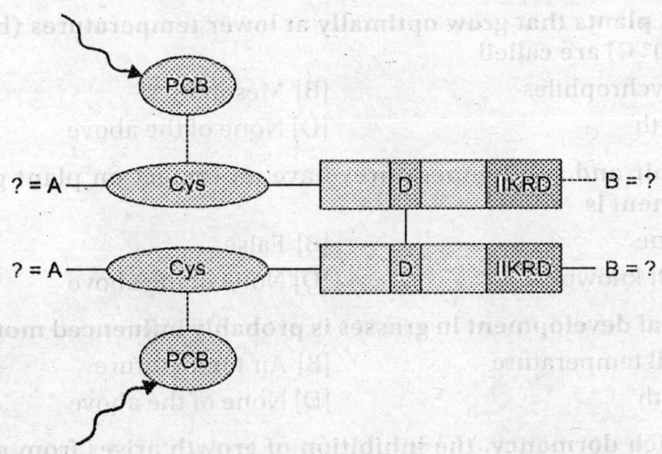

This block diagram of two phytochrome molecule (a dimer) shows the principle domains and subdomains. The A terminal photosensory domain binds the phytochromobilin chromophore (PCB) and is linked by a hinge region to the B terminal regulatory domain.

Find the missings i.e. A and B

[A] A= N- terminal, B= C-terminal [B] A= C- terminal, B= N-terminal

[C] Both [D] None of the above

6 TEMPERATURE: PLANT DEVELOPMENT AND DISTRIBUTION

1. **Plants are the machine of chemical metabolism, and one universal characteristic of chemical metabolism machines is their sensitivity to ....**
   [A] Temperature [B] Light
   [C] Water [D] All of the above
2. **Green plants probably first evolved in the .... region, not so much because of warmer temperatures, but because the temperatures there were relatively stable.**
   [A] Tropical [B] Temperate
   [C] Both [D] None of the above
3. **Those plants that grow optimally at lower temperatures (between 0° C and 10° C) are called**
   [A] Psychrophiles [B] Mesophiles
   [C] Both [D] None of the above
4. **Both air and soil temperatures have an impact on plant growth. This statement is**
   [A] True [B] False
   [C] Not known [D] None of the above
5. **The leaf development in grasses is probably influenced more by**
   [A] Soil temperature [B] Air temperature
   [C] Both [D] None of the above
6. **In which dormancy, the inhibition of growth arises from another part of the plant (e.g. apical dominance)?**
   [A] Paradormancy [B] Ecodormancy
   [C] Endodormancy [D] All of the above
7. **In which dormancy, the growth is imposed by limitation in environment?**
   [A] Paradormancy [B] Ecodormancy
   [C] Endodormancy [D] All of the above

8. In which dormancy, the dormancy is an inherent property of the dormant structure itself?

[A] Paradormancy [B] Ecodormancy

[C] Endodormancy [D] All of the above

9. Bud dormancy is an example of ....

[A] Paradormancy [B] Ecodormancy

[C] Endodormancy [D] All of the above

10. The onset of dormancy in buds is a typical .... response

[A] Short-day [B] Long-day

[C] Day-neutral [D] None of the above

11. Temperature near or just above freezing appear to be .... effective at breaking dormancy.

[A] More [B] Less

[C] Not [D] All of the above

12. In poplar species, ... gene has been implicated in bud dormancy.

[A] PtFLC [B] PtFCL

[C] PtLCF [D] All of the above

13. Many seeds have pores that are blocked with a plug called the ...., which must be mechanically dislodged before water and oxygen can enter.

[A] Strophiolar plug [B] Scarification plug

[C] Hilum [D] None of the above

14. A cocklebur contains two seeds: an upper .... seed and a lower .... seed.

[A] Dormant, non dormant [B] Non dormant, dormant

[C] Both of the above [D] None of the above

15. The effects of temperature on physiology and metabolism in turn influence plant distribution, it is called ....

[A] Biogeography [B] Biodistribution

[C] Biometry [D] None of the above

16. The decrease in temperature with elevation is called the ....

[A] Adiabatic lapse [B] Isothermal lapse

[C] Both [D] None of the above

17. The adiabatic lapse rate for dry air is constant at about

[A] 1° C/100 m elevation [B] 1° C/1000 m elevation

[C] 1° C/10 m elevation [D] 1° C/1 m elevation

[E] None of the above

18. .... has studied the distribution of C3 and C4 grasses along an elevational gradient in Hawaii.

[A] P.W. Rundel [B] M. Black

[C] Both [D] None of the above

19. *Tidestromia oblongifolia*, is a deciduous .... perennial native to Death Valley.

[A] C_3 [B] C_4

[C] C_2 [D] None of the above

20. Floral movements in members of Liliaceae such as *Tulip* and *Crocus* are the best examples of

[A] Effects of temperature differentials

[B] Slightly Effects of photoperiods

[C] Both

[D] None of the above

7 SECONDARY METABOLITES

1. **The total of all of the chemical reactions that takes place in an organism is called ....**
   [A] Catabolism [B] Metabolism
   [C] Anabolism [D] None of the above

2. **Lipids, proteins, nucleic acids and carbohydrates are called**
   [A] Primary metabolites [B] Secondary metabolites
   [C] Both [D] None of the above

3. **The unifying feature of Terpenes is that they are generally .... polymers.**
   [A] Lipophilic [B] Lipophobic
   [C] Both [D] None of the above

4. **Steroids with an alcohol group which is the case with practically all plant steroids are known as ....**
   [A] Phytoecdysones [B] Sterols
   [C] Gutta [D] None of the above

5. **Essential oils are produced in the .... and accumulated in a cavity that forms between the secretory cells and the overlying cuticle.**
   [A] Secretory cells [B] Epidermal cells
   [C] Mesophyll cells [D] None of the above

6. **The most abundant sterols in higher plants are .... and ...., which contribute up to 70% of the total sterols.**
   [A] Stigmasterol, sitosterol [B] Gutta, rubber
   [C] Latex, rubber [D] All of the above

7. **Steroids and sterols are synthesized from the ....**
   [A] Acyclic triterpense squalene [B] Cyclic triterpense squalene
   [C] Both [D] None of the above

8. **Sterols are .... molecules.**
   [A] Planar [B] Non planar
   [C] Both [D] None of the above

9. Which of the following sterol has protective function in plants?

[A] Ecdysones [B] Phytoecdysones
[C] Both [D] None of the above

10. Rubber is a polymer consisting of up to .... isopentenyl units.

[A] 15000 [B] 10000
[C] 9000 [D] None of the above

11. In the plant rubber occurs as small particles suspended in a milky-white emulsion called ....

[A] Latex [B] Chalcone
[C] Both [D] None of the above

12. Stigmasterol and sitosterol differs only in the presence or absence of a .... bond in ....

[A] Single, stigmasterol [B] Double, sitosterol
[C] Double, stigmasterol [D] Single, sitosterol
[E] None of the above

13. Latex contains about .... rubber and .... water

[A] 30 to 40%, 50% [B] 50%, 30 to 40%
[C] 12%, 20% [D] None of the above

14. The main commercial source of rubber is

[A] *Mentha piperita* [B] *Hevea brasiliensis*
[C] *Albizia* spp. [D] None of the above

15. In most plants, latex is produced in the ...., accumulating in a series of long, interconnected vessels called ....

[A] Phloem, laticifers [B] Xylem, laticifers
[C] Both [D] None of the above

16. The best known source of gutta is a desert shrubs, ..., which grows in the northern Mexico.

[A] *Parthenium argentatum* [B] *Agrimonies mexicana*
[C] Both [D] None of the above

17. Parthenium may contain as much as .... per cent latex by weight, which is not stored in laticifers but in the vacuole of roots and shoots.

[A] 2 [B] 20
[C] 11 [D] None of the above

18. Which of the following terpene is known generally as VOC (volatile organic carbon)?

[A] Monoterpens [B] Hemiterpenes
[C] Sesquiterpenes [D] All of the above

19. The sugar most commonly found in glycosides is ....

[A] Glucose [B] Sucrose
[C] Lactose [D] None of the above

20. Three interesting glycosides are the

[A] Saponins [B] Cardiac glycosides

[C] Cyanogenic glycosides [D] All of the above

21. Saponins may also occur as aglycones, which are known as

[A] Sapogenins [B] Saponaria

[C] Both [D] None of the above

22. The principle role of saponins appears to preform defense against attack by

[A] Fungi [B] Bacteria

[C] Virus [D] None of the above

23. Saponins are stored in the form of a biologically inactive molecule in the cells, called

[A] Monodesmosidic saponins

[B] Bisdesmosidic saponins

[C] Both

[D] None of the above

24. Bisdesmosidic saponins have .... sugar chains

[A] One [B] Two

[C] Three [D] None of the above

25. Saponins are .... toxic to fish.

[A] Highly [B] Lesser

[C] No [D] None of the above

26. Commercially, saponins from the bark of .... have been used as surfactants in photographic film, in shampoos, liquid detergents, toothpastes and beverages.

[A] *Quillaja saponaria* [B] *Glyscyrrhiza glabra*

[C] Both [D] None of the above

27. The cardiac glycosides are distinguished from other steroid glycosides by the presence of a .... ring and rare sugars that formed the glycoside.

[A] Lactone ring attached to the carbon 15

[B] Lactone ring attached to the carbon 17

[C] Lactone ring attached to the carbon 14

[D] None of the above

28. ...., a mixture of cardenolides extracted from the seeds, leaves and roots of purple foxglove, *Digitalis purpurea*.

[A] Digitalis [B] Digitoxin

[C] Both [D] None of the above

29. Common sources of cardenolides are the

[A] *Asclepias* [B] *Calotropis*

[C] Both [D] None of the above

30. The milkweed (*Asclepias* and *Calotropis*) are interesting because they are the principle host of ovipositing of

[A] Monarch butterflies [B] Monarch moth
[C] Blue jays [D] None of the above

31. In sorghum, the cyanogenic glycoside .... is synthesized and stored in epidermal cells.

[A] Dhurrin [B] Amygdalin
[C] - glycosidase [D] None of the above

32. A common cyanogenic glycoside is ...., which occurs in many representatives of family Rosaceae.

[A] Dhurrin [B] Amygdalin
[C] - glycosidase [D] None of the above

33. .... are found primarily in the mustard family (Brassicaceae) and related families in the order Capparales.

[A] Dhurrin [B] Amygdalin
[C] Glucosinolates [D] None of the above

34. Hydrolysis of Glucosinolates is catalyzed by an enzyme called ....

[A] Myrosinase [B] - glycosidase
[C] Both [D] None of the above

35. The shikimic acid pathway is common to bacteria, fungi and plants, but is not found in ....

[A] Animals [B] Viruses
[C] Both [D] None of the above

36. Opium, an exudate obtained from the .... seed capsule of the opium poppy (*Papaver somniferum*).

[A] Immature [B] Matured
[C] Both [D] None of the above

37. One of the better known and commonly used natural insect toxins is pyrethrin, a monoterpene ester that is produced in the .... of *Chrysanthemum cineraiifolium*.

[A] Petals of flower [B] Sepals of flower
[C] Ovary of flower [D] None of these

38. A saponin called yamogenin is obtained from .... and is used in making oral contraceptives.

[A] *Dioscorea* [B] *Digitalis lantana*
[C] Both [D] None of the above

39. Phytol side chain of chlorophylls is .... derivative.

[A] Monoterpene [B] Diterpene
[C] Both [D] None of the above

40. Carotenoids are ....

[A] Monoterpene [B] Diterpene
[C] Tetraterpens [D] None of the above

41. Alkaloids are usually optically active being

[A] Laevorotatory [B] Dextrorotatory
[C] Both [D] None of the above

42. In tobacco, nicotine is synthesized in .... but is translocated to and stored in ....

[A] Roots, leaves [B] Leaves, leaves
[C] Both [D] None of the above

43. An alkaloid which is known to inhibit mitotic spindle formation in cells is

[A] Morphine [B] Thebaine
[C] Colchicine [D] Quinine
[E] None of the above

44. An example of pyridine alkaloid is

[A] Nicotine [B] Stachydrine
[C] Coniine [D] None of the above

45. An example of pyrrolidine alkaloid is

[A] Nicotine [B] Stachydrine
[C] Coniine [D] None of the above

46. An example of piperidine alkaloid is

[A] Nicotine [B] Stachydrine
[C] Coniine [D] None of the above

47. An example of tropane alkaloid is

[A] Atropine [B] Quinine
[C] Papaverine [D] Lupinine
[E] Reserpine [F] None of the above

48. An example of quinoline alkaloid is

[A] Atropine [B] Quinine
[C] Papaverine [D] Lupinine
[E] Reserpine [F] None of the above

49. An example of isoquinoline alkaloid is

[A] Atropine [B] Quinine
[C] Papaverine [D] Lupinine
[E] Reserpine [F] None of the above

50. An example of quinolizidine alkaloid is

[A] Atropine [B] Quinine
[C] Papaverine [D] Lupinine
[E] Reserpine [F] None of the above

51. An example of indole alkaloid is

[A] Atropine [B] Quinine
[C] Papaverine [D] Lupinine
[E] Reserpine [F] None of the above

52. Which of the following is not an important monoterpene component of conifer resins?

[A] Menthol [B] Limonene
[C] Both [D] None of the above

53. An example of pyrrolizidine alkaloid is

[A] Heliotridine [B] Pilocarpine
[C] Both [D] None of the above

54. An example of tropolone alkaloid is

[A] Colchicine [B] Pilocarpine
[C] Both [D] None of the above

55. Canavanine, a non protein amino acid closely resembles in its structure to protein amino acid ....

[A] Argentine [B] Proline
[C] Both [D] None of the above

56. Azetidine-2-carboxylic acid, a non protein amino acid closely resembles in its structure to protein amino acid ....

[A] Argentine [B] Proline
[C] Both [D] None of the above

57. Major plant source of heterocyclic amino acid mimosine is

[A] *Mimosa pudica* [B] Corn cockle
[C] *Astragalus* spp. [D] None of the above

58. Major plant source of non protein aromatic amino acid, 2,4-Dihydroxy-6-methyl phenyl alanine is

[A] *Mimosa pudica* [B] Corn cockle
[C] *Astragalus* spp. [D] None of the above

59. Major plant source of non protein selenium containing amino acid, Se-Methyl-l-Cysteine is

[A] *Mimosa pudica* [B] Corn cockle
[C] *Astragalus* spp. [D] None of the above

60. Major plant source of non protein sulphur containing amino acid, S-Methyl-L-cysteine is

[A] *Phaseolus vulgaris* [B] *Crotalaria juncea*
[C] *Lathyrus* sp. [D] None of the above

61. Major plant source of non protein neutral aliphatic amino acid, 5-Hydroxynorleucine is

[A] *Phaseolus vulgaris* [B] *Crotalaria juncea*
[C] *Lathyrus* sp. [D] None of the above

62. Major plant source of non protein basic amino acid, γ-Hydroxyhomoarginine is

[A] *Phaseolus vulgaris* [B] *Crotalaria juncea*
[C] *Lathyrus* sp. [D] None of the above

63. Plant phenolics are biosynthesized in plants through

[A] Malonic acid pathways [B] Mevalonic acid pathways
[C] Shikimic acid pathways [D] None of the above

64. Except ...., all other plant phenolics are biosynthesized in plants from a common biosynthetic intermediate, phenylalanine or its close precursor shikimic acid through shikimic acid pathway.

[A] Flavonoids [B] Lignins
[C] Both [D] None of the above

65. The first step in the synthesis of phenolic compounds from phenylalanine in plants is .... of phenylalanine by the enzyme PAL.

[A] Deamination [B] Reduction
[C] Oxidation [D] None of the above

66. Lignin is highly complex and branched polymer of simple phenolic compounds with .... basic carbon skeleton.

[A] C_6-C_3 [B] C_3-C_6
[C] C_6-C_4 [D] None of the above

67. Which of the following is the phenolic compound?

[A] Flavonoids [B] Coumarins
[C] Lignin [D] All of the above

68. Terpenoid containing alkaloids (sterol alkaloids) occur in plants in combination with

[A] Protein [B] Carbohydrate
[C] Both [D] None of the above

69. Flavones and flavonols differ from anthocyanins in structure mainly in

[A] Ring-A [B] Ring-B
[C] Middle ring [D] None of the above

70. UV-radiations are absorbed by

[A] Flavones [B] Flavonols
[C] Both [D] None of the above

71. Gallic acid is the component of

[A] Hydrolysable tannins [B] Condensed tannins
[C] Both [D] None of the above

72. A widely used insecticide Rotenone is

[A] Sesquiterpene [B] Flavonols
[C] Isoflavonoid [D] None of the above

73. The colour of delphinidin is

[A] Deep red [B] Bright red
[C] Bluish-purple [D] None of the above

74. In anthocyanins, the sugars are mostly attached to

[A] 7th position of ring-A [B] 3rd position of middle ring

[C] 4th position of ring-B [D] None of the above

75. The main function of flavonoid in plants is

[A] Pigmentation [B] Defense

[C] Both [D] None of the above

76. The main accumulation site of flavonoids is

[A] Vacuole [B] Chloroplast

[C] Mitochondria [D] None of the above

77. Flavonoids are accumulated in vacuole although they are synthesized outside the vacuole.

This statement is

[A] True [B] False

[C] Not known [D] All of the above

78. Anthocyanins are found dissolved in cell sap in vacuole as ....

[A] Glycosides [B] Aglycones

[C] Both [D] None of the above

79. Gossypol, which is found in cotton plants and provides resistance to insects, fungal and bacterial pathogens is a

[A] Diterpene [B] Sesquiterpenes dimer

[C] Both [D] None of the above

80. Which of the following steroid has primary function in plants?

[A] Cholesterol [B] Ponasterone

[C] Sitosterol [D] None of the above

81. Digitoxin, a steroid glycoside obtained from *Digitalis*, is prescribed for

[A] Heart ailments [B] Kidney ailments

[C] Liver ailments [D] None of the above

82. Among the followings, which one is not a cyanogenic glycoside?

[A] Solanine [B] Lautostralin

[C] Amygdalin [D] None of the above

83. Which of the following alkaloid does not contain nitrogen in heterocyclic ring?

[A] Quinine [B] Morphine

[C] Ephedrine [D] Narcotine

84. Which of the following is not produced in the opium poppy?

[A] Atropine [B] Codeine

[C] Thebaine [D] Morphine

85. Natural products of mixed biosynthetic origins that are partially derived from terpenoids are often called ....

[A] Meroterpenes [B] Polyterpenes

[C] Both [D] None of the above

86. Systematic study of the monoterpenes led the German chemist Otto Wallach to recognize that many terpenoid compounds might be constructed by joining isoprene units, generally in a repetitive head-to-tail fashion. This concept is known as ....

[A] Isoprene rule [B] Biogenetic isoprene rule
[C] Both [D] None of the above

87. A compound is "isoprenoid" if it is derived biologically from an "isoprenoid" precursor, with or without rearrangements, it is called

[A] Isoprene rule [B] Biogenetic isoprene rule
[C] Both [D] None of the above

88. The menthol and menthyl acetate content of peppermint oil glands .... with leaf maturation.

[A] Increases [B] Decreases
[C] Not influenced [D] None of the above

89. Very few examples of the genetic engineering of terpenoids metabolism are available, although two notable successes have been achieved in the area of

[A] Terpenoid vitamin [B] Terpenoid protein
[C] Both [D] None of the above

90. Gallic acid is

[A] C6C1 [B] C5C1
[C] C4C1 [D] C3C1
[E] None of the above

91. Isoflavones (isoflavonoids) mostly occur in

[A] Leguminous plants [B] Palms
[C] Both [D] None of the above

92. Isoflavonoids are known to

[A] Act as insecticides [B] Act as phytoalexins
[C] Cause infertility in mammals [D] All of the above

93. The colour of tomato fruit is due to the presence of

[A] Carotenoids [B] Anthocyanins
[C] Flavonols [D] None of the above

94. Which of the following is not a simple plant phenolic?

[A] Salicylic acid [B] Abietic acid
[C] Ferulic acid [D] None of the above

95. Which of the following is the building block of lignin?

[A] Sinapyl alcohol [B] Coniferyl alcohol
[C] Both [D] None of the above

96. Flavonoids are phenolic compounds that contain

[A] 15-C [B] 10-C
[C] 30-C [D] None of the above

97. Condensed tannins are not easily hydrolyzed.

This statement is

[A] True [B] False

[C] Not known [D] None of the above

98. Rubber consists of a large number of isopentenyl units in which carbon-carbon double bounds have .... configuration

[A] Z-configuration [B] E-configuration

[C] Both [D] None of the above

99. Gutta consists of a large number of isoprene units in which carbon-carbon double bounds have .... configuration

[A] Z-configuration [B] E-configuration

[C] Both [D] None of the above

100. Alkaloids are .... in reaction.

[A] Acidic [B] Basic

[C] Neutral [D] None of the above

101. In plants, alkaloids usually accumulate in

[A] Epidermal and hypodermal cells

[B] Young actively growing part

[C] Bundle sheath cells and latex vessels

[D] All of the above

102. Approximately how many alkaloids have been isolated from plants?

[A] 1100 [B] 1500

[C] 3000 [D] 10000

8 RESPONSES OF PLANTS TO ENVIRONMENTAL STRESS

1. **According to ...., *biological stress is any change in environmental conditions that might reduce or adversely change a plant's growth or development (it's biological function).***

   [A] Levitt [B] Hans Selye
   [C] Both [D] None of the above

2. **According to ...., *stress was a syndrome of reactions within an organism in response to one or more stress factors and any environmental factors could act as an agent of stress when it deviated from its optimum level for the organism.***

   [A] Levitt [B] Hans Selye
   [C] Both [D] None of the above

3. **According to ...., *we should speak of a state of stress. When the stress factor is first experienced, there is the alarm reaction, in which the function of interest deviates markedly from the norm. Then comes the resistance stage (or restitution phase), in which the organism adapts to the stress factor and the function often returns toward its normal state factor increasing or continues for a long time, the stage of exhaustion may be reached, in which the function may again strongly deviate from the norm; this can eventually leads to death.***

   [A] Larcher [B] Levitt
   [C] Both [D] None of the above

4. **The study of functioning of plant under adverse environmental conditions is called**

   [A] Ecological physiology [B] Stress physiology
   [C] Environmental physiology [D] None of the above

5. **Among the vascular plants, the highest known temperature tolerance is found among .... and .... which can withstand temperature of 60 °C or more.**

   [A] Agaves, cacti [B] Coleus, Croton
   [C] Both [D] None of the above

6. ***Arabidopsis* is a chill- .... plant.**

   [A] Tolerant [B] Sensitive
   [C] Free [D] None of the above

7. Among ornamentals, *Coleus, Croton, Passiflora, Diffenbachia* are examples of chill- .... plants.

[A] Tolerant [B] Sensitive
[C] Free [D] None of the above

8. The cell membrane of chill-sensitive species has .... proportion of saturated fatty acid while those of chill-resistant species has .... proportion of unsaturated fatty acids in their lipid bilayers.

[A] Higher, higher [B] Higher, lower
[C] Lower, higher [D] None of the above

9. The beginning of ice crystal formation is called

[A] Ice nucleation [B] Water nucleation
[C] Both [D] None of the above

10. Thawing means ....

[A] Liquefying [B] Solidifying
[C] Both [D] None of the above

11. Example of drought escapers are

[A] Desert annuals [B] Mesquite
[C] Alfalfa [D] Palms
[E] None of these

12. Example of water spenders (deep tap roots) are

[A] Desert annuals [B] Mesquite
[C] Alfalfa [D] Palms
[E] More than one

13. Example of water collectors (succulents, have CAM photosynthesis, shallow root system, salt secretion, dew absorption) are

[A] Cacti [B] Mosses
[C] Lichens [D] None of the above

14. ...., measured as qP, reflects the capacity to utilize excess absorbed energy through metabolism and growth.

[A] Photochemical quenching [B] Non photochemical quenching
[C] Both [D] None of the above

15. .... reflects the capacity to dissipate excess absorbed energy as heat.

[A] Photochemical quenching [B] Non photochemical quenching
[C] Both [D] None of the above

16. Both photochemical quenching and non photochemical quenching capacity can be estimated *in vivo* using

[A] Pulse –amplitude –modulated chlorophyll fluorescence (PAM)
[B] Kautsky effects
[C] Both
[D] None of the above

17. If the samples are exposed to a light intensity sufficiently strong to initiate photosynthesis, a fluorescence signal characterized by complex transient can be detected. This is called the .... effects.

[A] Kautsky effect [B] PAM effects

[C] Photon counter [D] None of the above

18. One of the immediate plant responses to a chilling stress is the light-dependent inhibition of photosynthesis. Because low temperature inhibits the D1 repair cycle, this leads to chronic photoinhibition of .... and .... in cucumber.

[A] PSII, PSI [B] PSI, PSII

[C] Both [D] None of the above

19. In *Phaseolus vulgaris* (bean), the ratio of unsaturated/saturated fatty acids of membrane lipids of mitochondria isolated from chilling-sensitive and chilling-resistant tissues is

[A] 2.8 (shoot) [B] 1.7 (shoot)

[C] 2.1 (shoot) [D] 2.8 (shoot)

[E] None of the above

20. In *Ipomoea batatas* (sweet potato), the ratio of unsaturated/saturated fatty acids of membrane lipids of mitochondria isolated from chilling-sensitive and chilling-resistant tissues is

[A] 2.8 (tuber) [B] 1.7 (tuber)

[C] 2.1 (tuber) [D] 2.8 (tuber)

[E] None of the above

21. In *Zea mays*, the ratio of unsaturated/saturated fatty acids of membrane lipids of mitochondria isolated from chilling-sensitive and chilling-resistant tissues is

[A] 2.8 (shoot) [B] 1.7 (shoot)

[C] 2.1 (shoot) [D] 2.8 (shoot) [E] None of the above

22. In *Pisum sativum*, the ratio of unsaturated/saturated fatty acids of membrane lipids of mitochondria isolated from chilling-sensitive and chilling-resistant tissues is

[A] 2.8 (shoot) [B] 1.7 (shoot)

[C] 2.1(shoot) [D] 3.8 (shoot)

[E] None of the above

23. .... is an eight KDa protein involved in targeting other proteins for proteolytic degradation.

[A] Ubiquitin [B] HSP110

[C] HSP90 [D] None of the above

24. The photoreduction of oxygen by PSI is called the .... and results in the production of another toxic reactive oxygen species known as a superoxide radical.

[A] Mehler reaction [B] Asada-Halliwell pathway

[C] Both [D] None of the above

25. .... refers to heritable modifications in structure or function that increases the fitness of the organism in a stressful environment.

[A] Adaptation [B] Acclimation
[C] Both [D] None of the above

26. Adaptation is the consequence of

[A] Natural selection
[B] Complement of gene found in the genome
[C] Both
[D] None of the above

27. .... are adapted to survive extreme shade conditions and exhibit minimal capacity to acclimate to high-light conditions.

[A] Obligate shade plants [B] Facultative shade plants
[C] Both [D] None of the above

28. Leaves of .... plants exhibit higher chlorophyll contents than those of .... plants.

[A] Shade, sun [B] Sun, shade
[C] Both [D] None of the above

29. Shade adapted plant species exhibit .... ratio of chlorophyll a to chlorophyll b.

[A] Lower [B] Higher
[C] No [D] None of the above

30. In shade adapted plant species, the value of chlorophyll a to chlorophyll b is

[A] 2-2.5 [B] 5-5.5
[C] 7-7.5 [D] None of the above

31. In sun plant, the ratio of chlorophyll a to chlorophyll b is

[A] 3-3.6 [B] 5-5.5
[C] 7-7.5 [D] None of the above

32. High growth rates are supported by at least three factors

[A] A high shoot-root ratio [B] High stomatal conductance
[C] High photosynthetic rates [D] All of the above

33. The .... leaf generally thinner than the .... leaf.

[A] C4, C3 [B] C3, C4
[C] Both [D] None of the above

34. Plants growing in the shade of a canopy are frequently subjected to sudden transient fluctuations in fluence rate, known as ....

[A] Time-nested response [B] Absorption cross-section
[C] Sunflecks [D] All of these

35. The process whereby adjustments are made to the structure and function of the photosynthetic apparatus in response in changes in growth irradiance is called ....

[A] Photoacclimation [B] Thermal denaturation
[C] Transition temperature [D] More than one

36. The upper tree limit (tree line) occurs at the lower elevation when one moves from the equator until it reaches sea level some distance above the Arctic Circle. This phenomenon is known as

[A] Humboldt's law [B] Larcher's Law
[C] Both [D] None of the above

37. The ability of a single plant to produce morphologically and physiologically different seed is known as

[A] Heteroblasty [B] Monoblasty
[C] Both [D] None of the above

38. Several higher plants are able to grow and even flower under the snow, where the temperature is close to 0° C or sometimes below. Such plants are known as

[A] Geophytes [B] Spring ephemerals
[C] Both [D] None of the above

39. The halophytes which can resist a wide range of salt concentrations are called as ....

[A] Euryhaline [B] Stenohaline
[C] Both [D] None of the above

40. The halophytes which can resist a narrow range of salt concentrations are called as ....

[A] Euryhaline [B] Stenohaline
[C] Both [D] None of the above

41. Among the crop plants, maize, peas, beans, onion, citrus, lettuce, etc are ....to salts.

[A] Sensitive [B] Moderately salt tolerant
[C] Highly salt tolerant [D] None of the above

42. Cotton and barley are ....

[A] Sensitive to salt [B] Moderately salt tolerant
[C] Highly salt tolerant [D] None of the above

43. Sugarbeet and date palms are ....

[A] Sensitive to salt [B] Moderately salt tolerant
[C] Highly salt tolerant [D] None of the above

44. It is believed that on being wounded by insect or other herbivore, the host plant produces a short polypeptide (of 18 amino acids) called ....

[A] Systemin [B] Proline
[C] Leucine [D] None of the above

45. The transportation of systemin occurs through

[A] Xylem [B] Phloem

[C] Both [D] None of the above

46. The originator of elicitor molecules are

[A] Phytoalexins

[B] Cell walls of pathogen

[C] Plasmamembrane of plant cell

[D] All of the above

47. There is a accumulation of phytochelatins in vacuoles of plant cells, and it is the indication of

[A] Drought stress [B] Heat stress

[C] Heavy metal stress [D] None of the above

48. Among the followings, Creosote bush is a

[A] Cold tolerant plant [B] Salt tolerant plant

[C] Drought tolerant plant [D] None of the above

49. Among the followings, which of the following substance acts as a volatile air-born SAR inducing signal?

[A] Methyl salicylate [B] Jasmonate

[C] Salicylic acid [D] None of the above

50. Biotic stress in plant is caused by

[A] Bacteria and fungi [B] Insects

[C] Blue bull [D] All of the above

51. In which of the followings, Heat-shock proteins were first discovered ?

[A] *Arabidopsis* [B] *Drosophila*

[C] *Bacillus thruingensis* [D] All of the above

52. Which one of the following is the initial response of mesophyte plants to water stress?

[A] Beginning of leaf stimulation

[B] Declining of photosynthesis

[C] Declining of respiration

[D] Decreased cellular growth and leaf area

53. Jasmonate is a .... carbon unsaturated fatty acid.

[A] 12 [B] 15

[C] 17 [D] 20

54. Jasmonate is synthesized in plant from ....

[A] Linolenic acid [B] Diphosphatidylglycerol

[C] Both [D] None of the above

55. Medicarpin, glyceolin, rishitin and capsidiol are the examples of

[A] Phytoalexins [B] Pathogen related protein

[C] PCD [D] None of the above

56. Heavy metal accumulator plant absorbs heavy metals and accumulate them into their ....

[A] Nucleus [B] Chloroplast
[C] Mitochondria [D] More than one
[E] None of the above

57. *Thlaspi goesingenase* is a hyperaccumulator plant of heavy metal

[A] Cadmium [B] Nickel
[C] Chromium [D] More than one

58. *Viola calaminaria* is a hyperaccumulator plant of heavy metal

[A] Zinc [B] Lead
[C] Chromium [D] More than one

59. *Astragalus* and *Stanleya pinnata* are hyperaccumulator plants of heavy metal

[A] Selenium [B] Chromium
[C] Cadmium [D] More than one

60. Which of the following heavy metal forms complexes with histidine in *Alyssum lesbiacun*?

[A] Ni [B] Cd
[C] Pb [D] None of the above

61. The most abundant thiol (-SH) compound in plants is

[A] Glutathione [B] Glutamate
[C] Glycophytes [D] None of the above

62. Which of the following is considered to be the indicator of heavy metal stress?

[A] Phytochelatin accumulation in vacuole
[B] Phytochelatins accumulation in chloroplast
[C] Phytochelatins accumulation in nucleus
[D] None of the above

63. The halophytes which can resist a wide range of salt concentration are called as ....

[A] Euryhaline [B] Stenohaline
[C] Both [D] None of the above

64. .... is more sensitive to water stress than rate of ....

[A] Leaf expansion, photosynthesis
[B] Photosynthesis, leaf expansion
[C] Leaf expansion, respiration
[D] None of the above

65. Presence of salt glands on leaf surface is characteristic of

[A] *Tamarix pentandra* [B] *Suaeda fruticosa*
[C] *Atriplex spongiosa* [D] None of the above

66. In the choice given below, which of the plant(s) is example of resurrection plant?

[A] *Selaginella lepidophylla* [B] *Polypodium*
[C] Both [D] None of the above

67. Which of the following plant is most sensitive to frost injury?

[A] Temperate zone plants [B] Tropical zone plants
[C] Subtropical plants [D] None of the above

68. Among the choices, which of the following amino acid chiefly accumulates in cells of water stressed plants and also halophytes to maintain osmotic relations of the cells?

[A] Proline [B] Leucine
[C] Glycine [D] More of the above

69. What are the external symptoms of chilling injury to plants?

[A] Lesser growth [B] Lesion of leaves
[C] Chlorosis of leaves [D] All of the above

70. Morphological symptom of plants suffering from salt injury is

[A] Bleaching of chlorophyll [B] Necrotic spot on leaves
[C] Burning of leaf edge [D] All of the above

71. Ice nucleation means

[A] Formation of ice under the influence of nucleus
[B] Formation of ice in chloroplast
[C] Formation of ice crystals around large polysaccharides and proteins in cells walls
[D] None of the above

72. Which of the following acts as ice nucleators?

[A] Polysaccharides [B] Protein
[C] Both [D] None of the above

73. *Tidestromia oblongifolia*, a native of very hot climate of death valley, California, belongs to family

[A] Amaranthaceae [B] Fabaceae
[C] Brassicaceae [D] None of the above

74. .... is considered to be the pathway of SAR signal.

[A] Phloem [B] Xylem
[C] Both [D] None of the above

75. The transmission rate of SAR signal from infection site to other parts of plants is

[A] 3 cm/h [B] 10 cm/h
[C] 6 cm/h [D] None of the above

76. Which of the following is antidigestive proteins?

[A] Lectins [B] Proteolytic inhibitors
[C] -Amylase inhibitors [D] None of the above

77. After acclimation in plants, what will be the tolerance level of plants against a particular stress?

[A] Increased [B] Decreased

[C] Not affected [D] None of the above

78. Drought resistance in plant is related to

[A] +ve water potentials [B] –ve water potentials

[C] Both [D] None of the above

79. Among the followings which is a chill-tolerant plants?

[A] *Croton* [B] *Coleus*

[C] *Dieffenbachia* [D] *Arabidopsis*

[E] None of the above

80. Phytoalexins are

[A] Secondary metabolites [B] Synthetic compounds

[C] Plant hormones [D] None of the above

81. Hypersensitive response of plants includes

[A] Formation of pathogen related protein

[B] Formation of phytoalexins

[C] Synthesis of phytohormones

[D] All of the above

82. Cellular location of HSP (Heat-Shock Protein) class100 is

[A] Cytosol [B] Mitochondria

[C] Chloroplast [D] All of the above

83. Cellular location of HSP (Heat-Shock Protein) class 90 is

[A] Cytosol [B] ER

[C] Both [D] None

84. Cellular location of HSP (Heat-Shock Protein) class70 is

[A] Cytosol [B] Mitochondria

[C] Chloroplast [D] Nucleus

[E] All of the above

85. Cellular location of HSP (Heat-Shock Protein) class 60 is

[A] Mitochondria [B] Chloroplast

[C] Both [E] None of the above

86. Size of HSP class 100 is

[A] 100-114 kDa [B] 80-94 kDa

[C] 69-71 kDa [D] 57-60 kDa

[E] 15-30 kDa [F] None of the above

87. Size of HSP class 90 is

[A] 100-114 kDa [B] 80-94 kDa

[C] 69-71 kDa [D] 57-60 kDa

[E] 15-30 kDa [F] None of the above

88. Size of HSP class 70 is

[A] 100-114 kDa [B] 80-94 kDa
[C] 69-71 kDa [D] 57-60 kDa
[E] 15-30 kDa [F] None of the above

89. Size of HSP class 60 is

[A] 100-114 kDa [B] 80-94 kDa
[C] 69-71 kDa [D] 57-60 kDa
[E] 15-30 kDa [F] None of the above

9 GERMINATION AND DORMANCY

1. **Which of the following dye is used for testing the viability of seeds?**

[A] Safranine
[B] 2, 6 dichlorophenol indophenols
[C] 2, 3, 5, triphenyl tetrazolium chloride
[D] None of the above

2. **Those seeds whose life span varies from 15 years to 100 years or more are called**

[A] Macrobiotic [B] Mesobiotic
[C] Microbiotic [D] None of the above

3. **The dormancy of seeds of families such as Leguminosae, Convovulaceae, Solanaceae, Chenopodiaceae are due to**

[A] Seed coats impermeable to water
[B] Seed coats impermeable to oxygen
[C] Mechanically resistant seed coats
[D] Immaturity of embryo
[E] Germination inhibitors [F] None of the above

4. **In *Xanthium* and many grasses and some members of Compositae, seed dormancy is due to**

[A] Seed coats impermeable to water
[B] Seed coats impermeable to oxygen
[C] Mechanically resistant seed coats
[D] Immaturity of embryo
[E] Germination inhibitors [F] None of the above

5. **Seed dormancy of certain weeds such as *Amaranthus, Capsella and Alisma* are due to**

[A] Seed coats impermeable to water
[B] Seed coats impermeable to oxygen
[C] Mechanically resistant seed coats
[D] Immaturity of embryo
[E] Germination inhibitors [F] None of the above

6. **Seed dormancy of certain orchid, *Ginkgo biloba, Anemone nemorosa, Fraxinus excelsior* are due to**
   [A] Seed coats impermeable to water
   [B] Seed coats impermeable to oxygen
   [C] Mechanically resistant seed coats
   [D] Immaturity of embryo
   [E] Germination inhibitors
   [F] None of the above

7. **In Oat, the inhibitors of seed dormancy is present in**
   [A] Endosperm [B] Embryo
   [C] Testa [D] Glumes
   [E] None of the above

8. **Which of the following causes dormancy in seeds?**
   [A] Seed coats impermeable to water
   [B] Immaturity of the embryo
   [C] Mechanically resistant seed coats
   [D] Only [A] and [B]
   [E] All of the above

9. **Which of the following is germination stimulatory substance?**
   [A] Thiourea [B] KNO_3
   [C] Both [D] None of the above

10. **Which of the following is the common germination inhibitor?**
   [A] Coumarin [B] Phthalides
   [C] Ferulic acid [D] ABA
   [E] Dehydracetic acid [F] Prasorbic acid
   [G] All of the above

11. **In certain plants, seeds do not germinate immediately after harvest even under favorable conditions for seed germination due to**
   [A] Scotoperiodisim [B] Dormancy of seeds
   [C] Both [D] None of the above

12. **The seeds of *Lactuca sativa, Capsella bursa pastoris, Lepidium vigrinicum, Nicotiana tabacum, Lycopersicum esculentum,* are .... photoblastic.**
   [A] +ve [B] –ve
   [C] Non [D] None of the above

13. **The seeds of *Helleborus niger, Nigella damascene, Silene armeria* are .... photoblastic.**
   [A] +ve [B] –ve
   [C] Non [D] None of the above

14. The seeds of many plants are capable of germinating immediately after harvest become dormant if kept in environment in which at least one of the factors essential for germination is unfavorable. This induced dormancy is known as

[A] Declining secondary dormancy of seeds

[B] Secondary dormancy of seeds

[C] Primary dormancy of seeds

[D] None of the above

15. The process of rupturing or weakening the seed coats by mechanical or other different means for breaking of dormancy is called as

[A] Stratification [B] Scarification

[C] Both [D] None of the above

16. A process of breaking seed dormancy of some plants in which seeds are treated in moist medium at low temperature (5-10° C) for sufficient period of time is known as

[A] Stratification [B] Scarification

[C] Both [D] None of the above

17. .... is done by alternating the layers of seeds with layers of wet *Sphagnum*, sand or some other suitable material and keeping them at low temperatures.

[A] Artificial stratification [B] Artificial scarification

[C] Both [D] None of the above

18. The process of seed germination starts with

[A] Imbibition [B] Diffusion

[C] Osmosis [D] None of the above

19. Seeds which have life span of a few weeks to 4 years are known as

[A] Microbiotic [B] Mesobiotic

[C] Macrobiotic [D] None of the above

20. Seeds which have life span ranging from 3 years to 15 years are known as

[A] Microbiotic [B] Mesobiotic

[C] Macrobiotic [D] None of the above

21. Seeds which have life span ranging from 15 years to 100 years or more are known as

[A] Microbiotic [B] Mesobiotic

[C] Macrobiotic [D] None of the above

22. The seeds which just wait for suitable environmental conditions to germinate are said to be

[A] Quiescent [B] Dormant

[C] Dry seed [D] All of the above

23. Bewley and Black classified seeds relative to storage structure into three types

[A] Endosperm (grasses, castor bean, tomato, buckwheat)

[B] Embryo (legumes, lettuce)

[C] Perisperm (beet, yucca, coffee)

[D] All of the above

24. Based on solubility and method of separation, Osborne divided proteins into four groups

[A] Albuminus [B] Globulins

[C] Glutelins [D] Prolamins

[E] All of the above

25. In general, cereals are high in

[A] Albumins [B] Globulins

[C] Glutelins [D] Prolamins

[E] More than one

26. Legume seeds are generally rich in

[A] Albuminus [B] Globulins

[C] Glutelins [D] Prolamins

[E] More than one

27. Oat is an exception, with a protein composition of 80% .... instead of prolamins.

[A] Albuminus [B] Globulins

[C] Glutelins [D] Prolamins

28. The best example of important prolamins in cereals includes

[A] Zein in maize [B] Gliadin in wheat

[C] Hordenin in barly [D] All of the above

10

VERNALIZATION

1. **Vernalization term was coined by**
   [A] Klippart [B] Lysenko
   [C] Chouard [D] None of the above
2. **Which of the following is the best example of physiological preconditioning?**
   [A] Vernalization [B] Photoperiodism
   [C] Both [D] None of the above
3. **Which part of plant perceives cold stimulus?**
   [A] Apical meristems [B] Tuber
   [C] Bulb [D] None of the above
4. **Most suitable temperature for vernalization in plants ranges from**
   [A] 1-6° C [B] 1-10° C
   [C] 7-9° C [D] None of the above
5. **The Vernalization is an .... process**
   [A] Aerobic [B] Anaerobic
   [C] Both [D] None of the above
6. **Vernalization of dry seeds is possible? This statement is**
   [A] True [B] False
   [C] Not known [D] None of the above
7. **Phasic development theory for Vernalization was given by**
   [A] Lysenko [B] Melchers
   [C] Both [D] None of the above
8. **Hormonal theory for Vernalization was given by**
   [A] Lysenko [B] Melchers
   [C] Both [D] None of the above
9. **Which of the following is the practical utility of Vernalization?**
   [A] Vernalization increases the cold resistance of plant
   [B] Vernalization shortens the vegetative period of plant
   [C] Vernalization increases the resistance of fungal infection
   [D] All of the above

10. Which of the following hormone replaces the low temperature requirement in certain plant?

[A] Auxin [B] Gibberellins

[C] Both [D] None of the above

11. Vernalization plant can be devernalized by

[A] Treatment with high temperature

[B] Treatment with low temperature

[C] Treatment with high wind

[D] Treatment with very high wind

[E] None of the above

12. Vernalin is a

[A] Hormone for root elongation

[B] Hormone for shoot elongation

[C] Hormone for flowering

[D] None of the above

13. Which of the following statement is correct?

[A] Vernalization increases the flowering period

[B] Vernalization shortens the vegetative period

[C] Both

[D] None of the above

11 TRANSLOCATION OF ORGANIC SOLUTES

1. **The movement of organic food materials or the solutes in soluble form from one place to another in higher plants is called as**

   [A] Translocation of organic solute

   [B] Transformation of organic solute

   [C] Both   [D] None of the above

2. **Translocation of organic solute always takes place from the region of .... concentration of soluble form to .... concentration of soluble form.**

   [A] Higher, Lower   [B] Lower, Higher

   [C] No flow   [D] None of the above

3. **Translocation of organic solutes in higher plants takes place mainly through**

   [A] Xylem vessels   [B] Phloem sieve tubes

   [C] Both   [D] None of the above

4. **The example of sink tissue or organs in higher plants is mainly**

   [A] Seed   [B] Fruits

   [C] Stem   [D] Roots

   [E] All of the above

5. **In girdled plant, which die first?**

   [A] Roots   [B] Shoots

   [C] Leaves   [D] All of the above

6. **In symplastic loading the transport sugars are mainly**

   [A] Sucrose and other oligosaccharides

   [B] Sucrose

   [C] Only oligosaccharides   [D] None of the above

7. **In which form of sugar the symplastic loading takes place?**

   [A] Glucose   [B] Sucrose

   [C] Maltose   [D] None of the above

8. **In Apoplastic loading, type of companion cells in the small veins is**

   [A] Intermediary cells   [B] Ordinary or transfer cells

   [C] Both   [D] None of the above

9. **In Symplastic loading, type of companion cells in the small veins is**
   [A] Intermediary cells [B] Ordinary or transfer cells
   [C] Both [D] None of the above
10. **In Apoplastic loading, number of plasmadesmata connecting the sieve tubes (including companion cells) to surrounding cells is**
   [A] Fewer [B] Abundant
   [C] Both [D] None of the above
11. **In Symplastic loading, number of plasmadesmata connecting the sieve tubes (including companion cells) to surrounding cells is**
   [A] Fewer [B] Abundant
   [C] Both [D] None of the above
12. **Phloem loading and unloading occurs in higher plant through**
   [A] Symplastic pathways [B] Apoplastic pathways
   [C] Both [D] None of the above
13. **Assimilate partitioning is .... distribution of photoassimilates in different sinks of plant.**
   [A] Differential [B] Uniform
   [C] No [D] None of the above
14. **Which of the following is the controlling factor in assimilate partitioning and translocation of assimilates?**
   [A] Rate of photosynthesis and sink demand
   [B] Competition among sink tissues for available translocated assimilates
   [C] Long distance signals between sources and sinks and plasmadesmata
   [D] All of the above
15. **Movement of photosynthetic assimilates from sieve tube elements to receiver cells of sink is called**
   [A] Phloem loading [B] Phloem unloading
   [C] Both [D] None of the above
16. **The translocation of sugars from mesophyll cells to sieve tube elements in leaf is called as**
   [A] Phloem unloading [B] Phloem loading
   [C] Both [D] None of the above
17. **In higher plants, among the followings which is not related to translocation of organic solutes in higher plants?**
   [A] Protoplasmic streaming theory [B] Interfacial flow hypothesis
   [C] Electro-osmotic hypothesis [D] All of the above
18. **The ability of sink to mobilize photosynthates or assimilates toward it is known as**
   [A] Sink size [B] Sink strength
   [C] Sink activity [D] None of the above

19. **Sink strength=**
[A] Sink size x Sink activity
[B] Sink size + Sink activity
[C] Sink size - Sink activity
[D] None of the above

20. **Total weight of sink tissue is known as**
[A] Sink size
[B] Sink activity
[C] Both
[D] None of the above

21. **The rate of uptake of assimilates per unit weight of the sink is known as**
[A] Sink size
[B] Sink activity
[C] Both
[D] None of the above

22. **Rate of photosynthesis .... when sink/source ratio is decreased.**
[A] Declines
[B] Increases
[C] Not affected
[D] None of the above

23. **Rate of photosynthesis .... when sink/source ratio is decreased.**
[A] Declines
[B] Increases
[C] Not affected
[D] None of the above

24. **The signal between source and sink may be**
[A] Physical
[B] Chemical
[C] Both
[D] None of the above

25. **Electro-Osmotic theory for translocation of organic solutes was given by**
[A] Fenson and Spanner
[B] Mason and Phillis
[C] Cany and Thaine
[D] Van den Honert
[E] De Varies and Curtis

26. **Activated Diffusion Hypothesis for translocation of organic solutes was given by**
[A] Fenson and Spanner
[B] Mason and Phillis
[C] Cany and Thaine
[D] Van den Honert
[E] De Varies and Curtis

27. **Interfacial Flow Hypothesis for translocation of organic solutes was given by**
[A] Fenson and Spanner
[B] Mason and Phillis
[C] Cany and Thaine
[D] Van den Honert
[E] De Varies and Curtis
[F] All of the above
[G] None of the above

28. **Protoplasmic theory has been re-emphasized by**
[A] Fenson and Spanner
[B] Mason and Phillis
[C] Cany and Thaine
[D] Van den Honert
[E] De Varies and Curtis

29. Protoplasmic Streaming Theory for translocation of organic solutes was given by

[A] Fenson and Spanner [B] Mason and Phillis
[C] Cany and Thaine [D] Van den Honert
[E] De Varies and Curtis

30. According to ...., phloem unloading in potato tubers from sieve elements to cortical cells is a symplastic passive process

[A] Fenson and Spanner [B] Mason and Phillis
[C] Cany and Thaine [D] Van den Honert
[E] Oparka

31. According to ...., combinations of plant growth regulators may have additive, synergistic or inhibitory effects on assimilates partitioning.

[A] Fenson and Spanner [B] Mason and Phillis
[C] Cany and Thaine [D] Van den Honert
[E] Gifford and Evans

32. Pressure flow hypothesis is also known as

[A] Munch's Hypothesis [B] Fenson and Spanner hypothesis
[C] Both [D] None of the above

33. Munch's hypothesis accounts for translocation of organic solutes only in

[A] Downward direction [B] Upward direction
[C] Both [D] None of the above

34. The quantity of material passing through a given cross section of sieve tubes per unit of time is known as

[A] Mass transfer rate [B] Mass flow rate
[C] Both [D] None of the above

35. In most of the species, the maximum velocity of transport is

[A] 500 to 1500 mmh^{-1} [B] 700 to 900 mmh^{-1}
[C] Both [D] None of the above

36. Typically, the mesophyll cells of trees have an osmotic potential of .... Mpa, whereas sieve elements in leaves have an osmotic potential of about .... Mpa.

[A] -1.3 to -1.8, -2.0 to -3.0 [B] -2.0 to -3.0, -1.3 to -1.8
[C] Both [D] None of the above

12 ABSORPTION OF MINERAL SALTS

1. **Mineral salts are absorbed from the soil solution in the form of ....**

   [A] Cation ions [B] Anion

   [C] Both [D] None of the above

2. **Which part of the root is involved in absorption of mineral salts in higher plants?**

   [A] Root cap [B] Meristematic region

   [C] Zone of root hair [D] None of the above

3. **First step in the absorption of mineral salts is the process of**

   [A] Ion-exchange [B] Diffusion

   [C] Imbibition [D] None of the above

4. **Ion-Exchange does not require metabolic energy but it generally facilitates mineral salt absorption. This statement is**

   [A] True [B] False

   [C] Not known [D] None of the above

5. **There is requirement of metabolic energy in**

   [A] Active absorption of mineral salts

   [B] Passive absorption of mineral salts

   [C] Contact exchange of ions

   [D] None of the above

6. **In contact exchange theory, the ions adsorbed on clay particle may be exchanged with the ions adsorbed on root-surface directly without first being dissolved in soil solution. This statement is true /false**

   [A] True [B] False

   [C] Not known [D] None of the above

7. **When a plant is transferred from water to a salt solution, the rate of respiration increases. This is called**

   [A] Ions respiration [B] Salt respiration

   [C] Activated respiration [D] None of the above

8. In alga .... the cell sap accumulated K^+ and phosphate ions to such an extent that their concentration was thousands time greater than in the pond water in which the plant was growing.

[A] *Nitella* [B] *Chlorella*
[C] Both [D] None of the above

9. Cytochrome pump theory for absorption of mineral salts was first proposed by

[A] Lundegarth & Burstrom [B] Bennet-Clark
[C] Donnan [D] All of the above

10. Protein-lecithin theory for absorption of mineral salts was first proposed by

[A] Lundegarth & Burstrom [B] Bennet-Clark
[C] Donnan [D] All of the above

11. According to Carrier Concept, the plasma membrane is .... to free ions.

[A] Impermeable [B] Permeable
[C] Both [D] None of the above

12. Which of the following observation strongly supports the carrier concept of active absorption of mineral salts?

[A] Isotopic exchange [B] Saturation effects
[C] Specificity [D] All of the above

13. Lundegardh's Cytochrome Pump theory believed that there was a definite correlation between .... and .... absorption.

[A] Photosynthesis, anion [B] Respiration, anion
[C] Both [D] None of the above

14. The inhibition of salt respiration and accompanying absorption of anion by CO and cyanide .... the Lundegardh's Cytochrome Pump Theory.

[A] Supports [B] Declines
[C] Not known [D] None of the above

15. Which of the following assumption is a basis of Lundegardh's Cytochrome Pump Theory?

[A] Anion are absorbed through cytochrome chain by active process
[B] Mechanism of anion and cation absorption is different
[C] Cations are absorbed passively
[D] All of the above

16. The main defect of Lundegardh's Cytochrome Pump theory is

[A] It envisages active absorption of only anions
[B] It does not explain selective uptake of ions
[C] Both
[D] None of the above

17. **Protein associated with phosphatide is called**
[A] Lecithin [B] Proline
[C] Glycine betaine [D] None of the above

18. **Which of the following enzyme has no role in Bennet-Clark's protein-lecithin theory?**
[A] Lecithinase [B] Choline esterase
[C] Choline acetylase [D] None of the above

19. **Facilitated diffusion of solute through transport proteins across the membrane is**
[A] Active transport [B] Passive transport
[C] Both [D] None of the above

20. **The mineral salts in their ionic form move from one cell to another by**
[A] Apoplastic pathways [B] Trans membrane pathways
[C] Symplastic pathways [D] All of the above

21. **If a cell contains certain fixed NH_4^+ in it and placed in solution of cation and anion both, then Donnan's equilibrium will result in**
[A] Accumulation of cation inside the cells
[B] Accumulation of anion inside the cells
[C] Accumulation of both cation and anion inside the cells
[D] None of the above

22. **If a cell contains certain fixed COO^- in it and placed in solution of cation and anion both, then Donnan's equilibrium will result in**
[A] Accumulation of cation inside the cells
[B] Accumulation of anion inside the cells
[C] Accumulation of both cation and anion inside the cells
[D] None of the above

23. **As we know the ions carry a charge and are mobile, their diffusion across the membrane channel established an electric current, which can be detected by a special technique known as**
[A] Patch-clamp technology
[B] Voltage gate technology
[C] Both
[D] None of the above

24. **Patch-clamp technology was invented by**
[A] Neher and Sakmann [B] R. Mackinnon
[C] Both [D] None of the above

25. **The rate of ion diffusion through open channel is**
[A] $10^8\ s^{-1}$ [B] $10^2\ s^{-1}$
[C] $10^3\ s^{-1}$ [D] None of the above

26. **Anion channels are always**
[A] Outward rectifying [B] Inward rectifying
[C] Both [D] None of the above

27. Ca^{2+} channels are always

[A] Outward rectifying [B] Inward rectifying
[C] Both [D] None of the above

28. K^+ channels are

[A] Outward rectifying [B] Inward rectifying
[C] Both [D] None of the above

29. The rate of carrier mediated transport of solute is

[A] 10^4-10^5 [B] 10^8-10^9
[C] 10^1-10^2 [D] None of the above

30. Which of the following is an example of electroneutral pumps?

[A] H^+-PPase [B] H^+-ATPase
[C] Ca^{2+}-ATPase [D] H^+/K^+-ATPase
[E] None of the above

31. Among the followings, which ion can diffuse both outward and inward through channels across the plasmamembrane of plants?

[A] K^+ [B] Ca^{2+}
[C] Both [D] None of the above

32. In primary active transport, the required energy derived mainly from

[A] Hydrolysis of GTP [B] Hydrolysis of ATP
[C] Both [D] None of the above

33. In plant cell membrane the direction of electrogenic pump such as H^+-ATPase is

[A] Inward [B] Outward
[C] Both [D] None of the above

34. Na+/K+ -ATPase of animal cells is an example of

[A] Electrogenic pump [B] Electroneutral pump
[C] Both [D] None of the above

35. H^+-ATPase, H^+-PPase, and Ca^{2+}-ATPase are most common electrogenic pump in plant cells and direction of these pumps are

[A] Outward [B] Inward
[C] Lateral [D] None of the above

36. H^+-ATPases pumps are also known as

[A] P-type ATPase [B] L-type ATPase
[C] Q-type ATPase [D] None of the above

37. H^+-ATPases of mitochondria and chloroplast are also known as

[A] P-type ATPase [B] L-type ATPase
[C] F-type ATPase [D] None of the above

38. Na^+/K^+-ATPase, first discovered by

[A] Jens C. Skou [B] Neher and Sakmann
[C] R. Mackinnon [D] None of the above

39. H^+-ATPases of plasma membrane and tonoplast are

[A] Different [B] Same

[C] Not known [D] None of the above

40. H^+-ATPases of plasma membrane and tonoplast are inhibited by

[A] Vandate ions [B] Nitrate ion

[C] Cd ion [D] None of the above

41. H^+-PPase mainly found in

[A] Tonoplast [B] Plasmamembrane

[C] Chloroplast membrane [D] None of the above

42. Secondary active transport is also known as

[A] Cotransport mechanism [B] Primary mechanism

[C] Both [D] None of the above

43. Which of the following is most important in transport of ionisable solute across the membrane?

[A] Chemical potential gradient

[B] Concentration of solutes

[C] Electric potential gradient

[D] Electrochemical potential gradient

44. Which of the following is an example of secondary active transport of solutes in plants?

[A] Symport [B] Antiport

[C] Cotransport [D] All of the above

13 MINERAL NUTRITION OF PLANTS

1. **According to Epstein, there are .... main criteria for judging essentiality of an element in plant**
[A] Three [B] Two
[C] Four [D] None of the above

2. **Radioactive isotopes are measured quantitatively by an apparatus called**
[A] Scintillation Counter [B] Geiger-Muller Counter
[C] Mass spectrometer [D] More than one

3. **Stable (or heavy) isotopes are measured quantitatively by**
[A] Mass spectrometer [B] Photo spectrometer
[C] Both [D] None of the above

4. **In nitrogen deficiency which of the following leaves are affected first?**
[A] Older [B] Younger
[C] Both [D] None of the above

5. **Due to P deficiency**
[A] Leaves may change dark to blue-green in colour
[B] Dead necrotic area may be developed on leaves or fruits
[C] Premature leaf falls
[D] All of the above
[E] None of the above

6. **In S deficiency .... leaves are affected first**
[A] Older [B] Younger
[C] Both [D] None of the above

7. **Due to S deficiency**
[A] Tips and margins of the leaf roll inward
[B] Stem become hard due to the development of sclerenchyma
[C] Both
[D] None of the above

8. **Which of the following element act as second messenger in metabolic regulations?**
   [A] Calcium [B] Potassium
   [C] Iron [D] None of the above

9. **Due to Calcium deficiency**
   [A] There is degradation of meristematic region of plant
   [B] Yellowing occurs along the margins of the younger leaves.
   [C] Malformation of younger leaves
   [D] All of the above

10. **Magnesium deficiency causes .... of the leaves.**
    [A] Interveinal [B] Marginal
    [C] Both [D] None of the above

11. **Mottled chlorosis of leaves occurs due to deficiency of**
    [A] Potassium [B] Magnesium
    [C] Calcium [D] None of the above

12. **Die-back of citrus and reclamation disease of cereals and leguminous plants are due to deficiency of**
    [A] Copper [B] Zinc
    [C] Boron [D] None of the above

13. **Which of the following element has role in auxin biosynthesis?**
    [A] Copper [B] Zinc
    [C] Boron [D] Molybdenum

14. **...................... deficiency causes chlorosis of the older leaves which starts from tips and the margins**
    [A] Copper [B] Zinc
    [C] Boron [D] Molybdenum

15. **Mottle leaf disease in apple, citrus, walnut and other fruit trees are due to deficiency of**
    [A] Copper [B] Zinc
    [C] Boron [D] Molybdenum

16. **Which of the following element facilitates translocation of sugar?**
    [A] Boron [B] Zinc
    [C] Copper [D] None of the above

17. **Leaves become coppery in texture due to deficiency of**
    [A] Boron [B] Zinc
    [C] Copper [D] None of the above

18. **Specific role of boron in plant is**
    [A] In respiration [B] In photosynthesis
    [C] In hormonal synthesis [D] None of the above

19. Whip tail disease in cauliflower occurs due to

[A] Nickel deficiency [B] Chloride deficiency
[C] Mo deficiency [D] None of the above

20. Which of the following element is the cofactor of the enzyme urease in higher plants?

[A] Nickel [B] Chloride
[C] Mo [D] None of the above

21. Which of the following enzyme contains Mo in its prosthetic groups?

[A] Nitrogenase [B] Nitrate reductase
[C] Both [D] None of the above

22. Excluding sodium and silicon, how many different elements are considered essential for normal growth and development of majority of plants?

[A] 20 [B] 19
[C] 17 [D] None of the above

23. Among the followings, which of the following substance is used in hydroponic growth system?

[A] Gravel [B] Vermiculite
[C] Sand [D] All of the above

24. Soil less culture is also known as

[A] Hydroponics [B] Aeroponics
[C] Both [D] None of the above

25. Which of the following contains copper?

[A] Cytochrome oxidase [B] Plastocyanin
[C] Both [D] None of the above

26. Iron is an important component of

[A] Ferredoxin [B] Cytochromes
[C] Siroheme [D] All of the above

27. Magnesium is an important component of

[A] Cytochrome [B] Phaeophytin
[C] Chlorophylls [D] All of the above

28. Calcium is an important constituent of

[A] Secondary cell wall [B] Primary cell wall
[C] Middle lamella [D] All of the above

29. Equisetaceae (Scouring rushes) requires silicon to complete their life cycle where it may contain up to .... percent of the dry matter.

[A] 10 [B] 16
[C] 20 [D] None of the above

30. Deficiency of which element makes plant more susceptible to fungal infection and lodging?

[A] Silicon [B] Nickel
[C] Both [D] None of the above

31. Which of the following is the sulphur containing amino acid?

[A] Methionine [B] Cysteine
[C] Cystine [D] All of the above

32. Most important role of manganese is in

[A] Reduction of nitrates
[B] Phosphate transfer reaction
[C] Photolysis of water and oxygen evolution in photosynthesis
[D] None of the above

33. Which of the following enzyme depends on zinc for their activity?

[A] Carbonic anhydrase
[B] Alcohol dehydrogenase
[C] Pyridine nucleotide dehydrogenase
[D] All of the above

34. .... have built a strong case for the involvement of boron in carbohydrate transport within the plant.

[A] Gauch [B] Dugger
[C] Both [D] None of the above

35. In general, it is considered that symptoms of boron deficiency are symptoms of .... deficiency.

[A] Sugar [B] Protein
[C] Lipid [D] None of the above

36. .... reported a decrease in leaf epidermal cell size due to nitrogen deficiency in millet and buckwheat.

[A] Lutmen [B] Florell
[C] Lyon [D] Garcia
[E] None of the above

37. Phosphorus is found in plants as a constituent of

[A] Nucleic acid
[B] Phospholipids
[C] The coenzymes NAD and NADP
[D] ATP
[E] All of the above

38. Many symptoms of phosphorus deficiency can be confused with deficiency of?

[A] Nitrogen [B] Potassium
[C] Iron [D] None of the above

39. Purple and Red anthocyanin pigmentation of leaf can occur due to deficiency of?

[A] Nitrogen [B] Potassium
[C] Phosphorus [D] None of the above

40. According to Eaton, deficiency of phosphorus in sunflower, soyabean and black mustard causes accumulation of ....

[A] Carbohydrate [B] Protein
[C] Lipid [D] None of the above

41. The partial removal of calcium from the middle lamella of the cell can be done through the application of

[A] EDTA [B] Pectic acid
[C] Sulphuric acid [D] All of the above

42. Abnormal mitosis occurs due to deficiency of

[A] Calcium [B] Potassium
[C] Magnesium [D] None of the above

43. Among the followings, which element is the activator of many enzymes such as argenin kinase, adenosine triphosphate, adenyl kinase and potato apyrase ?

[A] Calcium [B] Potassium
[C] Both [D] None of the above

44. Calcium deficiency in .... plants results in increased level of carbohydrate in the .... and decreased levels in the .... and ....

[A] Cotton, leaves, stems, roots [B] Cotton, stems, leaves, roots
[C] Sugarcane, leaves, stems, roots [D] None of the above

45. Carbohydrate metabolism in plants is affected by deficiency of

[A] Magnesium [B] Calcium
[C] Potassium [D] None of the above

46. Among the following enzymes involved in carbohydrate metabolism, which has requirement of Mg^{2+} as an activator?

[A] Glucokinase [B] Fructokinase
[C] Galactokinase [D] Hexokinase
[E] Triosekinase [F] Gluconolactonase
[G] 6-phosphogluconic dehydrogenase
[I] Phosphopentokinase [J] Enolase
[K] Pyruvic kinase [L] Carboxylase
[M] Phosphoglyceric kinase [N] All of the above
[O] None of the above [P] Only A, B, C, D and E
[Q] Only F, G, H, I, J, K, L, M [R] Only A and B
[S] Only C and D [T] Only E and F

47. According to Lyon and Garcia, abundant supply of magnesium caused a depression of internal .... development and an .... of parenchymatous cells adjacent to the endodermis.

[A] Xylem, increase [B] Phloem, increase
[C] Xylem, decrease [D] None of the above

48. The highest concentration of potassium is found in which region of plants?

[A] Meristematic region [B] Zone of maturation
[C] Both [D] None of the above

49. Apical dominance in several plants appears to be lacking or is weak due to deficiency of

[A] Potassium [B] Magnesium
[C] Sulphur [D] None of the above

50. PAPS (3'-phosphoadenosine-5'-phosphosulfate), an active compound is involved in the reduction of

[A] SO_4^{2-} [B] CO_3^{2-}
[C] Both [D] None of the above

51. Which of the following element is involved in the metabolic activity of biotin, thiamine and co-enzyme A?

[A] Sulphur [B] Magnesium
[C] Calcium [D] None of the above

52. Under sulphur deficiency .... leaves showed chlorosis first.

[A] Younger [B] Older
[C] No any [D] None of the above

53. Which of the following form of iron is considered as metabolically active form of iron in plant?

[A] Ferric form [B] Ferrous form
[C] Both [D] None of the above

54. .... is an essential element in nitrogen metabolism and respiration, in both processes it functions as an enzyme activator.

[A] Iron [B] Manganese
[C] Sulphur [D] None of the above

55. Magnesium is most frequent replacer of

[A] Co [B] Zn
[C] Fe [D] Mn
[E] None of the above

56. Manganese acts as an activator of which of the following enzyme?

[A] Nitrate reductase
[B] Hydroxylamine reductase
[C] Both
[D] None of the above

57. Which of the following element is involved in the destruction of Indole -3- acetic acid?

[A] Manganese [B] Iron
[C] Sulphur [D] None of the above

58. The sensitivity of chlorophyll to light destruction .... under condition of .... deficiency.

[A] Increases, manganese [B] Decreases, manganese
[C] Increases, iron [D] Decreases, iron
[E] None of the above

59. Copper acts as an activator of which of the following enzyme?

[A] Phenolases [B] Laccase
[C] Ascorbic acid oxidase [D] None of the above
[E] All of the above

60. .... is the activator of Carbonic anhydrase.

[A] Zinc [B] Potassium
[C] Calcium [D] None of the above

61. Sodium is generally essential as a micronutrient to plants utilizing specifically the .... photosynthetic pathway.

[A] C_4 [B] C_3
[C] CAM [D] None of the above

62. Silicon may comprises of .... percent of the dry matter of maize.

[A] 1-2 [B] 4-5 [C] 9-10
[D] 12-16 [E] None of the above

63. Silicon may comprises of .... percent of the dry matter of *Equisetum arvensis*.

[A] 10 [B] 16
[C] 21 [D] None of the above

64. Silicon is accumulated in cell walls, especially of ....

[A] Epidermal cells [B] Cortex cell
[C] Pith cells [D] None of the above

65. Astragalus is known to tolerate high concentration of

[A] Selenium [B] Nickel
[C] Sodium [D] None of the above

66. The concentration of the nutrient measured in the tissue just below the level that gives maximum growth is defined as the

[A] Critical concentration [B] Adequate concentration
[C] Deficient concentration [D] None of the above

67. The concentration above the critical concentration where additional increment in nutrient content will have no beneficial effect on growth and the nutrient content is said to be

[A] Critical concentration [B] Adequate concentration
[C] Deficient concentration [D] None of the above

68. Boron is toxic to plant at concentration above than

[A] 75 $\mu g\ g^{-1}$ dry weight [B] 100 $\mu g\ g^{-1}$ dry weight
[C] 200 $\mu g\ g^{-1}$ dry weight [D] None of the above

69. Zinc is toxic to plant at concentration above

[A] 75 µg g^{-1} dry weight [B] 100 µg g^{-1} dry weight

[C] 200 µg g^{-1} dry weight [D] None of the above

70. Excess nitrogen normally favours a .... shoot/root ratio in plants.

[A] High [B] Low

[C] No [D] None of the above

71. At the soil pH lesser than 6.8, the predominant form of phosphorus and the form most readily absorbed by roots is the

[A] Monovalent orthophosphate anion

[B] Divalent orthophosphate anion

[C] Both

[D] None of the above

72. Potassium is frequently deficient in .... soils because of its high solubility and the ease with which K^+ leached out.

[A] Sandy [B] Clay

[C] Loamy [D] None of the above

73. The sulfur-containing .... are responsible for the pungent flavors of mustards, cabbage, turnips, horseradish and other plant of family Brassicaceae.

[A] Thiocyanates [B] Isothiocuanates

[C] Both [D] None of the above

74. According to .... , there are two principle criteria by which an element can be judged essential or nonessential to any plant. ***First, an element is essential if the plant cannot complete its life cycle in the absence of that element. Second, an element is essential if it forms part of any molecule or constituent of the plant that is itself essential in the plant.***

[A] Arnon and Stout [B] Epstein

[C] Both [D] None of the above

75. EDTA is a synthetic acid in common commercial use. It is sold under the trade name

[A] Versenate [B] Versant

[C] Sequestrene [D] More than one

[E] None of the above

76. Phytosiderophores have a high affinity for

[A] Fe^{3+} [B] Fe^{2+}

[C] Both [D] None of the above

77. 'Stem Crack' in celery and 'Heart rot' in sugar beet is due to deficiency of

[A] Boron [B] Iron

[C] Calcium [D] None of the above

78. **The browning of freshly cut apple and potato surface is due to the activity of copper – containing**

[A] Polyphenoloxidase [B] Phenolases

[C] Both [D] None of the above

79. **SOD (Superoxide dismutase), which detoxifies superoxide radicals, is an important .... containing enzyme.**

[A] Fe [B] Cu

[C] Mn [D] Ca

[E] Both Fe and Cu [F] Both Mn and Cu

[G] More than one [I] None of the above

80. **'Summer dieback' in citrus plants occurs due to deficiency of**

[A] Fe [B] Cu

[C] Mn [D] Ca

[E] Both Fe and Cu [F] Both Mn and Cu

[G] More than one [I] None of the above

81. **...................... readily forms a chelate with humic acid in the organic fraction of soil and may be involved in providing copper to the surface of the root.**

[A] Cu^{+} [B] Cu^{2+}

[C] Both [D] None of the above

82. **'Gray speck' of cereal grains, a disorder characterized by the appearance of greenish-gray, oval-shaped spots on the basal regions of young leaves, occurs due to the deficiency of**

[A] Mn [B] Ca

[C] B [D] More than one

83. **The Molybdenum concentration in dry matter (mmol/kg) considered adequate for normal growth is**

[A] 0.001 [B] 0.05

[C] 0.1 [D] 0.3

[E] 1.0

84. **The Nickel concentration in dry matter (mmol/kg) considered adequate for normal growth is**

[A] 0.001 [B] 0.05

[C] 0.1 [D] 0.3

[E] 1.0

85. **The Copper concentration in dry matter (mmol/kg) considered adequate for normal growth is**

[A] 0.001 [B] 0.05

[C] 0.1 [D] 0.3

[E] 1.0

86. The Zinc concentration in dry matter (mmol/kg) considered adequate for normal growth is

[A] 0.001 [B] 0.05 [C] 0.1
[D] 0.3 [E] 1.0 [F] 2.0

87. The Manganese concentration in dry matter (mmol/kg) considered adequate for normal growth is

[A] 0.1 [B] 0.3
[C] 1.0 [D] 2.0 [E] 3.0

88. The Iron concentration in dry matter (mmol/kg) considered adequate for normal growth is

[A] 0.001 [B] 0.05
[C] 0.1 [D] 1.0 [E] 2.0

89. The Boron concentration in dry matter (mmol/kg) considered adequate for normal growth is

[A] 0.001 [B] 0.05 [C] 0.1
[D] 0.3 [E] 1.0 [F] 2.0
[G] 3.0 [H] 30 [I] 60
[J] 80 [K] 125 [L] 250
[M] 1,000 [N] None of the above

90. The Chlorine concentration in dry matter (mmol/kg) considered adequate for normal growth is

[A] 1.0 [B] 2.0 [C] 3.0
[D] 4.0 [E] None of the above

91. The Sulfur concentration in dry matter (mmol/kg) considered adequate for normal growth is

[A] 30 [B] 60 [C] 80
[D] 125 [E] 250 [F] 1,000
[G] None of the above

92. The Phosphorus concentration in dry matter (mmol/kg) considered adequate for normal growth is

[A] 30 [B] 60 [C] 80
[D] 125 [E] 250 [F] 1,000
[G] None of the above

93. The Magnesium concentration in dry matter (mmol/kg) considered adequate for normal growth is

[A] 30 [B] 60 [C] 80
[D] 125 [E] 250 [F] 1,000
[G] None of the above

94. The Calcium concentration in dry matter (mmol/kg) considered adequate for normal growth is

[A] 30 [B] 60 [C] 80
[D] 125 [E] 250 [F] 1,000
[G] None of the above

95. The Potassium concentration in dry matter (mmol/kg) considered adequate for normal growth is

[A] 30 [B] 60 [C] 80

[D] 125 [E] 250 [F] 1,000

[G] None of the above

96. The Nitrogen concentration in dry matter (mmol/kg) considered adequate for normal growth is

[A] 30 [B] 60

[C] 80 [D] 125

[E] 250 [F] 1,000 [G] None of the above

97. In Vacuole, Ca is frequently precipitated as insoluble

[A] Oxalates [B] Carbonates

[C] Phosphate [D] Sulfate

[E] All of the above

98. According to Cheung and Roberts, in cytosol Ca becomes .... bound to a small protein called calmodulin

[A] Reversibly [B] Irreversibly

[C] Both [D] None of the above

99. Anion of organic acids, especially citric acid, appears to be most important for transport of .... through the ....

[A] Iron, xylem [B] Zinc, phloem

[C] Manganese, xylem [D] All of the above

100. Among the followings, which is the most studied phytosiderophore?

[A] Avenic acid [B] Mugineic acid

[C] Both [D] None of the above

101. Which of the following technique is used for measurement of concentration of elements in plants, soils and nutrient?

[A] Atomic absorption spectrometer

[B] Optical emission spectrometer

[C] Both

[D] None of the above

102. Which of the following plant seedling is used for proof of evidence that nickel is an essential element for plants?

[A] Barley [B] Wheat

[C] Rice [D] None of the above

103. Inorganic nutrients are present in the soil as

[A] Atoms [B] Molecules

[C] Electrically charged ions [D] None of the above

104. Sand culture is a component of

[A] Green house effect [B] Soil-less culture

[C] Photorespiration [D] None of the above

105. Permeability of protoplasm is accelerated by

[A] Sodium [B] Potassium
[C] Calcium [D] None of the above

106. Enzyme catalysed reaction can be inhibited by

[A] Copper ion [B] Zinc ion
[C] Mercury ion [D] None of the above

107. Bidirectional translocation of minerals takes place through

[A] Cambium [B] Phloem
[C] Xylem [D] None of the above

108. Element essential for the photolysis of water is

[A] Chlorine [B] Carbon
[C] Oxygen [D] Nitrogen
[E] None of the above

109. Role of inorganic nutrient in plant growth and metabolism was first discovered by

[A] Woodward [B] Knop
[C] De Sassure [E] Steward [F] None of the above

110. Element responsible for maintaining turgor is

[A] Na [B] K
[C] Ca [D] Hg

111. Mineral absorption is mostly

[A] Chemical process [B] Physical process
[C] Active process [D] Passive process
[E] None of the above

112. Nitrogen fixing bacteria were discovered by

[A] Winogradsky [B] Lister
[C] Koch [D] Pasture [E] None of the above

113. White bud in maize occurs due to deficiency of

[A] Zn [B] Fe
[C] K [D] Ca
[E] None of the above

114. Which of the following scientist was the first to record minerals/ elements in plant ash?

[A] Goldacre [B] Liebig
[C] Von Helmont [D] None of the above

115. Which of the following scientist first developed Hydroponics?

[A] Geriche [B] Bennet-Clark
[C] Both [D] None of the above

116. Mass Flow for the movement of ions was first given by

[A] Hylmno and Kramer [B] Neher and Sakmann
[C] Both [D] None of the above

117. Which of the following element is necessary for translocation of sugars in plants?

[A] Iron [B] Mo

[C] Manganese [D] More than one

[E] None of the above

118. Which of the following essential element land plants normally obtain directly from the air?

[A] Carbon [B] Phosphorus

[C] Nitrogen [D] Hydrogen

[E] None of the above

119. Importance of microelements was recognized late due to

[A] Presence as contaminants in macronutrients

[B] Their toxicity

[C] Their absence in plant ash

[D] Leaching

120. Framework elements are

[A] C, H and O

[B] Cu, Co and Fe

[C] Mg, Cu and Fe

[D] Mn, Ca and N

121. Plant ash indicates

[A] Mineral salts absorbed by plants

[B] Both mineral salts and organic matter

[C] Organic matter

[D] None of the above

122. Theory suggesting that carbon dioxide produced in respiration helps in mineral absorption is called

[A] Carbonic acid exchange theory

[B] Contact exchange theory

[C] Active mineral absorption

[D] Donnan equilibrium

123. Which of the following observation shows that ion uptake in plants is energy dependent?

[A] Ion uptake rates are different for different ions

[B] Ion uptake shows saturation kinetics

[C] Some of the ions accumulate against a concentration gradient

[D] Some ions enter the simplest before reaching the endodermis

[E] More than one

[F] None of the above

124. How does a plant accumulate ions from the soil which has a lower concentration than that in the plant?

[A] Applies tension to pull the ions into the cell

[B] Spends energy to transport ions against the concentration gradient

[C] Roots grow around soil particle trapping the ions in a cell

[D] More than one

[E] None of the above

125. Evidence for Na^+/H^+ antiport at the tonoplast emanated first from the experiments of .... with tonoplast vesicles

[A] Blumwald [B] Poole

[C] Both [D] None of the above

126. Scientist credited with development of formula for balanced solution of mineral elements is

[A] Went [B] Emerson

[C] Knop [D] Hill

[E] More than one

127. Which of the following essential mineral is not a constituent of any enzyme but stimulates the activity of many enzymes?

[A] Zn [B] Mg

[C] Mn [D] K

[E] More than one

128. Passage of mineral from top soil to subsoil through seepage of water is known as

[A] Percolation [B] Transpiration

[C] Leaching [D] Conduction

129. Neher and Sakmann proposed

[A] Patch-clamp technique [B] Single ion channels

[C] Ion vesicles [D] Ion carriers

[E] None of the above

14 TRANSPIRATION & GUTTATION

1. **The loss of water in form of vapour from aerial parts of plants is known as**
   [A] Transpiration [B] Guttation
   [C] Both [D] None of the above
2. **Loss of water in form of water from aerial parts of plants is known as**
   [A] Transpiration [B] Guttation
   [C] Both [D] None of the above
3. **Living cells are not required for**
   [A] Evaporation [B] Guttation
   [C] Transpiration [D] None of the above
4. **In higher angiosperm, water is lost from aerial parts of plants mainly through**
   [A] Cuticular transpiration [B] Stomatal transpiration
   [C] Lenticular transpiration [D] None of the above
5. **Evaporation is a**
   [A] Physical process [B] Physiological process
   [C] Chemical process [D] None of the above
6. **Transpiration is a**
   [A] Vital physiological Process [B] Vital Physical Process
   [C] Both [D] None of the above
7. **Living tissue is essential for**
   [A] Vital physiological Process [B] Vital Physical Process
   [C] Both [D] None of the above
8. **Transpiration is similar to evaporation in the loss of water vapours but differs from the same in**
   [A] Formation of water vapours internally
   [B] Being controlled by a number of plant factors
   [C] Occurrences of even when the air is saturated
   [D] Influence of light
   [E] All of the above

9. **.................... of water absorbed by a plant is lost through transpiration.**

[A] 98-99% [B] 90-91%
[C] 93-95% [D] None of the above

10. **A minimum amount of transpiration occurs through**

[A] Lenticular transpiration [B] Stomatal transpiration
[C] Cuticular transpiration [D] All of the above

11. **The amount of water transpired by a plant for the synthesis of a unit dry matter is known as**

[A] Transpiration Ratio [B] Water requirement
[C] Efficiency of transpiration [D] All of the above

12. **Stomata are usually confined in more number on the .... side of the leaves.**

[A] Lower [B] Upper
[C] Both [D] None of the above

13. **In .... stomata are distributed equally on both sides of the leaves.**

[A] Monocots [B] Dicots
[C] Both [D] None of the above

14. **In aquatic plants with floating leaves stomata are present on the .... surface.**

[A] Upper [B] Lower
[C] Both [D] None of the above

15. **Amount of transpiration by sunflower is**

[A] 0.5-2.5 g/dm²/hr [B] 2 litre/day
[C] 36-45 litres/day [D] 1 tonne/day

16. **Amount of transpiration by apple is**

[A] 0.5-2.5 g/dm²/hr [B] 2 liters/day
[C] 36-45 litre/day [D] 1 tons/day

17. **Amount of transpiration by Elm is**

[A] 0.5-2.5 g/dm²/hr [B] 2 litre/day
[C] 36-45 litres/day [D] 1 tons/day

18. **Transpiration ratio for CAM plants is**

[A] 50 [B] 100-200
[C] 300-500 [D] None of the above

19. **Transpiration ratio for C_4 plants is**

[A] 50 [B] 100-200
[C] 300-500 [D] None of the above

20. **Transpiration ratio for C_3 plants is**

[A] 50 [B] 100-200
[C] 300-500 [D] None of the above

21. Stomata remain open at relative humidity above than .... and close below relative humidity ....

[A] 70%, 50% [B] 50%, 70%

[C] Both [D] None of the above

22. .... was the first to point out that stomatal opening and closing is due to turgor changes in guard cells.

[A] Schwendener [B] Milbarrow

[C] Mouravieff [D] None of the above

23. Which of the following scientist proposed that water stress produces ABA which induces the closer of stomata?

[A] Schwendener [B] Milbarrow

[C] Mouravieff [D] None of the above

24. SPAC is

[A] Soil-plant-atmosphere continuum

[B] Soil-plant-animal continuum

[C] Both

[D] None of the above

25. The term Guttation was coined by

[A] Bergerstein [B] Steward

[C] Curtis [D] None of the above

26. Water coming out of guttation contains

[A] Inorganic substance [B] Organic substance

[C] Both [D] None of the above

27. The phenomenon of guttation is associated with the presence of special types of stomata at the margins of the leaves which are called as

[A] Water stomata [B] Hydathodes

[C] Both [D] None of the above

28. Each hydathode consists of a water pore which is .... open.

[A] Permanently [B] Temporarily

[C] Not [D] None of the above

29. Below the water pore of the hydathodes, there is a small cavity followed by a loose tissue called as

[A] Epithem [B] Epiblem

[C] Perithem [D] Periblem

[E] None of the above

30. The epithem is in close association with the

[A] Tracheids [B] Vessels

[C] Both [D] None of the above

31. Number of stomata/cm^2 Upper/lower leaf surface of *Oxalis acetosella* is

[A] 0/4500 [B] 5200/6800

[C] 2100/4600 [D] 1200/13000 [E] None of the above

32. Number of stomata/cm^2 Upper/lower leaf surface of *Zea mays* is

[A] 0/4500 [B] 5200/6800
[C] 2100/4600 [D] 1200/13000 [E] 101002/21600

33. Number of stomata/cm^2 Upper/lower leaf surface of *Triticum aestivum* is

[A] 0/4500 [B] 5200/6800
[C] 2100/4600 [D] 1200/13000 [E] 101002/21600

34. Number of stomata/cm^2 Upper/lower leaf surface of *Lycopersicon esculentum* is

[A] 0/4500 [B] 5200/6800
[C] 2100/4600 [D] 1200/13000 [E] 101002/21600

35. Number of stomata/cm^2 Upper/lower leaf surface of *Pisum sativum* is

[A] 0/4500 [B] 5200/6800
[C] 2100/4600 [D] 1200/13000 [E] 101002/21600

36. Number of stomata/cm^2 Upper/lower leaf surface of *Helianthus annuus* is

[A] 0/4500 [B] 5200/6800
[C] 2100/4600 [D] 8527/15504 [E] 101002/21600

37. Number of stomata/cm^2 Upper/lower leaf surface of *Cucurbita pepo* is

[A] 0/4500 [B] 5200/6800
[C] 2791/27132 [D] 1200/13000 [E] 101002/21600

38. Number of stomata/cm^2 Upper/lower leaf surface of *Solanum tuberosum* is

[A] 5100/16600 [B] 5200/6800
[C] 2100/4600 [D] 1200/13000 [E] 101002/21600

39. Number of stomata/cm^2 Upper/lower leaf surface of *Morus alba* is

[A] 0/48000 [B] 5200/6800
[C] 2100/4600 [D] 1200/13000 [E] 101002/21600

40. Number of stomata/cm^2 Upper/lower leaf surface of *Quercus eiccinea* is

[A] 0/4500 [B] 5200/6800
[C] 2100/4600 [D] 0/103800 [E] 101002/21600

41. Higher Root/Shoot ratio has .... rate of transpiration

[A] Higher [B] Lower
[C] Very less [D] None of the above

42. Transpiration is regarded as necessary evil. Who made this statement?

[A] Curtis [B] Steward
[C] Both [D] None of the above

43. Transpiration is regarded as unavoidable evil. Who proposed this?

[A] Curtis [B] Steward
[C] Both [D] None of the above

44. The instrument for measuring the rate of transpiration by shoots through measuring the rate of the water absorption is known as

[A] Potometer [B] Porometer
[C] Auxinometer [D] None of the above

45. An instrument that gives a rough idea about the degree of stomatal opening and closing is

[A] Potometer [B] Porometer
[C] Auxinometer [D] None of the above

46. In isobilateral leaf, the number of stomata per unit area is

[A] More on lower surface
[B] More on upper surface
[C] Approximately same on both the surface
[D] None of the above

47. In dorsiventral leaf, the number of stomata per unit area is generally

[A] More on lower surface
[B] More on upper surface
[C] Approximately same on both the surface
[D] None of the above

48. The stomata are called sunken when

[A] Guard cells occur in lower epidermis
[B] Guard cells are situated above epidermis
[C] Guard cells are situated below epidermis
[D] Guard cells are in line with epidermal cells

49. Stomatal frequency indicates

[A] Length of stomata
[B] Number of stomata per unit area
[C] Rate of water loss from per stomata
[D] None of the above

50. Porometer was developed by

[A] Darwin [B] Freeman
[C] Ganong [D] None of the above

51. Cobalt chloride paper test for transpiration was developed by

[A] Stahl [B] Curtis
[C] Ganong [D] None of the above

52. Margin of mature *Colocasia* leaves appear withered/necrotic due to

[A] Guttated salt [B] Excess of transpiration
[C] Excess of mineral [D] None of the above

53. The factors that favour guttation are

[A] High water absorption [B] High root pressure
[C] Reduced or no transpiration [D] All of the above

54. Which of the following plant shows phenomenon of guttation?

[A] *Tropaeolum* [B] *Garden nasturtium*

[C] *Colocasia* [D] All of the above

55. In plants, rate of transpiration is usually maximum in

[A] Little before noon [B] Morning

[C] Afternoon [D] Evening [E] All of the above

56. Which of the following substance is anti-transpirant?

[A] Phenylmercuric acetate

[B] Very high concentration of CO_2

[C] Low viscosits waxes and ABA

[D] Colour plastic and silicone oils

[E] All of the above

57. The colour of dry cobalt chloride paper when tightly pressed on the transpiring surface of a leaf changes from

[A] Green to blue [B] Pink to blue

[C] Blue to green [D] Blue to pink

[E] None of the above

58. Which of the following form of light has controlling influence on stomatal opening?

[A] Red [B] Blue

[C] Far-red [D] Green [E] None of the above

59. In plants, transpiration plays an important role in

[A] Translocation of mineral salts

[B] Regulation of temperature

[C] Ascent of sap [D] All of the above

60. Stomata open at night in

[A] C_3 plants [B] C_4 plants

[C] CAM plants [D] None of the above

61. Equisetum type stomata

[A] Always open [B] Always close

[C] Both [D] More than one

[E] None of the above

62. Who proposed the opening and closing of stomata is connected with the daily opening and closing ?

[A] Loft field [B] Lyoyd

[C] Von Mohl [D] None of the above

63. When half of the leaf is removed randomly, transpiration will show

[A] Lower magnitude but higher flux

[B] Both magnitude and flux increases

[C] Higher magnitude but lower flux or rate per unit

[D] Both magnitude and flux decreases

64. Non functional stomata can be seen in

[A] Lotus [B] *Hydrila*

[C] Mango leaf [D] Pea leaves

65. Hydathodes occur on

[A] Stem [B] Leaves

[C] Roots [D] All of the above

66. Which of the factor is most important in regulation of transpiration?

[A] Temperature [B] Humidity

[C] Wind [D] Light

67. At constant temperature, the rate of transpiration will be higher at

[A] Sea level [B] 1 km below sea level

[C] 1 km above sea level [D] 1.5 km above sea level

68. Elements involved in stomatal regulation is

[A] Zinc [B] Sodium

[C] Potassium [D] Iron

69. Low atmospheric pressure changes the rate of transpiration to

[A] Decrease slowly [B] Increase

[C] Remain unchanged [D] Decrease rapidly

70. Which one gives the most valid explanation for guard cell movements?

[A] Transpiration [B] Potassium influx and efflux

[C] Starch hydrolysis theory [D] Guard cell photosynthesis

71. The stomatal type of cereals which open only for a few hours during the day is

[A] Bean type [B] Alfalfa type

[C] Potato type [D] Barley type

[E] More than one [F] None of the above

72. In thin-leaved mesophytes, stomata open during the day and close during the night. They belong to

[A] Bean type [B] Alfalfa type

[C] Potato type [D] Barley type

[E] More than one [F] None of the above

73. Transpiring structure in plants is

[A] Phloem [B] Epidermis

[C] Xylem [D] Cortex

74. PMA (Phenyl Mercuric Acetate) results in

[A] Reduced photosynthesis [B] Reduced transpiration

[C] Reduced respiration [D] Killing of plants

75. Which wall of the guard cell is thick?

[A] Lateral [B] Inner

[C] Outer [D] All of the above

76. Water movement against gravity is caused by

[A] Imbibitions [B] Transpiration
[C] Osmosis [D] Diffusion

77. Most common type of transpiration is

[A] Stomatal [B] Cuticular
[C] Lenticular [D] Foliar

78. Transpiration differs from evaporation in

[A] Frequency of water loss
[B] Transpiration is a physical process and evaporation is physiological process
[C] Transpiration is a physiological process and evaporation is a physical process
[D] None of the above

79. In *Cucurbita*, Onion, *Portulaca*, Tulip and Banana, stomata open whole day and night except for a few hours during evening, it is known as

[A] Potato type [B] Barley/Cereal type
[C] Alfalfa/Leucerne type [D] None of the above

80. In Pea, Radish, Mustard, Turnip, Beet, Apple, Grape and Bean, stomata open whole day and close down during night, it is known as

[A] Potato type [B] Barley /Cereal type
[C] Alfalfa/Leucerne type [D] None of the above

81. The Stomatal pores open when

[A] H+ ions are pumped into the guard cells
[B] K+ ions are pumped out of the guard cells
[C] Water moves into the guard cells in response to a decrease in water potential in the guard cells.
[D] Water moves out of the guard cells in response to a decrease water potential of the guard cells
[E] None of the above

82. Root pressure is due to

[A] Increase in transpiration [B] Active absorption
[C] Low osmotic potential in soil [D] Passive absorption

83. The minimum light level for the opening of stomata in most plants is about

[A] 1/1000 to 1/30 of full sunlight [B] 1/500 to 1/20 of full sunlight
[C] 1/100 to 1/10 of full sunlight [D] None of the above

84. Stomata respond to light even in leaves in which photosynthesis had been reduced to zero by application of an inhibitor. This statement is true or false

[A] True [B] False
[C] Not known [D] None of the above

85. .... measured the wavelength of light that was most effective in causing stomata to open.

[A] Sharkey and Raschke [B] Outlaw and Tarczynski

[C] Both [D] None of the above

86. Blue light is nearly .... times as effective as red light in case of stomatal function.

[A] 10 [B] 20

[C] 30 [D] 40

[E] Not known

15 PLANT TISSUE CULTURE

1. **Protoplasts are the plant cells with a plasma membrane but no**

   [A] Cell wall [B] Ribosome

   [C] Mitochondria [D] None of the above

2. **Protoplast is a .... individual cell.**

   [A] Functional [B] Non functional

   [C] Not known [D] None of the above

3. **Which of the following scientist tried to isolate protoplast from plasmolysed higher plant cells by mechanically cutting the cells?**

   [A] Klerckel [B] Takabe

   [C] Carlson [D] None of the above

4. **Which of the following scientist, with the help of commercially available enzymes, isolated protoplast and able to regenerate whole tobacco plant from mesophyll protoplast?**

   [A] Klerckel [B] Takabe

   [C] Carlson [D] None of the above

5. **Which of the following scientist was successful in obtaining the first interspecific somatic hybrid using protoplast fusion between *Nicotiana glauca* and *Nicotiana longsdorffii*?**

   [A] Klerckel [B] Takabe

   [C] Carlson [D] None of the above

6. **The first intergeneric hybrid (pomato) between potato and tomato was demonstrated by**

   [A] Melchers [B] Zelcer

   [C] Fromm [D] None of the above

7. **...................... *et al.* obtained the first interspecific hybrid between protoplast of normal *Nicotiana sylvisteris* and X-ray irradiated protoplast of male sterile *N. tabacum*.**

   [A] Melchers [B] Zelcer

   [C] Fromm [D] None of the above

8. The first release of a commercial cultivar obtained from protoplast fusion was in

[A] 1990 [B] 1999
[C] 1980 [D] None of the above

9. Somatic embryogenesis involves many stages/phases that include

[A] Choice of plant material [B] Sterilization of the explants
[C] Enzymatic treatment [D] Protoplast fusion
[E] Culture of protoplast [F] Regeneration of plants
[G] Verification and control of fusion products and Testing the usefulness in plant bredding
[H] All of the above

10. Isolation of protoplast using mechanical methods involves isolation from vacuolated cells of storage tissue such as

[A] Onion [B] Radish
[C] Mesocarp of cucumber [D] All of the above

11. Protoplasts are counted by

[A] Haemocytometer [B] Protocytometer
[C] Gigermuller counter [D] None of the above

12. Protoplast suspension is centrifuged at speed

[A] 50-100 × g/5 min [B] 100-150 × g/5 min
[C] 200-250 × g/5 min [D] None of the above

13. Intact protoplasts form pellet. This statement is

[A] True [B] False
[C] Not known [D] None of the above

14. Protoplast Viability Test can be done through

[A] Exclusion of Evans Blue dye
[B] Presence of cytoplasmic streaming
[C] Change in protoplast size induced by changes in the level of osmoticum
[D] Measure of Photosynthetic and Respiratory activities
[E] All of the above

15. Presence of ammonium in culture medium is toxic to protoplast of

[A] Potato [B] Several fruit tree species
[C] Some species of compositae [D] All of the above

16. Protoplasts from actively growing cell culture require high

[A] Auxin/kinetin [B] Kinetin/auxin
[C] Auxin=kinetin [D] None of the above

17. Protoplasts from highly differentiated cell culture require high

[A] Auxin/kinetin [B] Kinetin/auxin
[C] Auxin=kinetin [D] None of the above

18. Suitable temperature for protoplast culture is

[A] 24-28°C [B] 20-28°C

[C] 19-28°C [D] None of the above

19. A short heat treatment at 45°C for 5 min improved protoplast division in

[A] Wheat [B] Rice

[C] *Pennisetum* [D] All of the above

20. The protoplast fusion requires .... contact between membrane surfaces.

[A] Direct [B] Indirect

[C] No contact [D] All of the above

21. The contact of protoplasts depends on surface

[A] Charge [B] Permeability

[C] Heat [D] None of the above

22. The outer membrane surface is negatively charged and varies from

[A] -10m V to -50 mV [B] -20 mV to -60 mV

[C] -10 mV to -20 mV [D] None of the above

23. The +vely charged plasma membrane can be obtained in the presence of

[A] Poly-L-lysin [B] Synthetic phospholipids

[C] Both [D] None of the above

24. There are several chemical and physical agents, known as fusogen or fusion agents, which help in fusion between protoplasts. These are

[A] Sodium nitrate [B] Polyethylene glycol

[C] Dextran sulphate [D] All of the above

25. High Ca^{2+} ion concentration and pH (10.5) of the incubation mixture .... fusion process.

[A] Facilitates [B] Decreases

[C] Do not affect [D] None of the above

26. The fusion and fusion product of protoplasts are observed in

[A] Inverted microscope [B] Phase contrast microscope

[C] Electron microscope [D] None of the above

27. The isolated protoplast regenerates new walls around themselves. This can be demonstrated by staining the reconstituted protoplast for 5 min with a fluorescent dye ....

[A] Calcofluor white [B] Evans blue

[C] Fluorescein diacetate [D] All of the above

28. In some species protoplasts can directly form somatic embryos, these are

[A] *Medicago* [B] *Brassica juncea*

[C] *Citrus* [D] All of the above

29. **In electric fusion of protoplast, fusion is initiated by field pulse of voltage range from .... for .... microseconds.**

[A] 500V-3 kV cm^{-1}, 10-1000
[B] 900V-5 kV cm^{-1}, 10-1000
[C] 100V-1 kV cm^{-1}, 10-1000
[D] None of the above

30. **The nuclei of two protoplasts may or may not fuse together even after fusion of cytoplasm. The binucleate cells are known as heterokaryon. The nuclei when get fused, the cells are called hybrids, and when only cytoplasm fuse and genetic information from one of the two nuclei is lost, it is called ....**

[A] Cybrid [B] Cytoplasmic hybrids
[C] Heteroplast [D] All of the above

31. **To isolate the fusion products and for selection of plant cells for desirable traits, various strategies have been developed**

[A] Direct isolation
[B] Resistance to anti-metabolites
[C] Metabolic inhibitors
[D] Fluorescence activated cell sorting
[E] All of the above

32. **Which of the following is the example of Inter-specific hybrid?**

[A] *Solanum pinnatisectum* + *S. tuberosum*
[B] *Brassica napus* + *B. nigra*
[C] *Cucumis melo* + *C. sativus*
[D] *Nicotiana tabacum* + *N. rustica*
[E] All of the above

33. **Which of the following is the examples of Inter-generic hybrid?**

[A] *Brassica napus* + *Raphanus sativus*
[B] *B. napus* + *Moricandia arvensis*
[C] *Citrus reticulata* + *Citropsis gilletiana*
[D] *Medicago sativa* + *Onobrychis vicifolia*
[E] All of the above

34. **Which of the following is the best example of Inter-familial hybrid?**

[A] *Nicotiana tabacum* + *Daucus carota*
[B] *Oryza sativa* + *Daucus carota*
[C] Both
[D] None of the above

35. **Which of the following is the best examples of Cybrid?**

[A] *Nicotiana tabacum* + *N. sylvestris*
[B] *Brassica oleracea* + *B. napus*
[C] Both
[D] None of the above

36. .... for the first time showed that mature pollen grains of *Ginkgo biloba* can be induced to proliferate in culture to produce haploid callus.

[A] Tulecke [B] Sipra guha

[C] C. Maheshwari [D] Both A and B

[E] Both B and C [F] None of the above

37. .... and .... reported the direct development of embryoids from microspores of *Datura innoxia* by the culture of excised anther.

[A] Tulecke [B] Sipra guha

[C] C. Maheshwari [D] Both A and B

[E] Both B and C [F] None of the above

38. .... and .... obtained the first haploid plants from isolated anthers of *Nicotiana tabacum.*

[A] Bourgin [B] Nitsch

[C] C. Maheshwari [D] Both A and B

[E] Both B and C [F] None of the above

39. Cause of Gynogenesis in plant is

[A] Development of unfertilized egg-cell

[B] Delayed pollination [C] Abortive pollen

[D] Alien pollen [E] All of the above

40. Which of the following species is highly responsive for haploids regeneration?

[A] Solanaceae [B] Cruciferae

[C] Gramineae [D] Ranunculaceaae

[E] All of the above

41. When a part of plant organ or tissue is put on a culture medium and maintained under optimum growth conditions, the meristematic cells de-differentiate and proliferate to form a mass of cells called

[A] Callus [B] Embryoids

[C] Both [D] None of the above

42. Embryogenesis from nuclei of unfertilized ovules of *Citrus sinensis* and *Citrus aurantifolia* is reported by

[A] Mitra & Chaturvedi [B] Kohn

[C] Chilton [D] Scowcroft [E] None of the above

43. Who coined term 'Somaclonal variation'?

[A] Larkin & Scowcroft [B] Krens

[C] Chilton [D] Kohn [E] Melchers

44. Shoot initiation from cryopreserved shoot apices of carnation was demonstrated by

[A] Larkin & Scowcroft [B] Krens

[C] Chilton [D] Seibert [E] Melchers

45. **Protoplast able to incorporate naked DNA was demonstrated by**
[A] Larkin & Scowcroft [B] Krens
[C] Chilton [D] Kohn [E] Melchers

46. **Production of transformed tobacco plants following single cell transformation or gene insertion was shown by**
[A] Larkin & Scowcroft [B] Krens
[C] Chilton [D] Kohn [E] Melchers

47. **Methods to double the chromosome number in microspores of *Datura* and *Nicotiana* were given by**
[A] Larkin & Scowcroft [B] Melchers
[C] Seibert [D] Nitsch [E] None of the above

48. **Discovery of Ti-plasmid was made by**
[A] Larkin & Scowcroft [B] Larebeke
[C] Chilton [D] Pierik [E] Melchers

49. **Somatic hybridization of tomato and potato was demonstrated by**
[A] Larkin & Scowcroft [B] Krens
[C] Chilton [D] Kohn [E] Melchers

50. **Publication of first extensive handbook on plant tissue culture was made by**
[A] Gautheret [B] Steward [C] Skoog and miller
[D] Philip White [E] Morel & Martin

51. **First successful isolation of protoplasts from a suspension culture of *Hapopappus grecilis* was by**
[A] Gautheret [B] Erikson & Jonassen
[C] Skoog and miller [D] Philip White
[E] Morel & Martin

52. **Selection of biochemical mutants *in vitro* was done by**
[A] Carlson [B] Tekebe
[C] Skoog and miller [D] Philip White

53. **First achievement of protoplast fusion was by**
[A] Carlson [B] Steward
[C] Power [D] Philip White
[E] Morel & Martin

54. **Successful culture of tomato roots was demonstrated by**
[A] Carlson [B] Steward
[C] Skoog and miller [D] Philip White
[E] Morel & Martin

55. **Among the followings, which of the culture is known as 'mericloning'?**
[A] Root culture [B] Meristem culture
[C] Stem culture [D] None of the above

56. The explants may carry microorganisms hence these are surface sterilized by

[A] Mercuric chloride [B] Sodium hypochlorite
[C] Both [D] None of the above

57. Microorganism may be present in nutrient medium at the time of its preparation. These microorganisms can be destroyed by proper plugging and autoclaving the culture flasks. Medium is sterilized completely by maintaining it at .... °C for about 20 minutes at pressure of .... kg cm^{-2} in autoclave.

[A] 121, 1.0-1.5 [B] 98, 2.0-2.5
[C] 121, 2.0-2.5 [D] None of the above

58. .... coined the term 'totipotency' to denote the capacity of cell to develop into an organism by regeneration.

[A] Morgan [B] Steward
[C] White [D] None of the above

59. The somatic embryo can be used for production of synthetic seeds for direct sowing in the field. This statement is true/false

[A] True [B] False
[C] Not known [D] None of the above

60. Which auxin is required for embryo induction as in case of cereal crops?

[A] 2,4-D [B] IAA
[C] NAA [D] IBA

61. Water soaking was described as 'Vitrification' by

[A] Debergh [B] Hussey
[C] Anonymus [D] None of the above

62. In hydrated synthetic seed, somatic embryos are encapsulated in

[A] Calcium alginate [B] Polyoxyethylene glycol
[C] Both [D] None of the above

63. Desiccated synthetic seeds are produced by coating somatic embryo in

[A] Calcium alginate [B] Polyoxyethylene glycol
[C] Both [D] None of the above

64. .... gave the concept of artificial seeds for the first time.

[A] T. Murashige [B] Ganapathy
[C] Both [D] None of the above

65. Shoot tips excised from the shoot cultures of banana are generally encapsulated in .... % sodium alginate solution.

[A] 3 [B] 2
[C] 1 [D] None of the above

66. Alginate artificial seeds are

[A] Spherical [B] Transparent
[C] Non-inhibitory [D] All of the above

67. An encapsulation machine can produce .... single embryos per hour.

[A] 6000 [B] 4000

[C] 2000 [D] None of the above

68. Which of the following propagules are mostly used for synthetic seeds production in *Actinidia deliciosa, Musa* sp., *Betula pendula, Brassica campestris, Zingiber officinale, Rubus* sp.?

[A] Shoot buds [B] Somatic embryos

[C] Protocorm-like bodies [D] Embryonic masses

[E] Axillary buds [F] All of the above

69. Which of the following propagules are mostly used for synthetic seeds production in *Arachis hypogaea, Asparagus cooperi, Camellia japonica, Daucus carota, Eucalyptus citriodora*?

[A] Shoot buds [B] Somatic embryos

[C] Protocorm-like bodies [D] Embryonic masses

[E] Axillary buds [F] All of the above

70. "Mingo" – the first licensed cultivar of .... was produced using haploid induced *in vitro* based on chromosome elimination, bulbosum method.

[A] Barley [B] Wheat

[C] Oat [D] None of the above

71. Which of the following method is employed for producing virus free plants?

[A] Meristem culture [B] Heat treatment

[C] Heat treatment followed by meristems culture

[D] Adventitious shoot formation followed by meristems culture

[E] Grafting of meristems on virus- free plants

[F] All of the above

72. The possible mechanisms of metal tolerance in plants are

[A] Binding of metal to cell wall

[B] Reduced transportation through cell

[C] Compartmentation of ions in vacuoles

[D] Metal-metal resistant enzymes

[E] Chelating by specific metal-binding proteins

[F] All of the above

73. The first report on mutant isolation through plant tissue cultures is on the selection of .... suspension cells tolerant to high temperature.

[A] *Antirrhinum* [B] Linum

[C] *Cicer* [D] None of the above

74. Hydroxyproline resistant lines of potato showed increased tolerance to

[A] Salinity [B] Frost

[C] Both [D] None of the above

75. The concept of cryopreservation is based on arrest of .... of biological material stored at low temperature (-196° C).

[A] Diffusion [B] Osmosis
[C] Imbibition [D] None of the above

76. All conventional methods of storage failed to prevent losses caused by

[A] Climate disturbance [B] Attack of pathogens and pests
[C] Natural disorders [D] Economic and political issues
[E] All of the above

77. Glycerol, DMSO and methanol are .... type of cryoprotectants.

[A] Permeating [B] Non permeating
[C] Both [D] None of the above

78. Sugars, sugars alcohols, polyvinyl pyrrolidone, polyethylene glycol (PEG), polyethylene oxide (PEO), hydroxystarch are the examples of .... type of cryoprotectants.

[A] Permeating [B] Non permeating
[C] Both [D] None of the above

79. In which plant mosaic virus was demonstrated to be eliminated by meristem culture?

[A] *Allium sativum* [B] *Brassica oleracea*
[C] *Chrysanthemum* sp. [D] *Dianthus barbatus*
[E] *Ipomoea batata*

80. In which plant Turnip mosaic virus and CaMV were shown to be eliminated by meristem culture?

[A] *Allium sativum* [B] *Brassica oleracea*
[C] *Chrysanthemum* sp. [D] *Dianthus barbatus*
[E] *Ipomoea batata*

81. In which plant, Virus B has been shown to be eliminated by meristem culture?

[A] *Allium sativum* [B] *Brassica oleracea*
[C] *Chrysanthemum* sp. [D] *Dianthus barbatus*

82. In which plant, Mottle virus, Vein mottle virus, Ring spot virus can be eliminated by meristem culture?

[A] *Allium sativum* [B] *Brassica oleracea*
[C] *Chrysanthemum* sp. [D] *Dianthus barbatus*
[E] *Ipomoea batata*

83. Ratio of light: dark period in environmentally controlled culture room normally is

[A] 16:8 [B] 11:8
[C] 14:6 [D] None of the above

84. What is the percentage of relative humidity that is maintained normally in environmentally controlled culture room?

[A] 60-70 [B] 30-40
[C] 20-30 [D] None of the above

85. What is the intensity of fluorescent light, in general, in environmentally controlled culture room?

[A] 1000 lux [B] 200 lux
[C] 3000 lux [D] None of the above

86. When embryogenesis occurs directly on the explants without production of callus, it is known as

[A] Direct embryogenesis [B] Indirect embryogenesis
[C] Both [D] None of the above

87. When explants produce callus and the callus forms embryos then it is called as

[A] Direct embryogenesis [B] Indirect embryogenesis
[C] Both [D] None of the above

88. In case of direct embryogenesis cells of explanted tissues are already determined for embryogenic development, and this is known as

[A] Pre-embryogenic determined cells
[B] Induced embryogenic determined cells
[C] Both
[D] None of the above

89. In case of indirect embryogenesis cells require redetermination through a period in culture and this is termed as

[A] Pre-embryogenic determined cells
[B] Induced embryogenic determined cells
[C] Both
[D] None of the above

90. When somatic embryos are transferred on induction medium they give rise to secondary somatic embryos. This method of obtaining embryos recurrently is termed as

[A] Repetitive embryogenesis [B] Cyclic embryogenesis
[C] Both [D] None of the above

91. Which of the following technology is used for Virus Indexing for biological material?

[A] TEM serology
[B] ELISA
[C] Sap of plants to test virus infection
[D] All of the above

92. "Ruseet Burbank" cultivars of potato have originated from .... protoplast.

[A] Leaf [B] Shoot
[C] Root [D] None of the above

93. Polyploid shows chromosomal abnormalities .... than diploids.

[A] More [B] Less
[C] Very less [D] None of the above

94. **..................... is done by removing the specimen from frozen state (liquid nitrogen stage) and transferring them in a water bath at 35-40°C for about 1-2 min or until specimen is warmed up.**

[A] Thawing [B] Regeneration

[C] Freezing [D] None of the above

95. **Which of the following institute is engaged in preservation of plant germplasm based on cryopreservation technology?**

[A] National Bureau of Plant Genetic Resources, New Delhi

[B] Department of Biotechnology, Govt. of India, New Delhi

[C] IRRI, Philippines

[D] International Plant Germplasm Research Institute

[E] All of the above

96. **Through suspension culture, Ojima and Ohira selected two cell lines of .... tolerant to Al and Mn by subculturing cells in excessive amount of $AlCl_3$ and $MnCl_3$ for several months.**

[A] Carrot [B] Cauliflower

[C] Cabbage [D] None of the above

97. **When normal protoplast of two species are fused, the resulting somatic hybrid plants have somatic chromosome complement of both the fusion parents. Such somatic hybrids are called**

[A] Symmetric hybrids [B] Asymmetric hybrids

[C] Both [D] None of the above

98. **Cybrids can be obtained by using any one of the following methods.**

[A] Fusion of normal protoplast from one parent with enucleated protoplast from the other parent

[B] Fusion of normal protoplast from one parent and protoplast containing non- viable nuclei from the other parent

[C] Selective elimination of one of the nuclei from the heterokaryon

[D] Selective elimination of chromosome of one parent at a later stage after fusion of the nuclei

[E] All of the above

99. **Young plant explants (i.e. soften) are more susceptible to vitrification. This statement is**

[A] True [B] False

[C] Not known [D] None of the above

100. **What is the medicinal value of *Artimisia annua* ?**

[A] Antimalarial [B] Antileukaemic

[C] Anticholinergic [D] Agglutination

[E] Anti-viral

101.

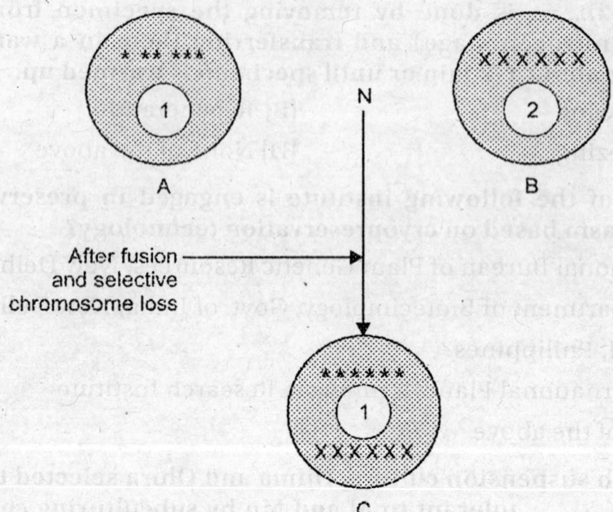

In above drawn figure, find C=?

[A] C= Heterokaryon [B] C= Cybrid

[C] Both [D] None of the above

102. What is the medicinal value of *Sanatalum album*?

[A] Antimalarial [B] Antileukaemic

[C] Anticholinergic [D] Agglutination

[E] Anti-viral [F] Hepatic stimulant

[G] Essential oil

103. What is the medicinal value of *Catharanthus roseus*?

[A] Antimalarial [B] Antileukaemic

[C] Anticholinergic [D] Agglutination

104. What is the medicinal value of *Datura innoxia*?

[A] Antimalarial [B] Antileukaemic

[C] Anticholinergic [D] Agglutination

105. What is the medicinal value of *Moringa oleifera*?

[A] Antimalarial [B] Antileukaemic

[C] Anticholinergic [D] Agglutination

[E] Anti-viral

106. What is the medicinal value of *Delonix regia*?

[A] Antimalarial [B] Antileukaemic

[C] Anticholinergic [D] Agglutination

107. What is the medicinal value of *Taraxacum officinale*?

[A] Antimalarial [B] Antileukaemic

[C] Anticholinergic [D] Agglutination

[E] Anti-viral [F] Hepatic stimulant

108. What is the medicinal value of *Hyoscyamus niger*?

[A] Antimalarial [B] Antileukaemic

[C] Anticholinergic [D] Antispasmolytic

109. What is the medicinal value of *Panax ginseng*?

[A] Antimalarial [B] Antileukaemic

[C] Anticholinergic [D] Antispasmolytic

[E] Tonic

110. Tissue culture medium (MS) is adjusted to a pH of .... before autoclaving.

[A] 5.6-6.0 [B] 6.0-7.0

[C] 3.5-5.6 [D] None of the above

16 ECO-PHYSIOLOGYA AND ENVIRONMENTAL BIOLOGY

1. **Biodiversity includes diversity of form right from molecular level through**

[A] Individual | [B] Organism
[C] Populations | [D] Communities
[E] Ecosystems | [F] Landscaping
[G] Biosphere | [H] All of the above

2. **....................... represents number of species in a specified area/region.**

[A] Point or -diversity | [B] -diversity
[C] -diversity | [D] None of the above

3. **....................... is represented by turnover of species across space.**

[A] Point or -diversity | [B] -diversity
[C] -diversity | [D] None of the above

4. **India is one of the .... mega-diversity countries in the world.**

[A] 12 | [B] 13
[C] 14 | [D] 15 | [E] None of the above

5. **The use of .... and .... help indicate about the rate of species extinction in given habitats/patches.**

[A] Species-area relations | [B] Endemics area relationship
[C] Both | [D] None of the above

6. **....................... refers to conservation of samples of genetic diversity away from their field habitats and done through botanic and zoological gardens, banks of germplasm, seed, pollen, seedlings, gene, DNA, tissue culture and culture collection.**

[A] *Ex situ* conservation | [B] *In situ* conservation
[C] Both | [D] None of the above

7. **................ refers to protection of habitat and ecosystem, conservation of plants and animals in their native habitat or even in man-made ecosystem where they naturally occur. e.g. National Parks, Sanctuaries, National Monuments, Native Reserve, Cultural Landscapes, Biosphere Reserves etc.**

[A] *Ex situ* conservation | [B] *In situ* conservation
[C] Both | [D] None of the above

8. Altitude zonation of vegetation on mountains is presented below. Write the different types of vegetation with increasing altitude.

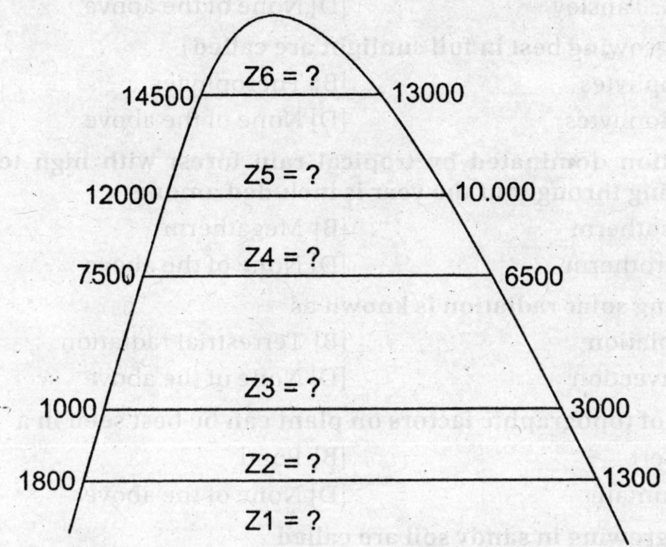

[A] Z1= Tropical rain forest, Z2= grassland or desert, Z3= Deciduous forest, Z4= Coniferous forest, Z5= Tundra, Z6 = Snow

[B] Z1= Tropical rain forest, Z2=Deciduous forest, Z3= grassland or desert, Z4= Coniferous forest, Z5= Tundra, Z6= Snow

[C] Z1= Tropical rain forest, Z2 and Z3= Deciduous forest, Z4= Coniferous forest, Z5= Tundra, Z6= Snow

[D] None of the above

9. In ...., the first patent for a higher category of animal was granted for the 'oncomouse'.

[A] 1987 [B] 1986

[C] 1985 [D] 1984

10. The most important strategy for the conservation of biodiversity is the establishment of

[A] Botanical gardens [B] Wild life sanctuaries

[C] Biosphere reserve [D] National park

11. Which of the following is known as ecology of individuals?

[A] Population ecology [B] Community ecology

[C] Autoecology [D] None of the above

12. Species content of plants growing in an area is referred to as

[A] Flora [B] Vegetation

[C] Community [D] None of the above

13. Bioenergetics approach for ecological studies was made by

[A] F.T. Clements [B] E. Haeckel

[C] A. G. Tansley [D] None of the above

14. Plants growing best in full sunlight are called

[A] Sciophytes [B] Therophytes

[C] Heliophytes [D] None of the above

15. Vegetation dominated by tropical rain forest with high temperature prevailing throughout the year is included among

[A] Mesotherm [B] Megatherm

[C] Microtherm [D] None of the above

16. Incoming solar radiation is known as

[A] Insolation [B] Terrestrial radiation

[C] Convection [D] None of the above

17. Effects of topographic factors on plant can be best seen in a

[A] Desert [B] Pond

[C] Mountain [D] None of the above

18. Plants growing in sandy soil are called

[A] Halophytes [B] Psammophytes

[C] Lithophytes [D] None of the above

19. Generally growth of plants is best supported by

[A] Sandy soil [B] Loam soil

[C] Silt [D] Clay

20. National Research Center on high altitude physiology is situated at

[A] Almora, Himachal Pradesh [B] Jaipur

[C] Jodhpur [D] None of the above

21. The chemical basis of competition between plants is referred to as

[A] Parasitism [B] Predation

[C] Allelopathy [D] None of the above

22. The microhabitat of leaf surface is known as

[A] Phyllosphere [B] Caulosphere

[C] Rhizosphere [D] None of the above

23. Which of the following is an example of commensalism?

[A] Zooxanthellae [B] Epiphytes

[C] Lichens [D] None of the above

24. Pneumatophores are characteristic of

[A] Psammophytes [B] Oxylophytes

[C] Heliophytes [D] None of the above

25. Sunken stomata are the characteristic feature of leaves of

[A] *Azolla* [B] *Nelumbo*

[C] *Nymphaea* [D] None of the above

26. Aerenchyma tissue develop on the leaves and stem of

[A] *Eichhornia* [B] *Nymphaea*
[C] *Potamogeton* [D] All of the above

27. Seasonal plants completing their life cycle in a single season are called

[A] Therophytes [B] Geophytes
[C] Chamaephytes [D] None of the above

28. Which of the following is suitable for study of forest community?

[A] Quadrat [B] Point-fram
[C] Belt transect [D] None of the above

29. Buds, organs or seeds are protected most in a

[A] Therophyte [B] Cryptophyte
[C] Chamaephyte [D] None of the above

30. Scientific study of seasonal changes in life cycle of plants is known as

[A] Phenology [B] Synecology
[C] Teratology [D] None of the above

31. Stratification is one of the characteristic features of a

[A] Population [B] Ecads
[C] Community [D] None of the above

32. Which of the following is earliest stage in plant succession?

[A] Invasion [B] Coaction
[C] Reaction [D] Nudation

33. The terminal stabilized stage of a plant succession is called

[A] Climax [B] Migration
[C] Aggregation [D] None of the above

34. Which of the following is concerned with international trade of endangered species?

[A] WWF [B] CITES
[C] IUCN [D] MAB

35. Which of the following is the earliest set up biosphere reserve in India?

[A] Nilgiri [B] Sunderbans
[C] Both [D] None of the above

36. What are the geographical scales for measuring the biodiversity?

[A] Alpha diversity [B] Beta diversity
[C] Gamma diversity [D] All of the above

37. Within the biological communities, some species may be important in determining the ability of large number of other species to persist in the community. These crucial species have been termed

[A] Keystone species [B] Species keystone
[C] Species diversity [D] None of the above

38. The first global conference on depletion of ozone layer was held in

[A] Japan [B] kathmandu

[C] Australia [D] Austria

39. Which of the following is concerned with environmental pollution?

[A] WWF [B] UNEP

[C] ITES [D] IUCN

40. Conservation biology rests on certain ethical principles. These are

[A] Biological diversity has intrinsic value

[B] Evolution is good

[C] Ecological complexity is good

[D] The diversity of organism is good

[E] All of the above

41. Alpha diversity refers to

[A] Number of species in single community

[B] Number of species in two communities

[C] Number of species in many communities

[D] None of the above

42. Beta diversity refers to

[A] Degree to which species composition changes along an environmental gradient

[B] Percent to which species composition changes along an environmental gradient

[C] Both

[D] None of the above

43. Species that may become endangered in the near future because populations of the species are decreasing in size throughout its range is known as

[A] Vulnerable [B] Extinct

[C] Endangered [D] None of the above

44. Species that have small total numbers of individuals often due to limited geographical ranges or low population densities are known as

[A] Rare [B] Critical

[C] Insufficiently known [D] None of the above

17 PHOTOSYNTHESIS

1. Photosynthesis is essentially the only mechanism of energy .... into the living world except chemosynthetic bacteria.

[A] Input [B] Output
[C] Both [D] None of the above

2. Photosynthesis is

[A] Anabolic process [B] Catabolic process
[C] Amphibolic process [D] None of the above

3. Chemosynthetic bacteria obtain energy by oxidizing .... substance.

[A] Inorganic [B] Organic
[C] Both [D] None of the above

4. Which of the following gas is called as 'dephlogisticated air' by Priestley?

[A] Carbon dioxide [B] Oxygen
[C] Argon [D] None of the above

5. In 1782, which of the following scientist showed that the presence of the noxious gas produced by animals and by plants in darkness stimulated production of "purified air" (O_2) in light?

[A] Jan Ingenhouse [B] Jean Senebier
[C] Von Hill [D] None of the above

6. The first quantitative measurement of photosynthesis was done by

[A] Jan Ingenhouse [B] Jean Saussure
[C] Von Hill [D] None of the above

7. The nature of the end product of photosynthesis- organic matter was demonstrated by

[A] Jan Ingenhouse [B] Jean Saussure
[C] Von Hill [D] Von Sachs
[E] None of the above

8. The light -driven split i.e. photolysis of water in the absence of CO_2 fixation became known as

[A] Hill reaction [B] Niel's hypothesis
[C] Arnon reaction [D] None of the above

9. End Product of photosynthesis is

[A] Organic compound [B] Inorganic compound

[C] Both [D] None of the above

10. Which of the following component is not required by plant during photosynthesis?

[A] Carbon dioxide [B] Water

[C] Light [D] None of the above

11. Most of the photosynthesis on earth is carried out by

[A] Phytosynthetic alga [B] Marine alga

[C] Forests [D] None of the above

12. Several observations leading to discovery of photosynthesis were made by

[A] Van Helmot [B] Stephan Hales

[C] Priestley [D] Ingenhouz

[E] Jean Senebier [F] Nicolas Theodore de Sassure Meyer

[G] All of the above

13. Who among the followings first showed that process of photosynthesis takes place in chloroplast?

[A] Julius Sachs [B] Jean Senebier

[C] Stephan Hales [D] None of the above

14. Which of the following scientist has contributed a lot in our understanding of electron transport system and photophosphorylation in chloroplast?

[A] Hill [B] Van Niel

[C] Kok *et al.* [D] None of the above

15. Water oxidizing clock model of photolysis of water was given by

[A] Hill [B] Bessel kok *et al*

[C] Emerson [D] None of the above

16. Main credit for investigation of sequence of dark reaction of photosynthesis goes to

[A] J. Bassham [B] Melvin Calvin

[C] A.A. Benson [D] None of the above

17. Radioactive isotope of Carbon (^{14}C) was discovered by

[A] A.A. Benson [B] Melvin Calvin

[C] Ruben and Kamen [D] None of the above

18. The idea that O_2 comes from water and not from CO_2 was first given by

[A] Van Niel [B] Arnon

[C] Ruben and Kamen [D] None of the above

19. Who among the followings first reported that 4-C containing dicarboxylic acids, malate and aspartate were the major labeled products when sugarcane leaves were allowed to photosynthesize for short periods in $^{14}CO_2$?

[A] Melvin Calvin [B] Kortschak, Harti and Burr
[C] Hatch and Slack [D] None of the above

20. The law of limiting factors was enunciated by
[A] F.F. Blackman [B] A.A. Benson
[C] Both [D] None of the above

21. The term photorespiration was coined by
[A] Krotkov [B] Warburg
[C] Hatch-Slack [D] None of the above

22. Photorespiration has been called as metabolic adjunct to calvin cycle by
[A] Hatch & Slack [B] A.A. Benson
[C] Krotokov [D] None of the above

23. The idea that air, light and leaves are important for nourishment of plants was given by
[A] Stephen Hales [B] Joseph Pristely
[C] Jan Ingen-Housz [D] Senebier

24. The idea that both green parts and sunlight are required for air purification and plant nourishment, was given by
[A] Stephen Hales [B] Joseph Pristely
[C] Jan Ingen-Housz [D] Senebier
[E] Lavosier

25. "Water is essential for photosynthesis along with carbon dioxide. Volume of oxygen evolved is equal to the volume of carbon dioxide absorbed", this concept was given by
[A] G.G. Stokes [B] Von Mayer
[C] Dutrochet [D] Pelletier and Caventon
[E] De Sassure [G] None of the above

26. Which of the following scientist established the importance of chlorophyll in photosynthesis?
[A] G.G. Stokes [B] Von mayer
[C] Dutrochet [D] Pelletier and Caventon
[E] De Sassure [G] None of the above

27. "In Photosynthesis, solar energy is changed to chemical energy", this concept was given by
[A] G.G. Stokes [B] Von Mayer
[C] Dutrochet [D] Pelletier and Caventon
[E] De Sassure [G] None of the above

28. "Starch is formed as reserve product of photosynthesis", this idea was given by
[A] G.G. Stokes [B] Von mayer
[C] Dutrochet [D] Pelletier and Caventon
[E] De Sassure [G] Sachs

29. "In photosynthesis the volume of O_2 released is equal to the volume of CO_2 absorbed", this concept was given by

[A] Bossingault [B] Engelmann
[C] Blackman [D] Willstatter and Stoll
[E] Warburg [E] Emerson and Arnold
[F] Hill [G] None of the above

30. Which of the following scientist obtained pure fractions of chlorophyll a and chlorophyll b?

[A] Bossingault [B] Engelmann
[C] Blackman [D] Willstatter and Stoll
[E] Warburg [E] Emerson and Arnold
[F] Hill [G] None of the above

31. Which of the following scientist obtained action spectra of photosynthesis?

[A] Bossingault [B] Engelmann
[C] Blackman [D] Willstatter and Stoll

32. Which of the following scientist studied structure of photosynthetic pigments?

[A] Bossingault [B] Engelmann
[C] Blackman [D] Willstatter and Stoll
[E] Warburg

33. Which of the following scientist first used light flashes for study of photosynthesis?

[A] Bossingault [B] Engelmann
[C] Blackman [D] Willstatter and Stoll
[E] Warburg

34. Which of the following scientist established the light and dark phase with the help of light flashes?

[A] Bossingault [B] Engelmann
[C] Blackman [D] Willstatter and Stoll
[E] Warburg [F] Emerson and Arnold
[G] Hill [H] None of the above

35. Which of the following scientist discovered Photophosphorylation?

[A] Arnon [B] Emerson
[C] Hatch and Slack [D] Huber
[E] None of the above

36. Which of the following scientist found that monochromatic light of more than 680 nm shows a sharp reduction in photosynthetic yield, which is known as red drop?

[A] Arnon [B] Emerson
[C] Hatch and Slack [D] Huber
[E] None of the above

37. Which of the following scientist found that carbon in plants comes from carbon dioxide in air?

[A] Liebig [B] Emerson

[C] Hatch and Slack [D] Huber

[E] None of the above

38. Chemi-osmotic theory was proposed by

[A] Mitchell [B] Emerson

[C] Hatch and Slack [D] Huber [E] None of the above

39. The guard cells would be unable to respond to light levels below the photosynthetic light compensation point. This statement is true or false?

[A] True [B] False

[C] Not known [D] None of the above

40. The pathway by which all photosynthetic eukaryotic organisms ultimately incorporate CO_2 into carbohydrate is known as

[A] Carbon fixation [B] Photosynthetic Carbon Cycle

[C] Calvin Cycle [D] All of the above

41. The time required to build up the necessary levels of PCR cycle intermediates in transition from dark to light is called the

[A] Photosynthetic induction time

[B] Photosynthetic incubation time

[C] Photosynthetic reaction time

[D] None of the above

42. Generally the chloroplast of higher plants are .... in shape

[A] Discoid [B] Ellipsoidal

[C] Both [D] None of the above

43. Which of the following pigment has open conjugated system of four pyrole rings in their molecules?

[A] Phycobillins [B] Chlorophylls

[C] Carotenoids [D] None of the above

44. Photosynthetic pigments are found in chloroplast in the form of pigment-protein complexes chiefly in

[A] Stroma [B] Thylakoids

[C] Stroma lamellae [D] None of the above

45. Which of the following is known as assimilatory power of dark reaction?

[A] RUDP and RUMP [B] ATP and NADPH

[C] Water and oxygen [D] None of the above

46. For synthesis of one hexose sugar, how many Calvin cycle takes place?

[A] 12 [B] 3

[C] 6 [D] None of the above

47. Photoinhibition

[A] Is the inhibition of photosynthesis due to the absorption of excessive light energy

[B] Is less common in C3 plants relative to C4 plants

[C] Results in damage tom proteins in the reaction centre

[D] Can result in the production of toxic oxygen molecules

48. Which of the following is not found in chloroplast?

[A] DNA [B] Ribosome

[C] Carotenoids [D] None of the above

49. Green alga and higher plants contain

[A] Chlorophyll-a and Chlorophyll-b

[B] Chlorophyll-a and Chlorophyll-c

[C] Chlorophyll-a and Chlorophyll-d

[D] All of the above

50. In chloroplast photo damage is prevented by

[A] Phycobillins [B] Carotenoids

[C] Chlorophyll [D] None of the above

51. Chlorophyll and carotenoids are .... in water and can be extracted only with .... solvents.

[A] Insoluble, organic [B] Soluble, inorganic

[C] Insoluble, water [D] None of the above

52. Phycobillins are .... in water

[A] Soluble [B] Insoluble

[C] Both [D] None of the above

53. .... are also known as carotenols.

[A] Xanthophylls [B] Carotenes

[C] Both [D] None of the above

54. Chlorophyll a is distributed in all plant kingdom except

[A] Bacteria [B] Algae

[C] Both [D] None of the above

55. Chlorophyll b is distributed in

[A] Higher plants [B] Green algae

[C] Both [D] None of the above

56. Chlorophyll c is distributed in

[A] Diatoms [B] Dinoflagellates

[C] Brown algae [D] All of the above

57. Chlorophyll e is distributed in

[A] *Tribonema* [B] Zoospores of *Vaucheria*

[C] Both [D] None of the above

58. Bacteriochlorophyll-a is distributed in

[A] Purple bacteria [B] Green bacteria

[C] Both [D] None of the above

59. Bacteriochlorophyll-g is distributed in

[A] Heliobacteria [B] Green bacteria

[C] Purple bacteria [D] None of the above

60. Mg-porphyrin 'head' is

[A] Hydrophilic [B] Lipophilic

[C] Both [D] None of the above

61. Phytol tail of chlorophyll molecule is

[A] Lipophilic [B] Hydrophilic

[C] Both [D] None of the above

62. Lycopene is most widely distributed carotenoid in

[A] Red alga [B] Higher plants

[C] Both [D] None of the above

63.

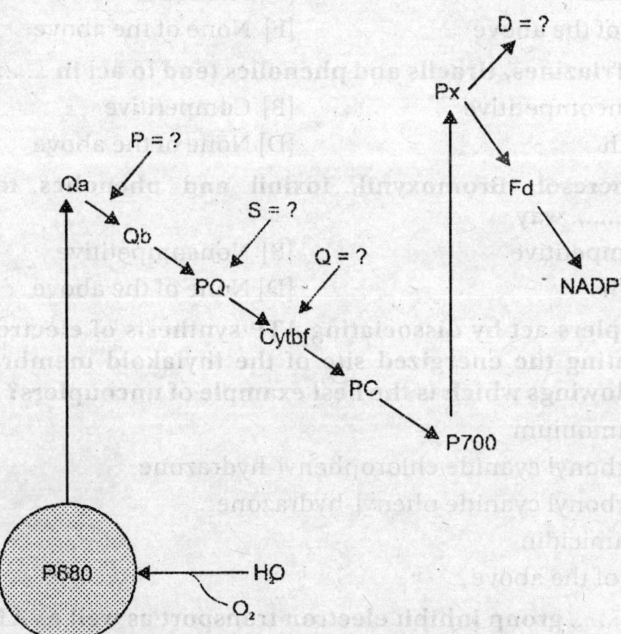

In the question, a scheme shows photosynthetic electron flow within the chloroplast from water to NADP⁺. There are number of sites of the action of different electron transport inhibitors in the reaction. So in this figure find the site of action of ureas, triazines and phenol-type herbicides

[A] P [B] S

[C] Q [D] D

[E] None of the above

64. In question number 63, find the site of action of bipyridyls

[A] P [B] S
[C] Q [D] D
[E] None of the above

65. In question number 63, find the site of action of analogues of plastoquinone

[A] P [B] S
[C] Q [D] D

66. Which of the following photosynthetic inhibitor is known as mobile herbicides?

[A] Urea [B] Triazines
[C] Uracils [D] Phenolics
[E] More than one [F] None of the above

67. Which of the following photosynthetic inhibitor is known as immobile herbicides?

[A] Dinitrocresol [B] Bromoxynil
[C] Ioxinil [D] Phenolics
[E] All of the above [F] None of the above

68. Urea, Triazines, Uracils and phenolics tend to act in .... way

[A] Noncompetitive [B] Competitive
[C] Both [D] None of the above

69. Dinitrocresol, Bromoxynil, ioxinil and phenolics tend to act in .... way

[A] Competitive [B] Noncompetitive
[C] Both [D] None of the above

70. Uncouplers act by dissociating ATP synthesis of electron transport by dissipating the energized site of the thylakoid membrane, so among the followings which is the best example of uncouplers?

[A] Ammonium
[B] Carbonyl cyanide chlorophenyl-hydrazone
[C] Carbonyl cyanide phenyl-hydrazone
[D] Gramicidin
[E] All of the above

71. .... group inhibit electron transport as well as ATP formation in coupled system.

[A] Energy transfer inhibitors [B] Inhibitory uncouplers
[C] Uncouplers [D] None of the above

72. Non herbicidal compounds that behave as energy transfer inhibitors include the antibiotics such as

[A] Dio-9 [B] Phlorizin
[C] DCCD [D] TBT
[E] All of the above

73. The unit name of quantity irradiance (energy) is

[A] Watts per square meter
[B] Moles per square meter per second
[C] Moles per square meter per second per nanometer
[D] None of the above

74. The unit name of quantity irradiance (moles of photons) is

[A] Watts per square meter
[B] Moles per square meter per second
[C] Moles per square meter per second per nanometer
[D] None of the above

75. The unit name of quantity spectral irradiance (moles of photons) is

[A] Watts per square meter
[B] Moles per square meter per second
[C] Moles per square meter per second per nanometer
[D] None of the above

76. Absorption spectra of allophycocyanin is

[A] In near red part of the spectrum
[B] In orange part of the spectrum
[C] In green part of the spectrum
[D] None of the above

77. Absorption spectra of phycocyanins is

[A] In near red part of the spectrum [B] In orange part of the spectrum
[C] In green part of the spectrum [D] None of the above

78. Absorption spectra of phycoerythrins is

[A] In near red part of the spectrum [B] In orange part of the spectrum
[C] In green part of the spectrum [D] None of the above

79. A leaf has an absorptivity coefficient of about

[A] 0.98 [B] 0.098
[C] 0.0098 [D] None of the above

80. According *to .... law, each molecule of a dissolved pigment absorbs the same fraction of light incident upon it. Thus, in a nonabsorbing medium, the light absorbed should be proportional to dissolved pigment concentration. The law often holds for dilute solution but fails as the light-absorbing properties of pigment molecules change at higher concentrations.*

[A] Lambert's law [B] Beer's law
[C] Stefan-Boltzmann law [D] None of the above

81. According to .... law, ***each layer of equal thickness absorbs an equal fraction of the light that traverses it.***

[A] Lambert's law [B] Beer's law
[C] Stefan-Boltzmann law [D] None of the above

82. A leaf emits radiation according to

[A] Lambert's law [B] Beer's law
[C] Stefan-Boltzmann law [D] Both A and B
[E] None of the above

83. According to .... law, ***all objects with temperature above absolute zero emit radiant energy.***

[A] Lambert's law [B] Beer's law
[C] Stefan-Boltzmann law [D] Both A & B [E] None of the above

84. Which of the following is known as universal precursor of tetrapyrroles?

[A] β-aminolevulinic acid [B] δ- aminolevulinic acid
[C] α- aminolevulinic acid [D] None of the above

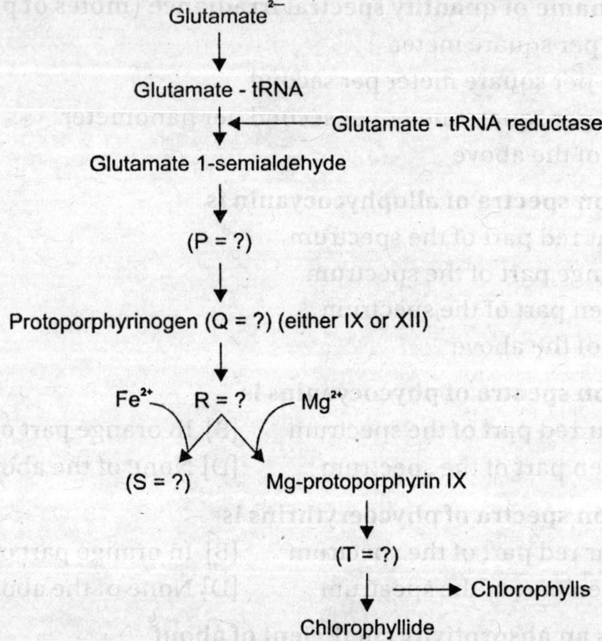

85. In question number 85, find the missing compounds in outline of the chlorophyll biosynthesis pathways.

[A] P= Beta ALA, Q=IX; R= Protoporphyrin; S= Hemes and open tetrapyrroles; T= Protochlorophyllide

[B] P= Delta ALA, Q=XII; R= Protoporphyrin; S= Hemes and open tetrapyrroles; T= Protochlorophyllide

[C] P= Protochlorophyllide; Q=IX; R= Protoporphyrin; S= Hemes and open tetrapyrroles; T= Delta ALA,

[D] P= Delta ALA, Q=VIII; R= Protoporphyrin; S= Hemes and open tetrapyrroles; T= Protochlorophyllide

[E] P= Alpha ALA, Q=VIII; R= Protoporphyrin; S= Hemes and open tetrapyrroles; T= Protochlorophyllide

[F] P= Delta ALA; Q=IX; R= Protoporphyrin; S= Hemes and open tetrapyrroles; T= Protochlorophyllide

86. Moll's Half Leaf experiment is the test for

[A] Photosynthesis

[B] Carbon dioxide is necessary for photosynthesis

[C] Oxygen is evolved during photosynthesis

[D] None of the above

87. The term assimilation number or photosynthetic number is coined by

[A] Warburg [B] Willstatter and Stoll

[C] Dicker and Tio [D] None of the above

88. Which one is the most efficient converter of sunlight?

[A] Rice [B] Wheat

[C] Sugarcane [D] None of the above

89. Absorption spectra of different chlorophylls differ *in vivo* and *in vitro*. This statement is true or false?

[A] True [B] False

[C] Not known [D] None of the above

90. The absorption band shown by the chlorophylls in .... regions is also called as soret band.

[A] Violet-blue [B] Red-blue

[C] Green –blue [D] None of the above

91. Molecular formula of chlorophyll a is

[A] $C_{55}H_{70}O_6N_4Mg$ [B] $C_{55}H_{72}O_5N_4Mg$

[C] $C_{40}H_{56}O_2$ [D] None of the above

92. Molecular formula of chlorophyll b is

[A] $C_{55}H_{70}O_6N_4Mg$ [B] $C_{55}H_{72}O_5N_4Mg$

[C] $C_{40}H_{56}O_2$ [D] $C_{55}H_{35}O_4N_5Mg$

93. Bacteriochlorophyll absorbs light rays heavily in

[A] UV-region of the spectrum

[B] Red region of the spectrum

[C] Near infra red region of the spectrum

[D] None of the above

94. CO_2 is necessary for photosynthesis. The chemical used to remove this gas most effectively from entering a control apparatus is

[A] Distilled water [B] Potassium hydroxide solution

[C] Calcium oxide [D] Sodium carbonate

95. During photosynthesis when PGA is changed into phosphoglyceraldehyde, which of the following reaction occurs?

[A] Hydrolysis [B] Reduction

[C] Oxidation [D] None of the above

96. Who first of all indicated that water is the electron donor in photosynthesis?

[A] Van Niel [B] Calvin

[C] Arnon [D] Blakeslee [E] None of the above

97. The size of the chlorophyll molecule is

[A] Head 15X15Å, tail 25Å [B] Head 20X20Å, tail 25Å

[C] Head 15X15Å, tail 20Å [D] None of the above

98. How many molecules of NADPH and ATP respectively are required for fixation of 1 CO_2 molecule in C4-pathway?

[A] 2 NADPH + 3ATP [B] 2 NADPH + 5 ATP

[C] 1 NADPH + 2 ATP [D] None of the above

99. $NADP^+$-Me Type C4-pathway is found in

[A] Sugarcane and maize [B] *Panicum maximum*

[C] *Atriplex* and *Amaranthus* [D] None of the above

100. How many molecules of NADPH and ATP respectively are required for the reduction of $6CO_2$ molecules to generate one molecule of hexose sugar in photosynthesis?

[A] 12 NADP+12ATP [B] 18 NADP+ 12 ATP

[C] 18 NADP+ 18 ATP [D] 12 NADP+ 18 ATP

101. Which of the following green alga was used by Melvin Calvin in his experiment to determine sequence of dark reaction of photosynthesis?

[A] *Chlorella* and *Chlamydomonas* [B] *Chlorella* and *Scenedesmus*

[C] *Chlorella* and *Ulothrix* [D] None of the above

102. Dimorphic chloroplasts are found in leaves of

[A] C4 plants [B] C3 plants

[C] CAM plants [D] None of the above

103. During phosphorylation, the proton electo-chemical gradient across the thylakoid membrane is from

[A] Stroma to Lumen of Thylakoids

[B] Lumen of Thylakoids to Stroma

[C] Lumen of one Thylakoids to Lumen of another Thylakoids

[D] None of the above

104. Chlorophyll is formed from protochlorophyll in light. The protochlorophyll lacks two hydrogen atoms one each at .... and .... C atoms in IV pyrole ring.

[A] 7th and 8th [B] 8th and 9th

[C] 10th and 11th [D] None of the above

105. The chlorophyll a and b differs from each other because in chlorophyll b there is a –CHO group instead of a $-CH_3$ group at the .... C atom in .... pyrole ring.

[A] 3rd, II [B] 3rd, I

[C] 2nd, I [D] 2nd, II [E] None of the above

106. Phycobilins consist of an open conjugated system of four pyrrol rings and lacks Mg and the phytol chain. This statement is true or false?

[A] True [B] False

[C] Not known [D] None of the above

107. An absorption spectrum is a measure of the extent to which a given substance absorbs the light of different colors or wavelengths. It is a function of the relationship between

[A] Absorption [B] Wavelength
[C] Both [D] None of the above

108. The immediate precursor of chlorophyll a is

[A] Protochlorophyll [B] Protoporphyrin IX
[C] Chlorophyllide a [D] None of the

109. Carotene, the first carotenoid to be named, was isolated from .... tissue by Wackenroder in 1831.

[A] Carrot root [B] Carrot leaf
[C] Carrot shoot [D] None of the above

110. Which of the following is considered as derivative of carotenoids?

[A] Lycopene, a blue pigment [B] Lycopene, a yellow pigment
[C] Lycopene, a purple pigment [D] None of the above

111. The chemical difference between α-carotene and β-carotene is that

[A] -carotene consists of the two -ionone rings and -carotene consists of one and one -ionone ring.

[B] -carotene consists of the one -ionone rings and -carotene consists of two and one -ionone ring

[C] -carotene consists of the two -ionone rings and -carotene consists of one and two -ionone ring

[D] None of the above

112. The carotenoids that consist of carbon and hydrogen are termed as ...., and carotenoids that contain oxygen are called ....

[A] Carotenes, xanthophylls [B] Xanthophylls, carotenes
[C] Xanthophylls, chlorophyll [D] None of the above

113. Inulin is an unbranched polymer of about .... units joined by β-2,1-linkages.

[A] 35 fructose [B] 25 glucose
[C] 20 glactose [D] None of the above

114. Emerson's enhancement effect and Red drop have been instrumental in the discovery of

[A] Photophosphorylation and Non –cyclic electron transport
[B] Two photosystems
[C] Photophosphorylation and cyclic electron transport
[D] None of the above

115. Which of the following has more energy per photon of light?

[A] UV-A [B] UV-B
[C] Red light [D] Blue light
[E] More than one [F] None of the above

116. The number of oxygen molecules released per photon of light in photosynthesis is called as

[A] Quantum yield [B] Quantum requirement

[C] Photosynthetic yield [D] None of the above

117. Rieske centre is

[A] Core complex I [B] Cytochrome b_6-f complex

[C] D1 protein [D] None of the above

118. The flow of electrons from donor to acceptor molecule in electron transport chain occurs along

[A] Increasing redox potential value

[B] Decreasing redox potential value

[C] Constant redox potential value

[D] None of the above

119. Starch synthesis in leaves

[A] Increases with a build up of phosphoglyceraldehyde in the stroma

[B] Is promoted in the light

[C] Occurs in the cytosol

[D] None of the above

120. C_4 plants are mostly abundant in

[A] Tropical region with more dry condition

[B] Tropical region with more humid condition

[C] Temperate region with more humid condition

[D] Temperate region with dry condition

121. Which of the following organism does not evolve oxygen in photosynthesis?

[A] *Funaria* [B] *Anabaena*

[C] Higher plant [D] *Rhodospirillium*

[E] None of the above

122. C_3 cycle represents which of the following reaction?

[A] Substrate level phosphorylation

[B] Salt respiration

[C] Oxidative carboxylation

[D] Reductive carboxylation

123. The larger subunit of Rubisco is coded by

[A] Nuclear genome [B] Chloroplast genome

[C] Mitochondrial genome [D] None of the above

124. Plastocyanin is a .... electron carrier between ....

[A] Mobile, PSII and cytochrome b complex

[B] Immobile, cytochrome b complex and PSI

[C] PSI and ferredoxin

[D] None of the above

125. Among the following components of chloroplast membranes, which one is the strongest reducing agent?

[A] Reduced cytochrome b6 [B] $NADPH_2$
[C] Reduced ferredoxin [D] None of the above

126. Dibromothymoquinone, an inhibitor of cytochrome bf complex of photosynthetic electron transport chain, was added to the green alga *Chlamydomonas*. After illumination it was found that in the alga

[A] P700 was in reduced state
[B] Plastocyanin was in reduced state
[C] Plastoquinone was in reduced state
[D] None of the above

127. In C_4 plants, phosphoenol pyruvate carboxylase is located in

[A] Chloroplast [B] Peroxisome
[C] Mitochondria [D] Cytosol

128. In organisms such as plants, algae and cyanobacteria, the reductant is

[A] H_2O [B] H_2S
[C] Lactate [D] None of the above

129. In green sulfur bacteria, the reductant is

[A] H_2O [B] H_2S
[C] Lactate [D] None of the above

130. The bacterial photosynthesis is different from that of the higher plants as

[A] Oxygen is not released [B] Light is not required
[C] Energy is not required [D] None of the above

131. During day time, the rate of CO_2 produced to that of CO_2 consumed is

[A] 1:1 [B] 1:5
[C] 1:10 [D] 1:50

132. The main limiting factor which limits the rate of photosynthesis on a clear day is

[A] Light [B] Carbon dioxide
[C] Chlorophyll [D] None of the above

133. CO_2 is necessary for photosynthesis; the chemical used to remove this gas most effectively from entering a control apparatus is

[A] Distilled water [B] Calcium oxide
[C] Sodium carbonate [D] Potassium hydroxide solution

134. It is advantageous to use Hydrilla to demonstrate photosynthesis than land plant because

[A] It respire slowly [B] It does not respire
[C] It photosynthesizes rapidly [D] None of the above

135. Decreased rate of photosynthesis at high concentration of oxygen is referred to as

[A] Pasture effects [B] Emerson effects
[C] Warburg effects [D] None of the above

136. The rate limiting step in photosynthesis is

[A] Light reaction [B] Oxygen evolution

[C] Dark reaction [D] None of the above

137. The Thylakoids are removed and kept in a culture medium containing CO_2 and H_2O. If the set up is exposed to light, hexose sugars are not formed as the end product. The most appropriate reason for this will be that

[A] The light –trapping device is nonfunctional

[B] Photosynthetic enzymes are not available

[C] The pigments(P700 and P680) are not linked

[D] Carbon assimilation can not takes place

138. In photosynthetic experiments, which plant would behave as a single chloroplast ?

[A] *Chlorella pyrenoidosa* [B] *Ulothrix*

[C] Both [D] *Pisum sativum*

139. The quantum yield in photosynthesis is

[A] 8% [B] 12% [C] 25% [D] 50%

140. Algae float in water during day time and sink during night time because

[A] They become light due to consumption of food materials in respiration

[B] They become buoyant in light due to accumulation of oxygen bubbles released in photosynthesis

[C] They come up to enjoy sunlight

[D] They lose weight at night

141. During day light hours, the rate of photosynthesis is higher than that of respiration, and the ratio of oxygen produced to that of consumed is

[A] 10:1 [B] 5:1

[C] 1:1 [D] 50:1

142. *"Plants very probably draw through their leaves some part of their nourishment from the air .... may not light, also by freely entering surfaces of leaves and flowers, contribute much to ennobling the principles of vegetables"*. **This famous statement was made by which famous scientist?**

[A] Hales [B] Priestley

[C] Ingenhousz [D] None of the above

143. *"Plants, instead of affecting the air in the same manner with animal respiration, reverse the effects of breathing and tend to keep the atmosphere sweet and wholesome when it has become noxious in consequence of animals either living and breathing or dying and putrefying in it"*. **This statement was made by**

[A] Woodward [B] Mayer

[C] Priestley [D] None of the above

144. *"Vegetables are not formed of water but of a certain peculiar terrestrial matter. It has been shown that there is considerable quantity of this matter contained in rain, spring and river water; that the greatest part of the fluid mass that ascends up into plants does not settle there but passes through their pores and exhales up into the atmosphere; that a great part of the terrestrial matter, mixed with water, passes up into the plant along with it and that the plant is more or less augmented in proportion as the water contains a greater or less quantity of that matter; from all of which we may reasonably infer, that earth, and not water, is the matter that constitutes vegetables".* **This famous statement was made by which famous scientist?**

[A] Woodward [B] Helmont
[C] De Saussure [D] None of the above

145. Which of the following element is essential for photolysis of water?

[A] Chlorine [B] Nitrogen
[C] Oxygen [D] None of the above

146. Phosphorylation in a chloroplast is most similar to which of the under mentioned mitochondrial reactions?

[A] Substrate level phosphorylation
[B] Oxidative phosphorylation
[C] Oxidative decarboxylation
[D] None of the above

147. RuBP carboxylase was crystallized in 1971 from

[A] Maize leaves [B] Rice leaves
[C] Chlorella [D] Tobacco leaves

148. Table sugar manufactured from sugarcane can be distinguished from that manufactured from sugarbeet by

[A] Carbon isotope ratio analysis
[B] Chemical analysis
[C] Hydrogen isotope ratio analysis
[D] None of the above

149. Choose the correct statement:

[A] C_3 plants are more enriched in ^{14}C than C_4 plants
[B] C_4 plants are more enriched in ^{14}C than C_3 plants
[C] C_4 plants are more enriched in ^{12}C than C_3 plants
[D] None of the above

150. Huber, Michel and Dissenhoffer analysed the structure of photosynthetic reaction centre of the bacterium *Rhodobacter* using

[A] Spectrophotometry [B] X-ray diffraction
[C] Autoradiography [D] Electron microscopy
[E] None of the above

151. The research on photosynthesis began

[A] In the 17th century [B] In the 18th century
[C] In the 19th century [D] In the early 20th century

152. The bulk fixation of carbon through photosynthesis takes place in
[A] Crop plants
[B] Tropical rain forest and crop plants
[C] Tropical rain forest
[D] Ocean

153. Who among the followings gave chemical composition of chlorophyll, carotenes and xanthophylls?
[A] Meyer and Calvin [B] Park and Biggins
[C] Willstatter and Stoll [D] Flemming and Waksman

154. The exact mechanism of photosynthesis is a process of converting light energy into chemical energy, was established by
[A] Robert Hill [B] Jean Senebier
[C] Ingen Housz [D] None of the above

155. The ultimate biological energy comes from
[A] Sunlight [B] Mitochondria
[C] Glucose [D] ATP [E] None of the above

156. The percentage of light energy fixed in photosynthesis is generally around
[A] 1 [B] 3
[C] 10 [D] 30

157. Chemical phase in photosynthesis was discovered for the first time by
[A] Blackman [B] Arnon
[C] Hill [D] Calvin [E] None of the above

158. Evidence for the existence of two photochemical systems in photosynthesis was proposed by
[A] Hill [B] Emerson
[C] Arnon [D] None of the above

159. Which of the following technique has been most useful for investigation of Calvin Cycle?
[A] Radioactive isotope technique [B] Flash light experiment
[C] X-ray technique [D] None of the above

160. The unit of photosynthesis is
[A] Quantasome [B] Angstrom
[C] Nanometer [D] None of the above

161. Quantasome contains
[A] 300 chlorophyll molecules [B] 250 chlorophyll molecules
[C] 230 chlorophyll molecules [D] 200 chlorophyll molecules
[E] None of the above

162. The plastids engaged in starch synthesis and storage are called
[A] Glucoplasts [B] Amyloplasts
[C] Chromoplasts [D] Leucoplasts [E] None of the above

163. In the process of photosynthesis, chlorophyll functions as

[A] A hydrogen acceptor [B] A raw material
[C] An end-product [D] An energy converter
[E] None of the above

164. Photo-oxidation of water in photosynthesis is in association of

[A] Pigment system I [B] Pigment system II
[C] Plastocyanin [D] Cytochrome B6
[E] None of the above

165. The experimental material which helped advances in photosynthesis research is

[A] *Chlorella* [B] *Hydrilla*
[C] *Clamydomonas* [D] None of the above

166. During photosynthesis, when PGA is changed into phosphoglyceraldehyde, which of the following reaction occurs?

[A] Hydrolysis [B] Reduction
[C] Oxidation [D] Electrolysis
[E] None of the above

167. Rate of photosynthesis is independent of

[A] Temperature [B] Quality of light
[C] Intensity of light [D] None of the above

168. The correlation between Kranz anatomy and C_4 photosynthesis was first established by

[A] Calvin [B] Dowton and Treguna
[C] Hill and Bendall [D] Arnold
[E] None of the above

169. Photosynthesis is an oxidation reduction process. The material that is oxidized is

[A] CO_2 [B] NADP
[C] PGA [D] None of the above

170. Which of the following plots P, Q, R and S in the given graph show the effect of light intensity on the rate of photosynthesis?

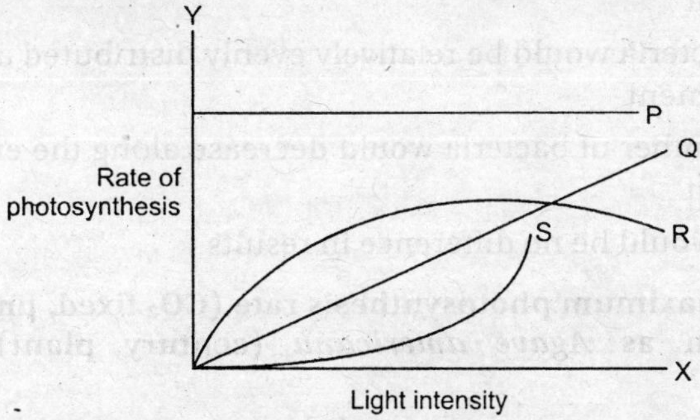

[A] P [B] Q
[C] R [D] S

171. The diagram given below shows the relation between light intensity and CO_2 concentration on rate of photosynthesis

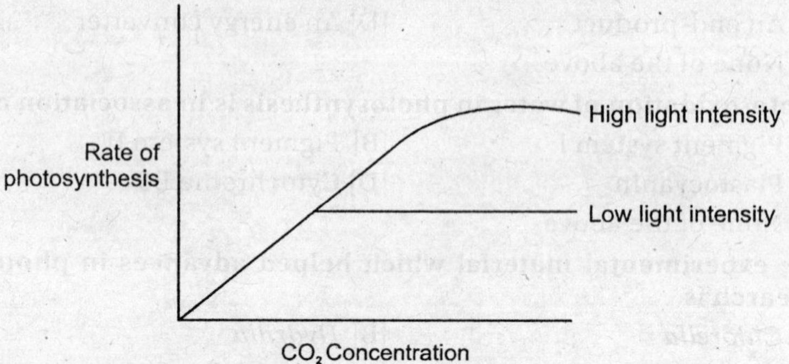

Which of the following explains the diagram correctly?

[A] The number of chlorophyll molecules prevent the rate of photosynthesis

[B] Photosynthesis is independent of light intensity

[C] Chlorophyll respond equally to light intensities and CO_2 concentration, thereby limiting the rate of photosynthesis

[D] At low intensity, the light dependent reaction is rate limiting but at high light intensity, the light independent reaction involving CO_2 is rate limiting

172. When DCPIP (Hill reagent) accepts electron from water, its colour changes to

[A] Blue [B] Red

[C] Colourless [D] None of the above

173. Thomas Engelmann illuminated a filament of algae with light that passed through a prism, thus exposing different segments of the algal filament to different wavelength of light. He added aerobic bacteria and found that these bacteria congregated in the areas illuminated by red and blue light. If you run the same experiment without passing light through a prism, what would you predict?

[A] The number of bacteria would increase along the entire length of the filament

[B] The bacteria would be relatively evenly distributed along the length of the filament

[C] The number of bacteria would decrease along the entire length of the filament

[D] There would be no difference in results

174. Find the maximum photosynthesis rate (CO_2 fixed, $\mu molm^{-2}s^{-1}$) of CAM plant such as *Agave americana* (century plant) under natural conditions

[A] 0.6-2.4 [B] 3.9- 5.6

[C] 3-9 [D] None of the above

175. Find the maximum photosynthesis rate (CO_2 fixed, $\mu molm^{-2}s^{-1}$) of tropical, subtropical and Mediterranean evergreen trees and shrubs and temperate zone evergreen conifers such as *Pinus sylvestris* under natural conditions

[A] 20-40 [B] 3.9- 5.6
[C] 3-9 [D] None of the above

176. Find the maximum photosynthesis rate (CO_2 fixed, $\mu molm^{-2}s^{-1}$) of temperate zone deciduous trees and shrubs such as *Fagus sylvatica* under natural conditions

[A] 3-12 [B] 10-20
[C] 10-24 [D] 20-24 [E] None of the above

177. Find the maximum photosynthesis rate (CO_2 fixed, $\mu molm^{-2}s^{-1}$) of temperate zone herbs and C3 pathway crop plants such as *Glycine max* under natural conditions

[A] 3-12 [B] 10-20
[C] 10-24 [D] 20-24 [E] None of the above

178. Photosynthetic efficiencies of crop plants never exceed the theoretical .... percent of absorbed PAR (Photosynthetic Active Radiation).

[A] 10 [B] 15
[C] 18 [D] None of the above

179. Theoretical energy requirement (CO_2: ATP: NADPH) for C_3 plant is

[A] 1:3:2 [B] 1:5:2
[C] 1:6.5:2 [D] None of the above

180. Theoretical energy requirement (CO_2: ATP: NADPH) for C_4 plant is

[A] 1:3:2 [B] 1:5:2
[C] 1:6.5:2 [D] None of the above

181. Theoretical energy requirement (CO_2: ATP: NADPH) for CAM plant is

[A] 1:3:2 [B] 1:5:2
[C] 1:6.5:2 [D] None of the above

182. Transpiration Ratio of C_3 plants is

[A] 400-950 [B] 250-350
[C] 18-125 [D] None of the above

183. Transpiration Ratio of CAM plants is

[A] 400-950 [B] 250-350
[C] 18-125 [D] None of the above

184. Transpiration Ratio of C_4 plants is

[A] 400-950 [B] 250-350
[C] 18-125 [D] None of the above

185. Leaf chlorophyll a to b ratio in C_3 plant is

[A] 2.8±0.4 [B] 3.9±0.6
[C] 2.5-3.0 [D] None of the above

186. Leaf chlorophyll a to b ratio in C_4 plant is

[A] 2.8±0.4 [B] 3.9±0.6

[C] 2.5-3.0 [D] None of the above

187. Leaf chlorophyll a to b ratio in CAM plant is

[A] 2.8±0.4 [B] 3.9±0.6

[C] 2.5-3.0 [D] None of the above

188. CO_2 compensation point (μ mol mol^{-1} CO_2) of C_3 plants is

[A] 30-70 [B] 0-10

[C] 0-5 in dark [D] None of the above

189. CO_2 compensation point (μ mol mol^{-1} CO_2) of C_4 plants is

[A] 30-70 [B] 0-10

[C] 0-5 in dark [D] None of the above

190. CO_2 compensation point (μ mol mol^{-1} CO_2) of CAM plants is

[A] 30-70 [B] 0-10

[C] 0-5 in dark [D] None of the above

191. Optimum temperature for photosynthesis in C_3 plants is

[A] 15-25°C [B] 30-47°C

[C] ~35°C [D] None of the above

192. Optimum temperature for photosynthesis in C_4 plants is

[A] 15-25°C [B] 30-47°C

[C] ~35°C [D] None of the above

193. Optimum temperature for photosynthesis in CAM plants is

[A] 15-25°C [B] 30-47°C

[C] ~35°C [D] None of the above

194. Which of the following plant is known as obligate shade plant?

[A] *Alocasia* [B] *Helianthus*

[C] *Solidago virgaurea* [D] None of the above

195. Which of the following plant is known as obligate sun plant?

[A] *Alocasia* [B] *Helianthus*

[C] *Solidago virgaurea* [D] None of the above

196. All gymnosperm, bryophytes and algae and most of pteridophyta that have been studied are .... plants.

[A] C_3 [B] C_4

[C] CAM [D] None of the above

197. Most of C_4 species are ....

[A] Monocots [B] Dicots

[C] Both [D] None of the above

198. In photosynthetic reaction, there is change of .... kJ per molecule of free energy, for formation of one molecule of hexose.

[A] +2840 [B] -2840

[C] +3840 [D] -3840 [E] None of the above

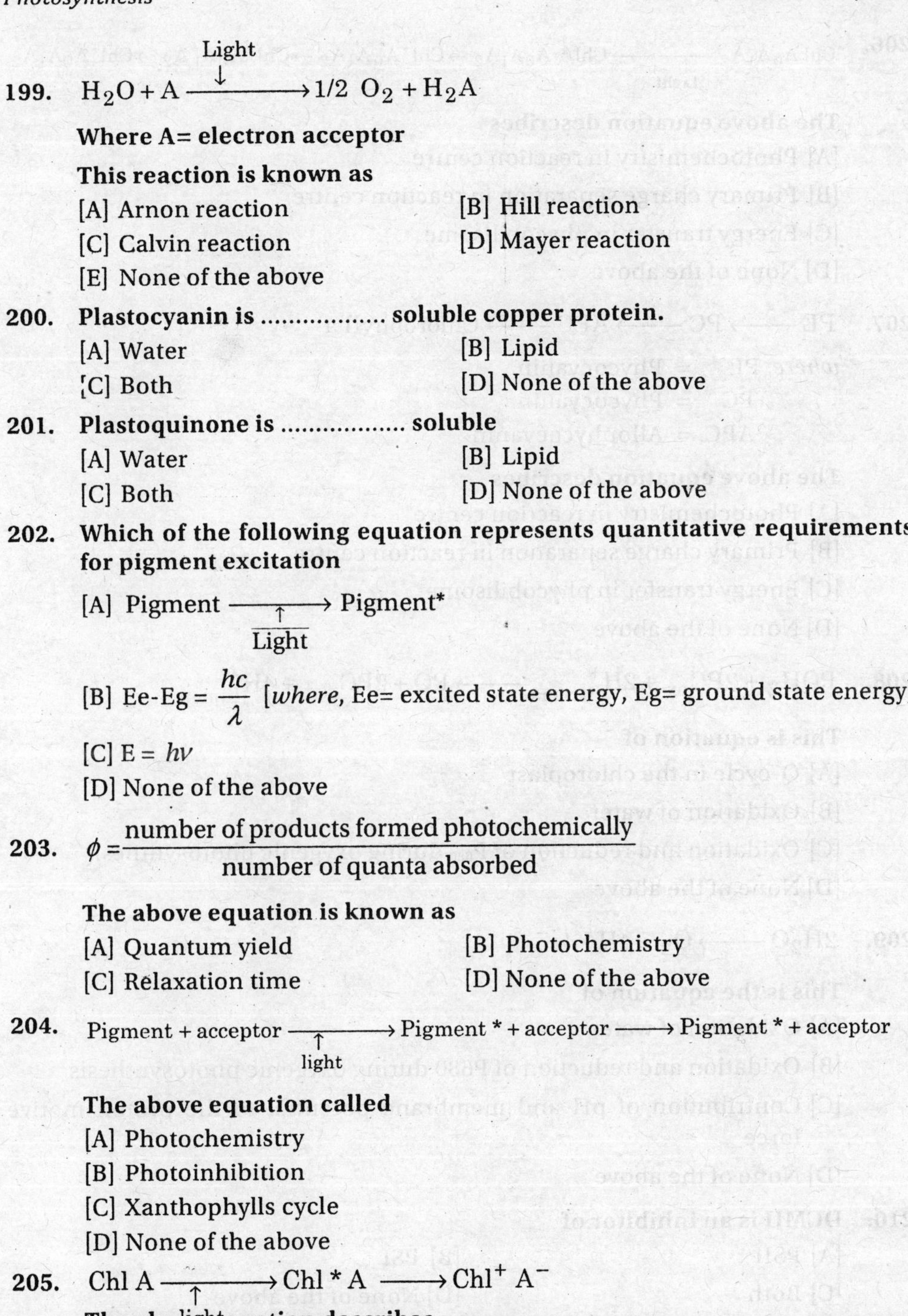

199. $H_2O + A \xrightarrow[]{\text{Light}} 1/2\ O_2 + H_2A$

Where A= electron acceptor

This reaction is known as

[A] Arnon reaction [B] Hill reaction

[C] Calvin reaction [D] Mayer reaction

[E] None of the above

200. Plastocyanin is .... soluble copper protein.

[A] Water [B] Lipid

[C] Both [D] None of the above

201. Plastoquinone is .... soluble

[A] Water [B] Lipid

[C] Both [D] None of the above

202. Which of the following equation represents quantitative requirements for pigment excitation

[A] Pigment $\xrightarrow[\text{Light}]{\uparrow}$ Pigment*

[B] Ee-Eg = $\dfrac{hc}{\lambda}$ [*where,* Ee= excited state energy, Eg= ground state energy]

[C] E = $h\nu$

[D] None of the above

203. $\phi = \dfrac{\text{number of products formed photochemically}}{\text{number of quanta absorbed}}$

The above equation is known as

[A] Quantum yield [B] Photochemistry

[C] Relaxation time [D] None of the above

204. Pigment + acceptor $\xrightarrow[\text{light}]{\uparrow}$ Pigment * + acceptor $\longrightarrow$ Pigment * + acceptor

The above equation called

[A] Photochemistry

[B] Photoinhibition

[C] Xanthophylls cycle

[D] None of the above

205. Chl A $\xrightarrow[\text{light}]{\uparrow}$ Chl * A $\longrightarrow$ $Chl^+\ A^-$

The above equation describes

[A] Photochemistry in reaction centre

[B] Primary charge separation in reaction centre

[C] Energy transfer in phycobilisome

[D] None of the above

206. $Chl\ A_0A_1A_2 \xrightarrow[\uparrow \text{Light}]{} Chl^*\ A_0A_1A_2 \rightarrow Chl^+A_0^-A_1A_2 \rightarrow Chl^+A_0A_1^-A_2 \rightarrow Chl^+A_0A_1A_2^-$

The above equation describes

[A] Photochemistry in reaction centre

[B] Primary charge separation in reaction centre

[C] Energy transfer in phycobilisome

[D] None of the above

207. $PE \longrightarrow PC \longrightarrow APC \longrightarrow \text{Chlorophyll a}$

where, PE = Phycocyanin
PC = Phycocyanin
APC = Allophycocyanin

The above equation describes

[A] Photochemistry in reaction centre

[B] Primary charge separation in reaction centre

[C] Energy transfer in phycobilisome

[D] None of the above

208. $PQH_2 + 2PC_{ox} + 2H^+_{Stroma} \longrightarrow PQ + 2PC_{red} + 4H^+_{lumen}$

This is equation of

[A] Q-cycle in the chloroplast

[B] Oxidation of water

[C] Oxidation and reduction of P_{680} during oxygenic photosynthesis

[D] None of the above

209. $2H_2O \longrightarrow O_2 + 4H^+\ 4e^-$

This is the equation of

[A] Oxidation of water

[B] Oxidation and reduction of P680 during oxygenic photosynthesis

[C] Contribution of pH and membrane potential to the proton motive force

[D] None of the above

210. DCMU is an inhibitor of

[A] PSII [B] PSI

[C] Both [D] None of the above

211. DBMIB is an inhibitor of the

[A] PSI [B] Cyt *b6f* complex

[C] Qa [D] Qb

[E] None of the above

212. SZP 680 $\xrightarrow[\text{light}]{\uparrow}$ SZP680$^+$ $\longrightarrow$ SZ$^+$P680 $\longrightarrow$ S$^+$ZP680

The above equation is of

[A] Q-cycle in the chloroplast

[B] Oxidation of water

[C] Oxidation and reduction of P_{680} during oxygenic photosynthesis

[D] None of the above

213. **In photosynthesis, assimilatory power is generated during**

[A] Light reaction [B] Dark reaction

[C] Photolysis of water [D] None of the above

214. **In photosynthesis, ultimate source of electron is**

[A] Water [B] Sunlight

[C] Phaeophytin [D] None of the above

215. **The photosynthetic process is a complicated**

[A] Hydrogenation process [B] Oxidation-reduction process

[C] Only reduction process [D] None of the above

216. **Which of the following is not a component of electron transport chain in chloroplast?**

[A] Water soluble Plastocyanin [B] Lipid soluble Plastoquinone

[C] Ubiquinone [D] None of the above

217. **The head of ATP-synthease (CF1) in thylakoids consists of how many different polypeptides?**

[A] Ten [B] Five

[C] Three [D] None of the above

218. **Which of the following does not form one electron redox system?**

[A] Plastocyanin [B] Plastoquinone

[C] Cytochrome [D] None of the above

219. **Non-cyclic electron transport in photosynthesis is also known by the name of**

[A] S-Scheme [B] Z-Scheme

[C] Q-Scheme [D] None of the above

220. **Ferredoxin (Fd) is a**

[A] Heme-iron protein [B] Non-heme iron-protein

[C] Copper containing protein [D] None of the above

221. **In photosynthesis, dark reaction or Blackman's reaction is**

[A] Enzymatic and occurs in absence of light

[B] Non-enzymatic and occurs in absence of light

[C] Enzymatic and independent of light

[D] None of the above

222. Which of the following does not occur in cyclic electron transport & phosphorylation?

[A] Oxygen evolution

[B] Photolysis of water

[C] Formation of reduced NADPH

[D] All of the above

223. First stable product of the dark reaction of photosynthesis is

[A] 3PGALD [B] 3 PGA

[C] 1, 3-bis PGA [D] None of the above

224. Chloroplast ATP synthase is a .... kDa enzyme that contains .... different subunits and includes both chloroplast and nuclear-encoded gene products.

[A] 400, nine [B] 700, five

[C] 800, four [D] None of the above

225. The overall stoichiometry of CF1 is

[A] α3β3γδƐ [B] α2β2γδ4Ɛ

[C] α1β9γδ9Ɛ5 [D] None of the above

226. The thylakoid ATP synthase consists of two segments: a transmembrane segment, called ...., and a hydrophilic segment on the stromal surface, called ....

[A] CF0, CF1 [B] CF1, CF0

[C] CF0, CF0 [D] CF1, CF1

227. Choose the correct alternative for polypeptide subunits of ATP synthase complex.

CF1

Protein	Gene	Location of gene	Mol. Mass (kDa)	Function
[A] α-subunit	atp A	Chloroplast	55	Catalytic
[B] β-subunit	atp B	Chloroplast	54	Catalytic
[C] γ-subunit	atp C	Nucleus	36	Proton gating
[D] Δ-subunit	atp D	Nucleus	20	Binding of CF1 to CF0, regulation
[E] Ɛ-subunit	atp E	Chloroplast	15	ATPase inhibition
[F] All of the above				

CF2

Protein	Gene	Location of gene	Mol. Mass (kDa)	Function
[A] I	*atp* F	Chloroplast	17	Binding of CF0 to CF1
[B] II	*atp* G	Nucleus	16	Binding of CF0 to CF1
[C] III	*atp* H	Chloroplast	8	Proton translocation
[D] IV	*atp* I	Chloroplast	27	Binding of CF0 to CF1
[E] All of the	above			

228. Find the correct matching of protein subunit of the PSII core complex

Protein	Gene	Location of gene	Mol. Mass (kDa)	Function
Hydrophobic subunit				
[A] D1	*psbA*	Chloroplast	32	Reaction centre protein
[B] D2	*psbB*	Chloroplast	34	Reaction centre protein
[C] CP47	*psbC*	Chloroplast	51	Antenna binding
[D] CP43	*psbD*	Chloroplast	43	Antenna binding
Cytb559				
[E] α-subunit	*psbE*	Chloroplast	9	Unknown
[F] β-subunit	*psbF*	Chloroplast	4	Unknown
[G] PsbH-PsbN	*psbH-psb N*	Chloroplast	3.8-10	Unknown
[H] 22 kDa	*psbS*	Nuclear	22	Photoprotection
Hydrophillic subunits				
[I] 33 kDa	*psbO*	Nucleus	33	Oxygen evolution
[J] 23 kDa	*psbP*	Nucleus	23	Oxygen evolution
[K] 17 kDa	*psbQ*	Nucleus	17	Oxygen evolution
[L] 10 kDa	*psbR*	Nucleus	10	Unknown
[M] All of the above				

229. Find the correct matching of polypeptide subunits of the cytochrome *b6f* complex.

Protein	Gene	Location of gene	Mol. Mass (kDa)	Function
[A] Cyt *f*	*petA*	Chloroplast	32	Cyt f apoprotein
[B] Cyt *b6*	*petB*	Chloroplast	24	Cyt b6 apoprotein
[C] RFeS	*petC*	Nucleus	19	Rieske Fe-S apoprotein
[D] Subunit IV	*petD*	Chloroplast	17	Quinone binding at Qp
[E] PetG, Pet M	*petG, petM*	Nucleus	4.0	Unknown
[F] Pet I	*petL*	Chloroplast	3.4	Unknown
[G] All of the above				

230. Find the correct matching of polypeptide subunit of PSI core complex

Protein	Gene	Location of gene	Mol. Mass (kDa)	Function
Hydrophobic subunits				
[A] PsaA	*psaA*	Chloroplast	83	Reaction centre protein
[B] PsaB	*psaB*	Chloroplast	82	Reaction centre protein
[C] PsaG	*psaG*	Nucleus	11	Unknown
[D] PsaI	*psaI*	Chloroplast	4	Unknown
[E] PsaJ, PsaM	*psaJ, psaM*	Chloroplast	5, 3.5	Unknown
[F] PsaK, PsaL	*psaK, psaL*	Nucleus	8.4, 18	Unknown
Hydrophilic subunits				
Stromal orientation				
[G] Psa C	*psaC*	Chloroplast	9	Fe-S apoprotein
[H] Psa D	*psaD*	Nucleus	18	Ferredoxin docking
[I] Psa E	*psaE*	Nucleus	10	Cyclic electron transport
[J] Psa H	*psaH*	Nucleus	10	LHC-I linker
Lumenal orientation				
[K] Psa F	*psaF*	Nucleus	17	PC docking
[L] Psa N	*psaN*	Nucleus	10	Unknown

[M] All of the above matchings are correct.

231.

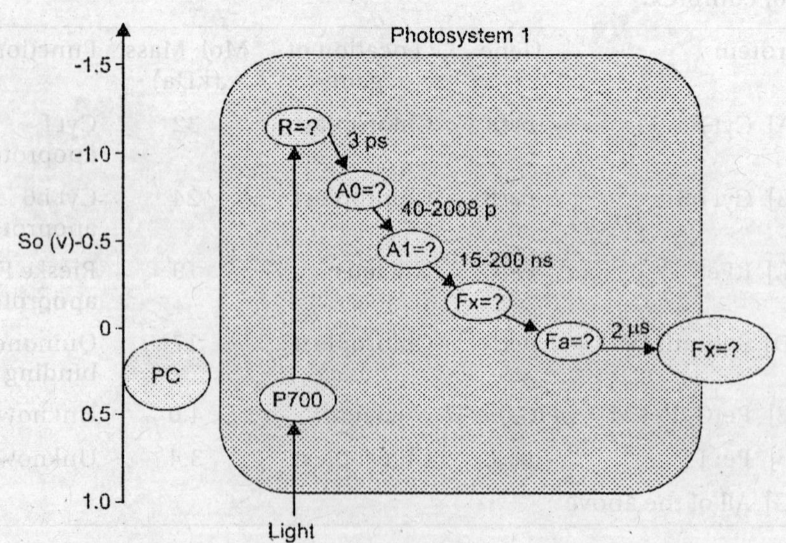

Find the missing in above figure which shows the PS I electron carriers and their kinetic behavior.

[A] R= P700*, Ao = Monomeric chlorophyll a, A1 = Phylloquinone, Fx and Fa = a series of additional electron carriers that include three different Fe-S centres.

[B] R= P700*, A1= Monomeric chlorophyll a, Ao = Phylloquinone, Fx and Fa = a series of additional electron carriers that include three different Fe-S centres.

[C] R= P700, Ao= Monomeric chlorophyll a, A1 = Phylloquinone, A1 and Fa = a series of additional electron carriers that include three different Fe-S centres.

[D] None of the above

232. In plants, Rubisco consists of .... larger and .... smaller subunits.

[A] Eight, eight [B] Nine, nine

[C] Seven, seven [D] None of the above

233. In most of the eukaryotes, larger subunits are encoded by the .... genome and smaller subunits are encoded by the .... genome.

[A] Chloroplast, nuclear [B] Nuclear, chloroplast

[C] Chloroplast, chloroplast [D] Nuclear, nuclear

[E] None of the above

234. How many enzymes are required in Calvin cycle?

[A] 12 [B] 13

[C] 14 [D] None of the above

235. How many enzymes are involved in regeneration phase of Calvin cycle?

[A] 10 [B] 11

[C] 12 [D] None of the above

236. Structural formula of Rubisco is

[A] L_8S_8

[B] L_8S_7 [where L= Larger subunit and S= Smaller subunit]

[C] L_8S_6

[D] None of the above

237. Photosynthetic green sulfur bacteria fix carbon via

[A] Reductive carboxylic acid cycle

[B] Hydroxyl propionate pathway

[C] Calvin cycle

[D] None of the above

238. Green non sulfur bacterium *Chloroflexus* fix carbon via

[A] Hydroxy propionate pathway [B] C4- cycle

[C] C3-cycle [D] None of the above

18 ENZYMES AND VITAMINS

1. **Which of the following represent Henri-Michaelis-Menten equation?**

[A] $V = \frac{Vmax\,[S]}{Km + [S]}$ [B] $V = \frac{Km + [S]}{Vmax\,[S]}$

[C] $V = \frac{Vmax + [S]}{Km + [S]}$ [D] None of the above

2. **Michaelis constants (Km) for enzyme-substrate pair of Catalase and H_2O_2 is**

[A] 25 mM [B] 0.4 mM

[C] 0.05 mM [D] None of the above

3. **Michaelis constants (Km) for enzyme-substrate pair of Hexokinase and ATP is**

[A] 25 mM [B] 0.4 mM

[C] 0.05 mM [D] None of the above

4. **Michaelis constants (Km) for enzyme-substrate pair of Hexokinase and D-Glucose is**

[A] 25 mM [B] 0.4 mM

[C] 0.05 mM [D] None of the above

5. **Michaelis constants (Km) for enzyme-substrate pair of Hexokinase and D-Fructose is**

[A] 25 mM [B] 0.4 mM

[C] 0.05 mM [D] None of the above

6. **Michaelis constants (Km) for enzyme-substrate pair of Chymotrypsin and Glycyltyrosinglycine is**

[A] 2.5 mM [B] 9 mM

[C] 108 mM [D] None of the above

7. **Michaelis constants (Km) for enzyme-substrate pair of Chymotrypsin and N-Benzoyltyrosinamide is**

[A] 2.5 mM [B] 26 mM

[C] 111 mM [D] None of the above

8. **Michaelis constants (Km) for enzyme-substrate pair of Carbonic anhydrase and HCO_3^- is**

[A] 7 mM [B] 8 mM
[C] 9 mM [D] None of the above

9. **Which of the following coenzyme is the carrier of hexose sugar molecules?**

[A] TPP [B] PAPS
[C] UDP [D] None of the above

10. **In higher plants which of the following is the rich source of enzymes and are widely used in experimental work?**

[A] Matured seeds [B] Germinating seeds
[C] Matured flowers [D] Immature flowers
[E] None of the above

11. **According to ...., an enzyme is "simple or combined proteins acting as specific catalysts".**

[A] Sumner and Myrback [B] Louis Pasture
[C] Traube [D] None of the above

12. **Which of the following is considered as good source for studies of enzymes?**

[A] Microorganism [B] Germinating seeds
[C] Both [D] None of the above

13. **The word enzyme was coined by**

[A] Willstate [B] Gobindjee
[C] F.W. Kuhne [D] None of the above

14. **The turnover number for Catalase is**

[A] 4×10^7 [B] 9×10^7
[C] 12×10^7 [D] 17×10^7
[E] None of the above

15. **E.C. number of the enzyme alcohol dehydrogenase is**

[A] 0.1.1.1 [B] 9.0.1.2
[C] 1.1.1.1 [D] None of the above

16. **First enzyme to be isolated in pure crystalline form is**

[A] Invertase [B] Diastase
[C] Zymase [D] Urease

17. **Which of the following is the common characteristic of any enzyme?**

[A] Enzymes exhibit enormous catalytic power
[B] Enzymes are highly specific
[C] Most (but not all) enzymes are proteins
[D] All of the above

18.

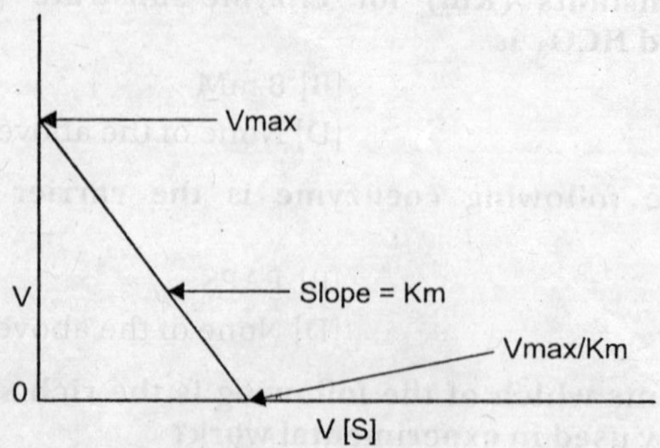

The graph V versus V/[S] is called

[A] Eadie-Hofstee plot [B] Double reciprocal plot

[C] Both [D] None of the above

19. Heterotrophic enzymes are

[A] Modified by their substrate concentration

[B] Different isoenzymes

[C] Stimulated and inhibited by an effectors or modulator molecule other than substrate

[D] All of the above

20. In a chemical reaction, transition state species have free energies

[A] Higher than reactant, but lower than products

[B] Higher than products but lower than reactant

[C] Higher than either the reactant or the products

[D] None of the above

21. A competitive inhibitor

[A] Decreases the km of an enzyme

[B] Increases the Km of an enzyme

[C] Decrease the Km but increases the Vmax of an enzyme

[D] Increases both the Vmax and the Km of an enzyme

22. Enzymes are basically made up of

[A] Proteins [B] Fat

[C] DNA [D] None of the above

23. Which of the following enzyme was first isolated and purified in the form of crystal?

[A] Ribonuclease [B] Pepsin

[C] Urease [D] Amylase

24. The enzymes which act normally within cells are called

[A] Ferment [B] Exoenzymes

[C] Endoenzymes [D] Apoenzymes

25. The non-protein part of enzyme is called

[A] Prosthetic group [B] Apoenzyme
[C] Holoenzyme [D] None of the above

26. Majority of enzymes belongs to .... category.

[A] Endoenzymes [B] Exoenzymes
[C] Both [D] None of the above

27. The enzymes which are always present in the organism in constant amounts regardless of its metabolic state are called as ....

[A] Constitutive enzymes [B] Inducible enzyme
[C] Isoenzymes [D] None of the above

28. An example of inductive or inducible enzyme in *E.coli* is

[A] Nitrate reductase [B] -amylase
[C] Both [D] None of the above

29. The lower value of Km means

[A] Lower affinity of enzyme with substrate
[B] Higher enzyme activity
[C] Higher substrate affinity of enzyme
[D] None of the above

30. Lactate dehydrogenase (LDH) which catalyses pyruvate to lactate is an example of

[A] Coenzyme [B] Isoenzyme
[C] Apoenzyme [D] None of the above

31. An excellent example of inducible enzyme in higher plant is

[A] Nitrate reductase [B] -amylase
[C] Both [D] None of the above

32. Multiple forms of enzymes with the same catalytic activity but different structures are known as

[A] Isoenzymes [B] Isozymes
[C] Both [D] None of the above

33. One of the first enzymes discovered in isozyme forms is

[A] Lactate dehydrogenase [B] Phosphatase
[C] Lipase [D] None of the above

34. How many forms of LDH have been discovered?

[A] Fours [B] Five
[C] Six [D] None of the above

35. Which of the following is not a zymogen?

[A] Trypsinogen [B] Pepsinogen
[C] Procarboxypeptidase [D] None of the above

36. The substrate for ribozyme is

[A] DNA [B] RNA
[C] Fat [D] None of the above

37. **.................... different types of peroxidase isoenzymes have been isolated from maize.**

[A] 18 [B] 20

[C] 22 [D] None of the above

38. **Some enzymes are synthesized in the cells in an inactive state and are called as**

[A] Zymogens [B] Proenzymes

[C] Pre-enzymes [D] All of the above

39. **Pepsinogen, trypsinogen, chymotrypsinogen and procarboxypeptidase are examples of**

[A] Zymogens [B] Ribozymes

[C] Isozymes [D] None of the above

40. **The self splicing group I introns and RNAse P are the examples of**

[A] Zymogens [B] Ribozymes

[C] Isozymes [D] None of the above

41. **Each enzyme has been assigned a specific code number consisting of .... digits. The first digit indicates the .... class, the second indicates the ...., the third digit indicates its .... while the fourth digit denotes the systematic specific name of the enzyme the first part of which indicates the name of the .... and the second part the ....**

[A] Four, major, sub-class, sub-sub class, substrate, nature of reaction

[B] Five, major, sub-class, sub-sub, substrate, nature of reaction

[C] Four, substrate, nature of reaction, sub-class, sub-sub class

[D] None of the above

42. **A non protein substance is required at the active centre for enzyme activity which is bound .... to the enzyme protein by .... linkage and is known as ....**

[A] Tightly, covalent, prosthetic group

[B] Loosely, covalent, prosthetic group

[C] Tightly, coordinate, cofactors

[D] None of the above

43. **According to the new system of classification the enzymes are grouped into**

[A] Six major classes [B] Five major classes

[C] Four major classes [D] None of the above

44. **Which of the following hydrolyzing enzyme does not hydrolyse ester bonds?**

[A] Lipases [B] Phosphatase

[C] Both [D] None of the above

45. **Which of the following organic compound acts as prosthetic groups in some enzyme reactions and as cofactors in some others?**

[A] FMN [B] FAD
[C] Heme-iron [D] Biotin
[E] All of the above

46. **Which of the following is the carrier of sugar molecules such as glucose, galactose and their derivatives?**

[A] UDP [B] PAPS
[C] TPP [D] None of the above

47. **Which of the following is the carrier of sulphate group?**

[A] UDP [B] PAPS
[C] TPP [D] None of the above

48. **Which of the following coenzyme is the carrier of acyl group?**

[A] Coenzyme II [B] Coenzyme A
[C] Coenzyme Q [D] None of the above

49. **The first ribozyme was discovered by**

[A] R. Cech [B] C. Woese
[C] F. Crick [D] L. Orgel

50. **Which of the following is the example of ribozyme?**

[A] Leadzyme [B] Hairpine
[C] Hammerhead [D] Hepatitis dalta virus
[E] Tetrahymena [F] All of the above

51. **Among the followings, which one is the accepted SI unit of enzyme activity?**

[A] Katal [B] Talka
[C] Latka [D] None of the above

52. **The amount of enzyme causing transformation of 1 molecule of substrate per minute at 25°C under optimal conditions of measurement is known as**

[A] Enzyme unit [B] The Katal (Kat)
[C] Rate constant [D] None of the above

53. **Transformation of 1 mol of substrate per second into product is known as**

[A] Enzyme unit [B] The Katal (Kat)
[C] Rate constant [D] None of the above

54. **The enzymes where catalysis involves transfer of electrons are named as**

[A] Transferases [B] Lyases
[C] Oxidoreductase [D] Isomerases

55. **The enzymes where catalysis involve group transfer are named as**

[A] Transferases [B] Lyases
[C] Oxidoreductase [D] Isomerases

56. The enzyme glutathione peroxidase which catalyzes destruction of H_2O_2 contains

[A] Mo [B] Fe
[C] Se [D] Zn

57. Sulpha drugs are effective antibiotics which inhibit the formation of folic acid by acting as a

[A] Non- competitive inhibitor [B] Competitive inhibitor
[C] Uncompetitive inhibitor [D] None of the above

58. Which one of the following vitamins does not act as a precursor for coenzymes?

[A] Thiamine [B] Biotin
[C] Folic acid [D] None of the above

59. Biotin is involved in which of the following type of reactions?

[A] Dehydrations [B] Carboxylations
[C] Hydroxylations [D] Dacarboxylations

60. Which of the following statement is true about non-competitive inhibition?

[A] Km decrease [B] Vmax increases
[C] Vmax decreases [D] Km increases

61. Abzymes are

[A] Enzymes that hydrolyze antibodies
[B] Antibodies that have catalytic activity
[C] Both
[D] None of the above

62. Induced fit theory of enzyme action was proposed by

[A] Koshland [B] Fischer
[C] Went [D] None of the above

63. Most of the enzymes show optimum activity in

[A] Alkaline medium [B] Neutral medium
[C] Acidic medium [D] Any of the above

64. Allosteric enzymes have

[A] Allosteric sites [B] Active site
[C] Both [D] None of the above

65. Among the followings, which enzyme has the highest turnover number?

[A] Catalase [B] Lysozyme
[C] Lipase [D] None of the above

66. Among the followings, which enzyme has lowest turnover number?

[A] Catalase [B] Lysozyme
[C] Lipase [D] None of the above

67. Malonic acid inhibits the activity of the enzyme

[A] Succinic dehydrogenase [B] Threonine dehydratase
[C] Hexokinase [D] None of the above

68. Pepsinogen (inactive form of enzyme pepsin) is converted into its active form by

[A] Hydrolysis [B] Isomerism
[C] Reduction [D] Oxidation

69. Most of the digestive enzymes belong to

[A] Oxidoreductase [B] Isomerases
[C] Lyases [D] Hydrolases

70. The active site of enzymes is formed by some of its

[A] Amino groups of amino acid
[B] Carboxyl groups of the amino acids
[C] Exposed sulfur bonds
[D] None of the above

71. The product of an enzyme-catalysed reaction can act as inhibitor of the reaction. This mechanism of control is known as

[A] Feedback inhibition [B] Competitive inhibition
[C] Metabolic [D] None of the above

72. The Km value of an enzyme is

[A] ½ of the Vmax
[B] The substrate concentration that gives ½ maximal velocity
[C] A dissociation constant
[D] None of the above

73. An enzyme that catalyzes the conversion of an aldose sugar to a ketose sugar would be classified as one of the

[A] Isomerases [B] Transferase
[C] Hydrolases [D] Oxidoreductase

74. An allosteric modulator influences enzyme activity by

[A] Binding to a site on the enzyme molecule distinct from the catalytic site
[B] Competing for the catalytic site with the substrate
[C] Changing the specificity of the enzyme for its substrate
[D] None of the above

75. Which one of the following coenzyme pair correctly with the group transferred?

[A] NADP+-phosphoryl [B] Biotin-carbon dioxide
[C] Thiamin-pyrophosphate [D] None of the above

76. Lactic dehydrogenase (LDH) is essential for lactic acid formation. This is an example of

[A] Isoenzymes [B] Abzyme
[C] Apoenzyme [D] Chemienzyme

77. Which of the following vitamin is a precursor of coenzyme that is required in enzymatic reactions involving transfer of acyl groups?

[A] Biotin [B] Riboflavin

[C] Lipoic acid [D] p-aminobenzic acid

78. Which metabolic process in bacteria is directly inhibited by the antibiotic penicillin?

[A] ATP hydrolysis [B] Synthesis of fats

[C] Cellular respiration [D] None of the above

79. Catalytic antibodies function as enzymes on the principle of

[A] Enzymatic conversion of antibodies

[B] Monoclonal antibodies with chemical capability

[C] Antigen-antibody affinity

[D] None of the above

80. An uncompetitive inhibitor of enzyme binds to

[A] Any other site and modifies part of an enzyme

[B] The active site of enzyme

[C] Site other than the active site

[D] None of the above

81. Which vitamin is necessary for coenzyme A synthesis?

[A] Pyridoxine [B] Pantothenic acid

[C] Biotin [D] None of the above

82. At what [S] the velocity[V0] of an enzyme –catalysed reaction is 25% of the Vmax?

[A] ¾ Km [B] 4Km

[C] $\frac{1}{3}$ Km [D] ¼ km

83. Km specifies affinity of enzyme with substrate when

[A] $K_2>>>K_{-1}$ [B] $K_2<<<K_{-1}$

[C] $K_2=K_{-1}$ [D] None of the above

84. Which of the following is a cofactor and not a coenzyme?

[A] Methylcobalamin [B] Copper

[C] Tetrahydrofolic acid [D] Biotin

85. Which of the following is the water soluble vitamin?

[A] Thiamines [B] Riboflavin

[C] Niacin [D] Biotin

[E] Panthothenic acid [F] Folic acid

[G] Cobalamin [H] Pyridoxine

[I] Ascorbic acid [J] All of the above

86. Which of the following is the fat soluble vitamin?

[A] Vitamins A [B] Vitamins D

[C] Vitamins E [D] Vitamins K

[E] All of the above

87. Deficiency of niacin causes

[A] Pellagra [B] Low vision

[C] Colorblindness [D] None of the above

88. Which of the following vitamin can act without phosphorylation?

[A] Thiamine [B] Lipoamide

[C] Niacin [D] Pyridoxine

89. Which of the following coenzyme acts as an electron sink to promote catalysis?

[A] Pyridoxal phosphate [B] Lipoamide

[C] CoA [D] Biotin

90. Panthothenic acid is a constituent of the coenzyme involved in which of the following event?

[A] Acetylation [B] Reduction

[C] Decarboxylation [D] Dehydrogenation

91. Most Vitamins are

[A] Covalently linked to enzymes [B] Synthesized by humans

[C] Precursor of coenzymes [D] None of the above

92. The enzyme acts by

[A] Lowering the energy of activation of reactions

[B] Increasing the energy of activation of reaction

[C] Conserving the energy of activation of reaction

[D] None of the above

93. If an enzyme substrate system has lower value of Km, then binding of substrate to its enzyme will be

[A] More stronger [B] Less stronger

[C] Inhibited [D] None of the above

94. Allosteric enzymes have

[A] Active sites [B] Allosteric sites

[C] Both [D] None of the above

95. Mitochondria account for ....% of a cell volume.

[A] 70 [B] 50

[C] 30 [D] None of the above

96. Which of the following enzyme is known as meat tenderizer because of its continued action even at the high temperature of cooking?

[A] Papain [B] Oxaloacetic acid

[C] Mutases [D] None of the above

97. When the velocity of the reaction is almost proportional to the substrate concentration, the rate of reaction is said to be

[A] First order reaction [B] Second order reaction

[C] Zero order reaction [D] None of the above

98. Induced fit model has sufficient experimental evidence from the

[A] Tracer technique [B] X-ray diffraction
[C] Auto radiography [D] None of the above

99. Km is not dependent on the concentration of ....

[A] Enzyme [B] Substrate
[C] Both [D] None of the above

100. For majority of enzymes, the Km values are in the range of

[A] 10^{-5} to 10^{-2} moles [B] 10^{-6} to 10^{-3} moles
[C] 10^{-9} to 10^{-6} moles [D] None of the above

101. Iodoacetate is an irreversible inhibitor of the enzyme

[A] Papain
[B] Glyceraldehyde 3-phosphate dehydrogenase
[C] Both [D] None of the above

102. All enzymes have a specific .... structure.

[A] Three dimensional [B] Two dimensional
[C] One dimensional [D] None of the above

103. Abzymes were first reported by

[A] Lerner and Tramontano [B] Northrop
[C] James B. Sumner [D] None of the above

104. A .... is obtained when velocity is plotted against the substrate concentration.

[A] Rectangular hyperbola [B] Rectangular parabola
[C] Liner [D] None of the above

105. .... enzymes are thermolabile.

[A] All [B] Few
[C] Most of [D] None of the above

106. Proenzyme is the .... precursor of an enzyme.

[A] Active [B] Inactive
[C] Both [D] None of the above

107. The essential chemical components of many co-enzyme are

[A] Vitamins [B] Carbohydrate
[C] Protein [D] None of the above

108. The modern system of enzyme classification was introduced by International Union of Biochemistry (IUB) in

[A] 1961 [B] 1962
[C] 1963 [D] None of the above

19 INSTRUMENTATION IN PLANT PHYSIOLOGY

1. **Autoclave is built to withstand the steam pressure of at least .... lb per sq. inch.**
   [A] 30 [B] 70
   [C] 110 [D] None of the above
2. **The common feature to all the chromatography methods is the use of two phases**
   [A] Stationary phase [B] Mobile phase
   [C] Both [D] None of the above
3. **In chromatography methods, if the stationary phase is solid, the process is called**
   [A] Adsorption [B] Partition chromatography
   [C] Both [D] None of the above
4. **In chromatography methods, if the stationary phase is liquid, the process is called**
   [A] Adsorption [B] Partition chromatography
   [C] Both [D] None of the above
5. **On the basis of stationary and mobile phase, chromatography system is of .... type.**
   [A] Liquid-solid [B] Gas-solid
   [C] Liquid-liquid [D] Gas-liquid [E] All of the above
6. **Classical adsorption chromatography, TLC and ion-exchange chromatography are the examples of**
   [A] Liquid-solid [B] Gas-solid
   [C] Liquid-liquid [D] Gas-liquid
7. **Classical partition chromatography and paper chromatography are the examples of**
   [A] Liquid-solid [B] Gas-solid
   [C] Liquid-liquid [D] Gas-liquid
8. **Capillary column chromatography is the example of**
   [A] Liquid-solid [B] Gas-solid
   [C] Liquid-liquid [D] Gas-liquid

9. The curve obtained by plotting the amount absorbed against concentration at constant temperature is called

[A] Sorption isotherm [B] Sorption isolate
[C] Sorption dekatherm [D] None of the above

10. The general chromatography method was introduced into biology by

[A] Tswett [B] A.T. James
[C] P. Martin [D] None of the above

11. The technique of gas chromatography was used for the first time by

[A] A.T. James [B] A.J. Martin
[C] P. Martin [D] All of the above

12. The use of radioactive radiations to obtain the photographic film of the test material, incorporated with the radioactive tracers, is called .... and the film obtained is called ....

[A] Autoradiography, autoradiograph
[B] Autoradiograph, autoradiography
[C] Replication, blanks
[D] None of the above

13. Autoradiography can be detected either directly with a .... or indirectly via their effect on ....

[A] Scintillation counter, photographic film
[B] Photographic film, scintillation counter
[C] Only spectrophotometer
[D] Only atomic absorption photometer

14. How many types of radiations are used in autoradiography?

[A] Three [B] Two
[C] One [D] None of the above

15. Scanning electron microscope combines the mechanism of electron microscopy and ....

[A] Television [B] Radio
[C] Telephone [D] None of the above

16. In bright field microscope, the microscope field (the area observed) is .... and the microorganism appears ....

[A] Bright, dark [B] Dark, bright
[C] Bright, bright [D] None of the above

17. Phase contrast principle was discovered by

[A] Fritz Zernike [B] Knoll
[C] Ruska [D] None of the above

18. Autoradiography can be directly measured by various counters like

[A] Giger Muller counter [B] Scintillation Counter
[C] Proportional counter [D] All of the above

19. Which of the following is the stage in the operation of Scintillation counter

[A] Absorption

[B] Scintillation

[C] Conversion of light into electronic impulsions

[D] All of the above

20. .... is made of a glass or metal tube containing a mixture of gases, an inert gas like helium or argon and an organic vapour or a gas e.g. isopropanol or isolutane or cooking gas. It has a thin window, usually made of mica at one end to enclose the gas. Depending upon the thickness of the Window and the energy of radiation of the radio-isotopes a fraction of the particles enter the counting tube and ionize the gas mixture inside. The small current generated is magnified at high voltage and measured with a scaler calibrated to record disintegrations per unit time. The counts recorded include counts due to ionization of the gas mixture by cosmic rays passing through the tube, any radioactivity present in the surrounding area.

[A] Geiger-Muller Counter [B] Tracer technique

[C] Scintillation Counter [D] None of the above

21. The result of Geiger-Muller counter is expressed in

[A] Count/min [B] Count/sec

[C] Count/hr [D] None of the above

22. .... developed the process of electrophoresis.

[A] Tiseleus [B] Tswett

[C] Went [D] None of the above

23. The .... is operated at a temperature of 160 to 180°C for a period of one and half hour. If the temperature goes above 180°C there is a danger of cotton being charred. It is used for sterilizing all kind of laboratory glassware, such as test tubes, Petri dishes, pipettes, flasks, bottles, etc.

[A] Gel filtration [B] Hot Air Oven

[C] Autoclave [D] None of the above

24. .... is a process of separation of particles where on the basis of difference in charges and molecular size and shape the particles under the influence of an electric field migrate.

[A] Electrophoresis [B] Centrifugation

[C] Chromatography [D] None of the above

25. To ensure uniform distribution of charged molecules, agents like .... is added during electrophoresis.

[A] SDS (Sodium dodecyl sulphate)

[B] KDS (Potassium dodecyl sulphate)

[C] Both

[D] None of the above

26. In angle head centrifuge, the sample is kept at an angle of .... to the horizontal.

[A] 30° [B] 50°
[C] 70° [D] None of the above

27. In differential centrifuge the components of a mixture are separated according to their

[A] Size [B] Shape
[C] Density [D] All of the above

28. .... is an equipment having an air blower in the rear side of the chamber which can produce air flow with uniform velocity along parallel flow lines. There is a special filter system-high efficiency particulate air filter which can remove particles as small as 0.3 μm.

[A] Laminar air flow [B] Incubator
[C] Autoclave [D] None of the above

29. .... is an equipment that consists of copper/steel chamber, around which warm water or air circulates either by electricity or by means of small gas flame.

[A] Incubator [B] Hot air oven
[C] Electrophoresis [D] None of the above

30. .... is a cylindrical vessel having double walls around all parts except the upper side. It is built to withstand the steam pressure of at least 30 lb per sq. inch.

[A] Autoclave [B] Hot air oven
[C] Both [D] None of the above

31. The most frequent used wavelength in the UV region is

[A] 340 nm [B] 280 nm
[C] Both [D] None of the above

32. Which of the following is the suitable fluorescence used in the fluorescent microscopy?

[A] Acridine orange [B] Acridine yellow
[C] Acriflavine auramine [D] Rhencine
[E] Fluorescence titan yellow [F] All of the above

33. .... chromatography is a noble form of column chromatography in which the molecule (principally proteins and nucleic acids) to be isolated from the sample under study are retarded in their passage through the column by their specific biological reaction with the column matrix.

[A] Affinity chromatography [B] Thin layer chromatography
[C] Paper chromatography [D] None of the above

34. Find the correct adsorbents used for thin layer chromatography.

[A] Silica gel [B] Alumina
[C] Kieselguir [D] Celite
[E] Cellulose powder [F] Hydroxylapatite
[G] Polyethlenimine [H] All of the above

35. In Transmission electron microscopy, the electron source is commonly a tungsten filament at .... KV potential.

[A] 30-150 [B] 200-250
[C] 300-350 [D] None of the above

36. Silica cuvette range is

[A] 360-800 nm [B] 200-800 nm
[C] 100-300 [D] None of the above

37. Glass cuvette range is

[A] 360-800 nm [B] 200-800 nm
[C] 100-300 [D] None of the above

38. Adsorption chromatography is widely used in the separation of

[A] Amino acids [B] Lipids
[C] Steroids [D] Sugars
[E] All of the above

39. Find the isoelectric point of lysozyme

[A] 11.0 [B] 10.6
[C] 9.5 [D] None of the above

40. Find the isoelectric point of pepsin

[A] 1.0 [B] 1.2
[C] 1.3 [D] 1.4

41. Find the isoelectric point of ribonuclease

[A] 9.5 [B] 9.7
[C] 9.9 [D] None of the above

42. .... is a dry air type sterilizer with three walls and two air spaces. The outer walls are made up of thick asbestos to reduce the radiation of heat. The hot air steriliser is operated electrically. In this case the heater coil is either be placed at the bottom of the instrument or on the side walls. A convection current travel a complete circuit through the wall space and interior of the instrument. The temperature inside the instrument is controlled by thermostat.

[A] Hot air oven [B] Autoclave
[C] Electro focussing [D] None of the above

43. Which of the following isotope is stable in nature?

[A] ^{2}H [B] ^{15}N
[C] ^{18}O [D] All of the above

44. Which of the following isotope is unstable in nature?

[A] ^{3}H [B] ^{14}C
[C] ^{32}P [D] ^{35}S [E] All of the above

45. One Curie (Ci) is equivalent to.................... disintegration per second.

[A] 3.7×10^{10} [B] 3.7×10^{9}
[C] 3.7×10^{8} [D] None of the above

46. One Bequerel (Bq) is

[A] 10^{10} disintegration per second
[B] 10^{9} disintegration per second
[C] 10^{8} disintegration per second
[D] None of the above

47. Biological half life of Iodine is

[A] 156 days [B] 104 days
[C] 8.1 days [D] None of the above

48. .... is a method for separating molecules on the basis of size (i.e. molecular weight) difference. The procedure is carried out using glass columns packed with nonionic, porous gel particles. Since the gel do not posses ionic groups, separation of a mixture of macromolecules does not involve the temporary formation of bound with the gel. The most commonly used gels are cross-linked dextrans produced by reacting with epichloronydrin.

[A] Gel filtration [B] Molecular sieving
[C] Both [D] None of the above

49. The sedimentation coefficient of a particle may be experimentally determined in an instrument known as .... This instrument is equipped with an optical system that permits visual observation of the sedimenting particles as one or more moving boundaries formed between regions of the particle suspension having different refractive indexes.

[A] Analytical ultracentrifuge [B] Differential centrifuge
[C] Dialysis and Ultra filtration [D] None of the above

50. Autoradiography is a special modification of the

[A] Radioisotopes tracer techniques
[B] Gel filtration
[C] Gas chromatography
[D] None of the above

20

THE PLANT CELL

1. **Credit for discovery of cell goes to**
[A] Robertson [B] Robert Hooke
[C] Robert Brown [D] None of the above

2. **Robert Hooke was basically a**
[A] Mathematician and physicist [B] Botanist
[C] Chemist [D] None of the above

3. **The branch which deals with the study of cell structure is known as**
[A] Ecology [B] Cytology
[C] Histology [D] None of the above

4. **Resolving power of human eye is**
[A] 150 μ [B] 50 μ
[C] 200 μ [D] None of the above

5. **The smallest size of cell visible to naked eye is**
[A] 100μ [B] 10μ
[C] 1000μ [D] None of the above

6. **The size of most cells is best expressed in**
[A] Inches [B] Micrometers
[C] Meters [D] None of the above

7. **The smallest measuring unit in cytology is**
[A] nm [B] Angstrom
[C] Micron [D] None of the above

8. **One micron is equal to**
[A] 1/10 of a millimeter [B] 1/1000 of a millimeter
[C] 1/100 of a millimeter [D] None of the above

9. **An angstrom is equal to**
[A] 10^{-3} meters [B] 10^{-10} meters
[C] 10^{-8} meters [D] 10^{-5} meters

10. **Angstrom, the unit of measurement in transmission electron microscopy is equal to**
[A] 0.01 μm [B] 0.001 μm
[C] 0.0001 μm [D] None of the above

11. Cell doctrine was given by

[A] Singer and Nicolson [B] Leeuwenhoek

[C] Both [D] None of the above

12. The branch of botany dealing with the internal structure of cell and its function is called

[A] Cell biology [B] Cytology

[C] Both [D] None of the above

13. The term cell was first of all used by

[A] Harvey [B] Flemming

[C] Hopkins [D] None of the above

14. Robert Hooke

[A] Invented the lens

[B] Constructed a microscope

[C] Discovered the cork cells

[D] Lived in the 17th century

15. The figure of the cork cells as seen by Robert Hooke was published in his book

[A] Micrographia [B] Genera Plantarum

[C] Origin of species [D] None of the above

16. What Robert Hooke had discovered in the thin section of the cork as a cell was really

[A] Nuclei [B] Cellulose

[C] Protoplasm [D] Cell walls

17. The idea of individuality of cells was expressed by

[A] Corti [B] Schwan

[C] Dutrochet [D] Dujardin

18. Plant cell differs from animal cell in having

[A] Cell wall [B] Ribosome

[C] Cytoskeleton [D] None of the above

19. Father of modern cytology is

[A] C.P. Swanson [B] R. Altman

[C] C. Benda [D] None of the above

20. Book "The Cell" was written by

[A] C.P. Swanson [B] R. Altman

[C] C. Benda [D] None of the above

21. Well defined nucleus is found in

[A] Prokaryotes [B] Eukaryotes

[C] Cyanobacteria [D] Bacteria

22. "All plants are made of cells", this theory was given by

[A] M.J. Schleiden [B] T. Schwann

[C] Both [D] None of the above

23. "Animal cell differs from plant cell in absence of cell wall", this statement was made by

[A] M.J. Schleiden [B] T. Schwann
[C] Both [D] None of the above

24. Robert Hooke used the word .... for the walls separating the chambers (or cells).

[A] Diaphragm [B] Object plate
[C] Cell wall [D] None of the above

25. "Micrographia" published by Robert Hooke is among the great classics of biology. The Chapter .... of the Micrographia entitled "of the Schematisme or Taxture of cork and of the cells and pores of some other such frothy bodies" contains the observation of Robert Hooke on cells.

[A] XVIII [B] XVII
[C] XVI [D] None of the above

26. Who defined that the cells are membrane enclosed nucleus containing structure?

[A] M.J. Schleiden [B] T. Schwann
[C] Both [D] None of the above

27. What was the main lacking point in cell theory of Schleiden and Schwann?

[A] They failed to explain how the new cell arises.
[B] They failed to explain how cells transport material from inner side of cells
[C] They failed to explain cell wall organization
[D] None of the above

28. Extension to cell theory was made by

[A] Mendel [B] Palade
[C] Virchow [D] None of the above

29. The statement that cells are produced by preexisting cells was made by

[A] Mendel [B] Palade
[C] Virchow [D] None of the above

30. Virchow discovered that the cells divide and all new cells arise from the pre-existing cells. His discovery

[A] Had no effect on the cell theory
[B] Resulted in the modification of cell theory
[C] Resulted in the acceptance of the cell theory
[D] None of the above

31. The first person to see a 'free' cell under the microscope was

[A] N. Grew [B] Robert Hooke
[C] Purkinje [D] None of the above

32. Which of the following is the exception of cell theory?

[A] Virus [B] Bacteria
[C] B.G. algae [D] Rhizopus
[E] All of the above

33. "All organisms are made of protoplasm", this statement was made by

[A] Max Schultz [B] H. W. Mohl
[C] Purkinje [D] None of the above

34. "Body of all multicellular organism is complete mass of protoplasm which is incompletely subdivided into many small units (as cell) for proper functioning". Who proposed this?

[A] Sachs [B] O. Hertwing
[C] Corti [D] None of the above

35. What is the mode of cytokinesis in plant cell?

[A] Centrifugal [B] Centripetal
[C] Both [D] None of the above

36. What is the mode of cytokinesis in animal cell?

[A] Centrifugal [B] Centripetal
[C] Both [D] None of the above

37. Outermost covering of plant cell is

[A] Living [B] Semi living
[C] Non living [D] All of the above

38. Write the exceptional example of plant in which cell wall is absent and only plasma membrane is outermost covering.

[A] Gamete of zoospore [B] Flagella of zoospore
[C] Cilia of zoospore [D] None of the above

39. Outermost covering of animal cell is

[A] Living [B] Semi living
[C] Non living [D] All of the above

40. What is the exceptional example of animal cell in which outermost covering is non-living structure?

[A] Hard shell of cledoic egg [B] Sperm of animal
[C] Ovum of female [D] None of the above

41. Protoplast term was given by

[A] Hanstein [B] Negilli
[C] Both [D] None of the above

42. Non-living part of cell is also known as

[A] Ergastic substance [B] Deutoplasm
[C] Cell inclusion [D] Metaplasm [E] All of the above

43. Organelle of cell remains embedded in

[A] Protoplasm [B] Metaplasm
[C] Deutoplasm [D] All of the above

44. Largest component of cell is

[A] Nucleus [B] Mitochondria
[C] Plastid [D] All of the above

45. Largest organelle of plant cell is

[A] Nucleus [B] Mitochondria
[C] Plastid [D] All of the above

46. In old plant cell largest component of plant cell is

[A] Nucleus [B] Mitochondria
[C] Plastid [D] None of the above

47. Smallest component of cell is

[A] Microfilament [B] Ribosome
[C] Plastid [D] None of the above

48. Who first observed the protoplasm?

[A] Corti [B] Dujardin
[C] Purkinjee [D] None of the above

49. Who first highlighted the significance of protoplasm and named it as sarcode?

[A] Corti [B] Dujardin
[C] Purkinjee [D] None of the above

50. .... gave the term protoplasm.

[A] Purkinjee [B] Von Mohl
[C] Huxley [D] None of the above

51. .... emphasized the importance of protoplasm

[A] Purkinjee [B] Von Mohl
[C] Huxley [D] None of the above

52. "Protoplasm is physical basis of life", this statement was made by

[A] Huxley [B] Reinke
[C] Both [D] None of the above

53. In protoplasm, what is the percentage of dry mass?

[A] 10-20 [B] 30-40
[C] 50-60 [D] None of the above

54. In dry mass of protoplasm, what is the percentage of protein?

[A] 10-20 [B] 30-40
[C] 50-60 [D] None of the above

55. In normal condition .... is most abundant constituent of protoplasm.

[A] Water [B] Protein
[C] Fat [D] None of the above

56. In dry condition .... is most abundant constituent of protoplasm.

[A] Water [B] Protein
[C] Fat [D] None of the above

57. What is the nature of protoplasm?

[A] Polyphasic crystello colloidal [B] Monophasic crystello colloidal
[C] Both [D] None of the above

58. On increasing number of cell wall the size of lumen of plant cell

[A] Decreases [B] Increases
[C] Remains unchanged [D] All of the above

59. Plasmodesmata was discovered by

[A] Strasburger [B] Brown
[C] Altman [D] None of the above

60. Cytoplasm is a part of

[A] Protoplasm [B] Plastids
[C] Nucleus [D] Cell wall
[E] None of above

61. The term metabolism was coined by

[A] Swanson [B] Schleiden
[C] Schwann [D] Jensen

62. On the basis of his own observations, as well as those of others, Schleiden in 1883 proposed that cell is the structural and functional unit of life. The idea was a

[A] Assumption [B] Observation
[C] Generalisation [D] None of the above

63. The concept of cell lineage was given by

[A] Schleiden [B] Schwann
[C] Jensen [D] None of the above

64. The cell wall does not totally (or, completely) isolated (or separate) from the plant cell due to

[A] Plasmodesmata [B] Tight junction
[C] Interdigitation [D] None of the above

65. .... is known as intercom system of plant cells.

[A] Plasmodesmata [B] Cell wall
[C] Ribosome [D] None of the above

66. Cell wall outside the protoplast constitutes

[A] Apoplast [B] Symplast
[C] Phragmoplast [D] None of the above

67. Continuity of protoplasm through plasmodesmata is called as

[A] Intercellular space [B] Symplast
[C] Endoplasmic reticulum [D] None of the above

68. Middle lamella consists of

[A] Lignin [B] Cellulose
[C] Calcium pectate [D] None of the above

69. Find the missing in the flow chart given below

Glucose molecule (1 lexose)
↓ P = ?
1 Cellulose chain
↓ Q = ?
1 cellulose micelle
↓ R = ?
1 cellulose microfibril
↓ S = ?
Cellulose macrofibril

[A] P= 3000 molecules of glucose, enzyme-polymerase, Q=100 molecule of cellulose chain, enzyme-polymerase, R=25 molecule of cellulose micelle, enzyme-polymerase, S=250 molecule of cellulose microfibril, enzyme-polymerase

[B] P= 3000 molecules of glucose, enzyme-polymerase, Q=150 molecule of cellulose chain, enzyme-polymerase, R=55 molecule of cellulose micelle, enzyme-polymerase, S=250 molecule of cellulose microfibril, enzyme-polymerase

[C] P= 3000 molecules of glucose, enzyme-polymerase, Q=100 molecule of cellulose chain, enzyme-polymerase, R=100 molecule of cellulose micelle, enzyme-polymerase, S=20 molecule of cellulose microfibril, enzyme-polymerase

[D] None of the above

70. Specific stain for lignin is

[A] Phloroglucinol hydrochloride
[B] Acidic phloroglucin
[C] Both
[D] None of the above

71. Which of the following is the characteristic of plant cell?

[A] Cell wall [B] Plastid
[C] Both [D] None of the above

72. Which of the following is the characteristic of plant cell?

[A] Cell wall [B] Plastid
[C] Mitochondria [D] None of the above

73. Outermost layer of the plant cell wall when the plant cells are in group (attached)

[A] Middle lamella [B] Primary cell wall
[C] Secondary cell wall [D] None of the above

74. Outermost layer of cell wall when cell is isolated (single)

[A] Middle lamella [B] Primary cell wall
[B] Secondary cell wall [D] None of the above

75. Which of the following is used to isolate the plant cell from its group

[A] Strong acid [B] Strong base
[C] Lipase [D] More than one
[E] None of the above

76. Which of the following has maximum percentage of cellulose

[A] Cotton [B] Wood
[C] Jute [D] None of the above

77. The term plasmamembrane was given by

[A] Pfeffer [B] J.Q. Plowe
[C] Cramer and Nageli [D] None of the above

78. The term plasmalemma was given by

[A] Pfeffer [B] J.Q. Plowe
[C] Cramer and Nageli [D] None of the above

79. The term cell membrane was given by

[A] Pfeffer [B] J.Q. Plowe
[C] Cramer and Nageli [D] None of the above

80. The thickness of plasma membrane ranges from

[A] 75Å to 215Å [B] 15Å to 25Å
[C] 5Å to 15Å [D] None of the above

81. Unit-Membrane model of cell membrane was proposed by

[A] Singer and Nicolson [B] Danielli-Devson
[C] Robertson [D] None of the above

82. Which of the following is a cell organelle?

[A] Vacuole [B] Mitochondrion
[C] Plasma membrane [D] Microtubule
[E] None of the above

83. The existence of cell membrane was proved by

[A] Overton [B] Nageli
[C] Plowe [D] None of the above

84. It is not possible to see a cell usually without the help of a microscope because

[A] Cells are mostly smaller than 100 μm in size.
[B] Resolving power of human eye is 0.1 nm
[C] Most cells are colourless
[D] All of the above

85. Strasburger coined the term

[A] Cytoplasm [B] Nucleoplasm
[C] Plastids [D] None of the above

86. Schleiden believed that new cells are formed by budding off from the

[A] Cell wall [B] Protoplasm
[C] Nucleus [D] All of the above

87. Plant cell walls contain

[A] Proteoglycan [B] Peptidoglycan

[C] Pectin [D] Glycosaminoglycan

[E] None of the above

88. Receptor mediated endocytosis from plasma membrane requires which one of the following coat protein?

[A] Adaptin [B] Glycophorin

[C] Arrestin [D] Clathrin

89. Plasma membrane is made of

[A] Phospholipoprotein [B] Phospholipoprotein and carbohydrate

[C] Lipoprotein [D] More than one

[E] None of the above

90. What is the amount of lipid in plasma membrane?

[A] 20-40% [B] 59-70%

[C] 1-5% [D] None of the above

91. What is the amount of protein in plasma membrane?

[A] 20-40% [B] 59-70%

[C] 1-5% [D] None of the above

92. What is the amount of carbohydrate in plasma membrane?

[A] 20-40% [B] 59-70%

[C] 1-5% [D] None of the above

93. "Iceberg of protein in sea of lipid", this statement was made by

[A] Singer and Nicholson [B] Darwin

[C] Dixon and jolly [D] None of the above

94. Protein can shift its position .... in phospholipids bilayers

[A] Laterally [B] Vertically

[C] Both [D] None of the above

95. Intrinsic protein extended throughout the thickness of phospholipidbilayer is known as

[A] Tunnel protein [B] Transmembrane protein

[C] Both [D] None of the above

96. Lipid molecule can shift its position by

[A] Flip-flop movement [B] Lateral movement

[C] Both [D] None of the above

97. Flip-flop movement is also known as

[A] Lateral diffusion [B] Transverse diffusion

[C] Both [D] None of the above

98. Long chain fatty acids have higher transition temperature than shorter chain fatty acids, which means that membrane enriched with long chain fatty acids tend to exhibit .... fluidity.

[A] Increased [B] Decreased

[C] Both [D] None of the above

99. **Unsaturated fatty acid has lower transition temperature than saturated fatty acids, giving membrane enriched with unsaturated fatty acids .... fluidity.**
[A] Increased [B] Decreased
[C] Same [D] None of the above

100. **...................... is the most valuable instrument used to study the behavior of the lipid films.**
[A] Langmuir trough [B] Centrifugation of lipid layer
[C] Gel electrophoresis [D] None of the above

101. **Which of the following is the cell adhesion molecule?**
[A] Lysine [B] Keratin
[C] Myosin [D] Integrin

102. **Which of the following lipid is commonly found in biological membranes?**
[A] Monoglycerides [B] Diglycerides
[C] Triglycerides [D] None of the above

103. **After membranes have been isolated and purified, their protein composition is determined by extraction with the detergent .... and fractionating the extracted polypeptides by one-or two-dimensional gel electrophoresis.**
[A] Sodium dodecyl sulfate [B] Calcium sulfate
[C] KCl [D] None of the above

104. **...................... is a mobile carrier diffusing back and forth through lipid bilayers.**
[A] Valinomycin [B] Gramicidin
[C] Both [D] None of the above

105. **...................... functions as a channel former, creating a fixed transmembrane channel through which ions can diffuse.**
[A] Valinomycin [B] Gramicidin
[C] Both [D] None of the above

106. **Gramicidin is selective in the ions it transports, the selectivity of the gramicidin channel follows the order**
[A] $H^+ > NH_4^+ > K^+ > Na^+ > Li^+$ [B] $Li^+ > H^+ > NH_4^+ > K^+ > Na^+$
[C] $Li^+ > H^+ > K^+ > Na^+ > NH_4^+$ [D] None of the above

107. **Valinomycin, a small cyclic molecule constructed from .... alternating amino acids.**
[A] 12 [B] 15
[C] 18 [D] None of the above

108. **Gramicidin, an inophore consisting of a linear chain of .... amino acids.**
[A] 12 [B] 15
[C] 18 [D] None of the above

109. Among the followings which one is transported most efficiently through plasma membrane?

[A] D-Glucose [B] D-Mannose

[C] D-Galactose [D] None of the above

110. Glycocalyx is

[A] Non living [B] living

[C] Not known [D] None of the above

111. Functions of glycocalyx are

[A] Antigenesis [B] As receptor

[C] Reorganization of other cells [D] All of the above

112. Due to the presence of glycocalyx (or cell coat) the plasma membrane becomes

[A] Asymmetrical [B] Symmetrical

[C] Both [D] None of the above

113. Plasma membrane remains holded primarily by or stability of plasma membrane is due to

[A] Hydrophilic attraction [B] Hydrophobic attraction

[C] Covalent bond [D] None of the above

114. Which of the following component of plasma membrane is amphipathic in nature?

[A] Phospholipid [B] Extrinsic protein

[C] Intrinsic protein [D] Both A and B

[E] None of the above

115. Which component of plasma membrane provides fluidity to the plasma membrane?

[A] Protein [B] Lipid

[C] Both [D] None of the above

116. Which component provides elasticity (or, flexibility) to the plasma membrane?

[A] Protein [B] Lipid

[C] Carbohydrate [D] None of the above

117. Biomembrane is

[A] Living [B] Nonliving

[C] Both [D] None of the above

118. What is the specialization due to evegination of plasma membrane?

[A] Flagella [B] Cilia

[C] Stereocilia [D] Microvilli

[E] All of the above

119. What is the specialization due to contact of plasma membrane?

[A] Tight junction [B] Intermediary junction
[C] Desmosome [D] Interdigitation
[E] Gap junction [F] All of the above

120. Tight junction is also known as

[A] Zonula occludens [B] Zonula adherence
[C] Terminal bar [D] None of the above

121. Intermediary junction is also known as

[A] Terminal bar [B] Zonula adherence
[C] Both [D] None of the above

122. Gap junction is also known as

[A] Nexus [B] Macula occludens
[C] Both [D] None of the above

123. Phagocytosis was discovered by

[A] Metchnikoff [B] Lewis
[C] Both [D] None of the above

124. Pinocytosis was discovered by

[A] Lewis [B] Nicolson
[C] Govindjee [D] None of the above

125. Endoplasmic reticulum was discovered and named by

[A] Porter [B] Palade
[C] Benda [D] None of the above

126. Ergastoplasm is .... in nature

[A] Basophilic [B] Acidic
[C] Neutral [D] None of the above

127. The ground phase of the cytoplasm is known as

[A] Cytosol [B] Cell-inclusion
[C] Both [D] None of the above

128. What are the main components of smooth endoplasmic reticulum?

[A] Tubule [B] Vesicle
[C] Both [D] None of the above

129. What are the main components of rough endoplasmic reticulum?

[A] Cisternae [B] Tubule
[C] Both [D] None of the above

130. Endoplasmic reticulum consists of following structure

[A] Cisternae [B] Tubule
[C] Vesicle [D] All of the above

131. Endoplasmic reticulum originated from

[A] Nuclear membrane [B] Chloroplast
[C] Mitochondria [D] All of the above

132. What is the function of endoplasmic reticulum?

[A] It provides endoskeleton to the cell
[B] It provides increased surface area for increased rate of reaction
[C] It helps in intra cellular transport of material
[D] It indirectly forms lysosome
[E] It helps in formation of nuclear membrane and cell wall
[F] It directly forms Golgi body/complex
[G] It helps in intra cellular conduction of impulse
[H] All of the above

133. After fractionation of cell which of the following new structures will be formed?

[A] Microcosm [B] Microsome
[C] Microtech [D] Microblast
[E] All of the above

134. Endomembrane system or cytoplasmic vacuolar system includes

[A] Nuclear membrane [B] Endoplasmic reticulum
[C] Golgi body [D] All of the above

135. Golgi complex or Golgi apparatus or Golgi bodies are discovered by Camillio Golgi by

[A] Silver metallic impregnation technique
[B] Sodium metallic impregnation technique
[C] Iron metallic impregnation technique
[D] None of the above

136. Collective name of all the Golgi bodies of a cell is known as

[A] Lipochondria [B] Mitochondria
[C] Dictyosomes [D] None of the above

137. Lipochondria termed was given by

[A] Baker [B] Fillip
[C] Benda [D] None of the above

138. Small and scattered form of Golgi body present in most of plant cells is known as

[A] Dictyosomes [B] Golgi-Body
[C] Lipochondria [D] None of the above

139. Dictyosomes are found in

[A] Plant cell [B] Animal cells
[C] Prokaryote [D] None of the above

140. Zone of exclusion is also present around

[A] Centriole [B] Centriole forming area
[C] Aster [D] Microtubule
[E] All of the above

141. .... is the area of cytoplasm around the Golgi body which is devoid of glycogen and organelles like ribosome, mitochondria and plastid.

[A] Zone of exclusion [B] Zone of Golgi
[C] Zone of condensation [D] None of the above

142. Golgi body shows structural polarity. This statement is

[A] True [B] False
[C] Not known [D] None of the above

143. Golgi body originated from

[A] Smooth endoplasmic reticulum
[B] Smooth portion of rough endoplasmic reticulum
[C] Both
[D] None of the above

144. What is the function of Golgi body?

[A] It is mainly associated with secretion
[B] It is concerned with condensation, packing, storage, transfer
[C] It perform Glycosidation of protein and lipid
[D] It form acrosome of sperm
[E] All of the above

145. In plant, Golgi bodies are concerned mainly in the formation of

[A] Plasma membrane [B] Cell wall
[C] Tonoplast [D] None of the above

146. Desmotubules are found in

[A] Endoplasmic reticulum [B] Plasmodesmata
[C] Cytoskeleton [D] None of the above

147. Lysosome was discovered by

[A] De Duve [B] R. Altman
[C] W. Flemming [D] None of the above

148. Periplasmetic space acts as lysosome in

[A] Prokaryote cell [B] Eukaryotic cell
[C] Both [D] None of the above

149. Lysosome shows polymorphism, this statement is

[A] True [B] False
[C] Not known [D] None of the above

150. Secondary lysosome is also known as

[A] Heterophagic vacuole [B] Heterophagosome
[C] Both [D] None of the above

151. Enzyme of lysosome acts on

[A] Acidic PH [B] Basic PH
[C] Neutral PH [D] None of the above

152. Enzyme of cell acts on
[A] Acidic pH [B] Neutral pH
[C] Basic pH [D] None of the above

153. Oxidative enzyme is found in
[A] Lysosome [B] Mitochondria
[C] Both [D] None of the above

154. Enzyme of lysosome is
[A] Hydrolytic [B] Catalytic
[C] Anabolic [D] None of the above

155. Enzyme of lysosome is
[A] Hydrolytic in nature [B] Catabolic in nature
[C] Both [D] None of the above

156. Catabolic enzyme is found in
[A] Lysosome [B] Mitochondria
[C] Both [D] None of the above

157. Function of lysosome is
[A] Intracellular digestion [B] Autophagy
[C] Extracellular digestion [D] All of the above

158. In plant cell ribosome was discovered by
[A] Robinson and Brown [B] Palade
[C] Both [D] None of the above

159. Ribosome is also called
[A] Palade particle [B] Protein factory
[C] Cell engine [D] Ribonucleoprotein (RNP) particle
[E] All of the above

160. Ribosome is
[A] Membrane less organelle [B] Smallest organelle of cell
[C] -vely charged [D] All of the above

161. 55 'S' ribosome found in
[A] Mammalian mitochondria [B] Animal mitochondria
[C] Fungal mitochondria [D] None of the above

162. 77 'S' ribosome is present in
[A] Mammalian mitochondria [B] Animal mitochondria
[C] Fungal mitochondria [D] None of the above

163. In 80 'S' ribosome what is the amount of r-RNA?
[A] 45% [B] 65%
[C] 75% [D] None of the above

164. In 70 'S' ribosome what is the amount of r-RNA?
[A] 45% [B] 65%
[C] 75% [D] None of the above

165. Which of the following subunit of ribosomes is attached with the membrane of rough endoplasmic reticulum?

[A] Smaller subunit [B] Larger subunit
[C] Both [D] None of the above

166. Which of the following antibiotic can stop protein synthesis on 80 'S' ribosome?

[A] Cyclohexamidin [B] Chloremphenical
[C] Both [D] None of the above

167. Who discovered mitochondria and named it as sarcosome?

[A] Kolliker [B] Altman
[C] Fleming [D] None of the above

168. .... observed mitochondria and named it as 'bioblast' and suggested their association with respiration.

[A] Kolliker [B] Altman
[C] Fleming [D] None of the above

169. .... named mitochondria as 'Fila'

[A] Kolliker [B] Altman
[C] Fleming [D] None of the above

170. .... gave the term 'mitochondria'

[A] Benda [B] Altman
[C] Went [D] None of the above

171. Mitochondrial DNA was discovered by

[A] Nass and Nass [B] Seekevitz
[C] Lahinger [D] None of the above

172. In mitochondria what is the amount of DNA?

[A] 0.02% [B] 0.05%
[C] 0.09% [D] None of the above

173. The mitochondria having only inner membrane and matrix is known as

[A] Mitoplast [B] Mitochondrion
[C] Protoplast [D] None of the above

174. Which of the following can perform oxidative phosphorylation?

[A] Mitochondria [B] Mitoplast
[C] Both [D] None of the above

175. After fractionation of cell which of the following can be formed?

[A] Microsome [B] Mitoplast
[C] Both [D] None of the above

176. What is the function of mitochondria?
[A] It is the site for cellular respiration
[B] It performed oxidative phosphorylation
[C] Conservation of energy
[D] Help in cell division
[E] It form nebenkern
[F] Thermogenesis
[G] All of the above

177. Mitochondria is also known as
[A] Power house of cell [B] Biochemical machine
[C] Both [D] None of the above

178. Colour of mitochondria is yellow due to presence of
[A] Vitamin B_2 [B] Vitamin B_1
[C] Vitamin B_6 [D] None of the above

179. Stain of mitochondria is
[A] Genus green A [B] Genus green B
[C] Genus green C [D] None of the above

180. Smallest mitochondria is found in
[A] Yeast [B] Oocyte of Amphibia
[C] Spinacia [D] None of the above

181. Largest mitochondria is found in
[A] Yeast [B] Oocyte of amphibian
[C] Spinacia [D] None of the above

182. Living mitochondria was first observed by
[A] Lewis and Lewis [B] Lewis and Naas
[C] Lewis and Altman [D] None of the above

183. Plastid term was used by
[A] Kolliker [B] Schimper
[C] Both [D] None of the above

184. The term chloroplast was given by
[A] Errera [B] Haeckel
[C] N. Pringsheim [D] None of the above

185. Which of the following does not contain DNA?
[A] Chloroplast [B] Mitochondria
[C] Nucleus [D] None of the above

186. In spirogyra, shape of chloroplast is
[A] Ribbon shaped [B] Satellite shaped
[C] Bell shaped [D] None of the above

187. What is the shape of chloroplast in *Chlorella*?
[A] Ribbon shaped [B] Satellite shaped
[C] Bell shaped [D] None of the above

188. **What is the shape of chloroplast in *Ulothrix*?**
[A] Girdle shape [B] Collar shape
[C] C-shape [D] Band shape [E] All of the above

189. **What is the diameter of chloroplast?**
[A] 5-10 μm [B] 10-20 μm
[C] 30-40 μm [D] None of the above

190. **What is the range of periplastidial space in chloroplast?**
[A] 100-300Å [B] 300-600Å
[C] 600-900Å [D] None of the above

191. **Outer membrane of chloroplast has**
[A] Low protein [B] High permeable
[C] Both [D] None of the above

192. **Inner membrane of chloroplast has**
[A] More protein [B] Low permeable
[C] Both [D] None of the above

193. **Plastidial DNA was discovered by**
[A] Ris and Plaut [B] Nass and Nass
[C] Benda [D] None of the above

194. **Size of each grana is**
[A] 0.3 to 2.7 μm [B] 0.9 to 3.7 μm
[C] Both [D] None of the above

195. **Number of grana per chloroplast is**
[A] 40 to 60 [B] 80 to 90
[C] 100 to 110 [D] None of the above

196. **Thylakoid membrane is supposed to be considered as ………………… membrane of chloroplast.**
[A] 3rd [B] 2nd
[C] 4th [D] None of the above

197. **What is the functional unit of chloroplast?**
[A] Quantasome [B] Chloroplast
[C] Cell [D] None of the above

198. **What is the molecular weight of Quantasome?**
[A] 2×10^6 [B] 4×10^6
[C] 6×10^6 [D] None of the above

199. **Length of quantasome is**
[A] 185Å [B] 155Å
[C] 100Å [D] None of the above

200. **Width of the quantasome is**
[A] 185Å [B] 155Å
[C] 100Å [D] None of the above

201. Thickness of quantasome is
[A] 185Å [B] 155Å
[C] 100Å [D] None of the above

202. Number of chlorophyll molecule/quantasome
[A] 230 to 300 [B] 350 to 450
[C] 500 to 600 [D] None of the above

203. Number of carotenoid molecule /quantasome
[A] 50 [B] 100
[C] 150 [D] None of the above

204. What is the amount of DNA in chloroplast molecule?
[A] 0.5% [B] 0.9%
[C] 0.8% [D] None of the above

205. Autonomic genomic systems are found in
[A] Mitochondria [B] Chloroplast
[C] Both [D] None of the above

206. DNA of mitochondria and chloroplast has
[A] More GC ratio [B] More AT ratio
[C] More AU ratio [D] None of the above

207. Plastidial system carrying the genetic information is called
[A] Plastidom [B] Plastogene
[C] Chondriogene [D] None of the above

208. Mitochondria and chloroplast are also called
[A] Semiautonomus body [B] Endosymbiont of cell
[C] Cell within the cell [D] All of the above

209. After careful breaking of cell which of the following organelle remain alive for some time?
[A] Mitochondria [B] Chloroplast
[C] Both [D] None of the above

210. Chlorophyll does not dissolve in water even after prolonged boiling. This statement is
[A] True [B] False
[C] Both [D] None of the above

211. Environment enriched with carbon dioxide prevents destruction of chlorophyll molecule. This statement is
[A] True [B] False
[C] Both [D] None of the above

212. Which of the following type of plastid other than amyloplast also store starch grain?
[A] Chloroplast [B] Chromoplast
[C] Elaioplast [D] None of the above

213. What is the chemical composition of chlorophyll 'a'?

[A] $C_{55}H_{72}O_5N_4Mg$ [B] $C_{56}H_{72}O_5N_4Mg$

[C] $C_{57}H_{72}O_5N_4Mg$ [D] None of the above

214. What is the chemical composition of chlorophyll 'b'?

[A] $C_{55}H_{70}O_6N_4Mg$ [B] $C_{59}H_{72}O_5N_4Mg$

[C] $C_{60}H_{72}O_6N_4Mg$ [D] None of the above

215. What is the chemical composition of chlorophyll 'c'?

[A] $C_{35}H_{32}O_5N_4Mg$ [B] $C_{37}H_{42}O_5N_4Mg$

[C] $C_{40}H_{52}O_6N_4Mg$ [D] None of the above

216. What is the chemical composition of chlorophyll 'd'?

[A] $C_{54}H_{70}O_6N_4Mg$ [B] $C_{55}H_{71}O_6N_4Mg$

[C] $C_{56}H_{72}O_6N_4Mg$ [D] None of the above

217. What is the chemical composition of bacteriochlorophyll?

[A] $C_{55}H_{74}O_6N_4Mg$ [B] $C_{55}H_{71}O_6N_4Mg$

[C] $C_{56}H_{72}O_6N_4Mg$ [D] None of the above

218. What is the chemical composition of xanthophylls?

[A] $C_{40}H_{56}O_2$ [B] $C_{40}H_{55}O_3$

[C] $C_{41}H_{57}O_2$ [D] None of the above

219. In prokaryotic cell, the nucleus is known as

[A] Central body [B] Central mass

[C] Central unit [D] None of the above

220. Nucleoplasmic index is equal to

[A] Volume of nucleus/volume of cytoplasm

[B] Volume of cytoplasm/volume of nucleus

[C] Volume of cytoplasm/volume of chloroplast

[D] None of the above

221. What is the position of nucleus?

[A] Present in such area of cell where metabolic activity is maximum

[B] Usually it occupies geometric central position in cell

[C] In old plant cells nucleus become excentric in position due to enlarged vacuole

[D] All of the above

222. What is the shape of nuclear pore?

[A] Octagonal [B] Circular

[C] Rectangular [D] None of the above

223. Which type of nuclear pore is found in plant cell?

[A] Direct pore [B] Annulated pore

[C] Both [D] None of the above

224. The term nucleolus was coined by

[A] Bowman [B] Brown

[C] Fontana [D] None of the above

225. .... ions are supposed to maintain intact organization of nucleus.

[A] Ca^{++} [B] Mg^{++}

[C] Na^{+} [D] None of the above

226. Usually one nucleolus is present in one nucleus except

[A] Onion cell [B] Oocyte of xonopus

[C] Both [D] None of the above

227. What is the function of nucleolus?

[A] Helps in biosynthesis of ribosome

[B] It is the site for assembly of ribosome

[C] Both

[D] None of the above

228. Telome theory was given by

[A] Zimmermann [B] W. Fleming

[C] Tejo and Laven [D] None of the above

21 PLANT MOLECULAR BIOLOGY

1. .... proposed the term chromosome.
   [A] W. Waldeyer [B] W. Fleming
   [C] Tejo and laven [D] None of the above
2. Chromosomal theory of heredity was given by
   [A] Sutton and Bovary [B] W. Fleming
   [C] Strasburger [D] None of the above
3. Central dogma of molecular biology was given by
   [A] F.H.C. Crick [B] F.A. Janssen
   [C] E. Wilson [D] None of the above
4. Chiasmatype or one-plain theory was given by
   [A] F.H.C. Crick [B] F.A. Janssen
   [C] E. Wilson [D] L.W. Sharp
5. Classical or two plane theory was proposed by
   [A] F.H.C. Crick [B] F.A. Janssen
   [C] E. Wilson [D] L.W. Sharp
6. Copy choice theory of recombination was proposed by
   [A] J. Belling [B] H.J. Muller
   [C] K.W. Wolff [D] None of the above
7. 'Dosage compensation' was proposed by
   [A] J. Belling [B] H.J. Muller
   [C] K.W. Wolff [D] None of the above
8. Epigenesis theory was proposed by
   [A] J. Belling [B] H.J. Muller
   [C] K.W. Wolff [D] None of the above
9. One gene one polypeptide hypothesis was proposed by
   [A] V.M. Ingram [B] M. Kimura
   [C] S. Ohno [D] None of the above
10. DNA differs from RNA in
    [A] Base [B] Sugar
    [C] Phosphate [D] More than one

11. DNA differs from RNA in

[A] Base [B] Sugar

[C] Phosphate [D] None of the above

12. Basis of naming of DNA and RNA is

[A] Sugar [B] Base

[C] Phosphate [D] More than one

[E] None of the above

13. How many water molecules will be released to make a nucleoside?

[A] One [B] Two

[C] Three [D] None of the above

14. How many water molecules will be released to make a nucleotide from a nucleoside?

[A] One [B] Two

[C] Three [D] None of the above

15. How many water molecules will be released to make a nucleotide?

[A] Two [B] One

[C] Three [D] None of the above

16. How many water molecules will be released to make a dinucleotide?

[A] Five [B] Six

[C] Seven [D] None of the above

17. How many water molecules will be released to make a dinucleotide from two nucleotides?

[A] Two [B] One

[C] Three [D] None of the above

18. How many water molecules will be released to unite two nucleotides?

[A] Two [B] One

[C] Three [D] None of the above

19. How many water molecules will be released to make a dinucleotide from two nucleosides?

[A] Three [B] Four

[C] Five [D] None of the above

20. How many water molecules will be released to make a polynucleotide chain by joining the n number of nucleotide.

[A] n-1 H_2O molecule [B] n-2 H_2O molecule

[C] n-3 H_2O molecule [D] None of the above

21. Transformation phenomenon was discovered by .... in

[A] Griffith, *Diplococcus pneumoniae*

[B] Griffith, *E. Coli*

[C] Griffith, *Bacillus subtilis*

[D] None of the above

22. Which of the following organism was used by Hershey and Chase for their experiment to demonstrate that DNA is the genetic material?

[A] T_4 phase and *E.coli* [B] T_2 phase and *E.coli*
[C] T_1 phase and *E.coli* [D] None of the above

23. The process of identification of the genetic material began in 1928 with the experiments by?

[A] Griffith [B] Bridges
[C] Castle [D] None of the above

24. If DNA of somatic cell of plant has 31% [A] and 19% [C] then what is the % of [G] in DNA of germ cell?

[A] 19% [B] 31%
[C] 9.5% [D] None of the above

25. Frankel-Conrat and Singer demonstrated that RNA is a genetic material in?

[A] M13 bacteriophage [B] Tomato leaf curl virus
[C] Tobacco mosaic virus [D] Cauliflower mosaic virus

26. If DNA of somatic cell of plant has 31% [A] then what is % of [G] in such cell?

[A] 19% [B] 31%
[C] 9.5% [D] None of the above

27. The transformation principle was demonstrated to be DNA by which of the following?

[A] Frankel-Conrat and Singer
[B] Hershey and Chase
[C] Avery, MacLeod and McCarty
[D] None of the above

28. If DNA of somatic cell of plant has 40[A] then what will be the number of [G] in such DNA ?

[A] 10 [B] 40 [C] 20
[D] Unpredictable [E] More than one

29. In a B-DNA total number of base is 200. If there is 80[C] then what is the number of [A] in such DNA?

[A] 40 [B] 80
[C] 20 [D] None of the above

30. In a B-DNA total number of base pair is 200. If there is 80 [C], then what is the number of [A] and [T]?

[A] 70 [B] 140
[C] 80 [D] 20

31. If a B-DNA has 400 base pairs then what is the total number of nucleotide in such DNA?

[A] 400 [B] 800
[C] 200 [D] None of the above

32. What is the total number of nucleotide in short length of a DNA double helix having 200 [A] and 200 [C] base?

[A] 400 [B] 220

[C] 440 [D] None of the above

33. If a B-DNA has 20 [A] then what is the total number of nucleotide in that DNA?

[A] 40 [B] 100

[C] 80 [D] None of the above

34. Which of the following is not consistent with DNA double helix?

[A] Density of DNA decreasing on heating

[B] A+G/T+C equal to one

[C] The ratio of A+G and G+C is not consistent in different organisms of same species

[D] According to X-ray diffraction pattern DNA is double helical structure

35. It is unreasonable to think that a DNA (D.S) has which of the following type of bases?

[A] 5 [B] 10

[C] 3 [D] More than one

36. Before the explanation of Avery, MacLeod and McCarty, which of the following was considered as genetic material?

[A] RNA [B] DNA

[C] Protein [D] More than one

37. Transformation experiment was explained by

[A] Frankel-Conrat and Singer

[B] Hershey and Chase

[C] Avery, MacLeod and McCarty

[D] None of the above

38. In most of the organisms genetic material is

[A] DNA [B] RNA

[C] Protein [D] More than one

39. RNA is genetic material in

[A] Most of organism [B] Most of animal virus

[C] Most of plant virus [D] More than one

40. In most of animal viruses genetic material is DNA except

[A] Polio virus [B] Toga virus

[C] Rabies virus [D] Mumps virus

[E] All of the above

41. A poly nucleotide chain has 20[A], 30[C] and 40[G], this is part of

[A] DNA [B] RNA

[C] Unpredictable [D] Both DNA and RNA

42. A polynucleotide chain has 20[A], 30[C] and 40[G] and deoxyribose sugar then it is part of

[A] DNA [B] RNA
[C] Both [D] None of the above

43. Wobble hypothesis of codon- anticodon base pairing was given by

[A] F.H.C. Crick [B] G. Gammow
[C] O. WInge [D] None of the above

44. The theory of polyploidy as basis of evolution was given by

[A] O. Winge [B] C. Darwin
[C] Moses [D] None of the above

45. Theory of organic evolution was given by

[A] O. Winge [B] C. Darwin
[C] Moses [D] J.H. Taylor

46. Precosity theory of synapsis was given by

[A] C.D. Darlington [B] Moses
[C] J.H. Taylor [D] None of the above

47. Genetic material of viroids consists of which of the following

[A] Nucleoprotein [B] Protein
[C] RNA [D] DNA

48. A nucleoside consists of which of the following?

[A] Phosphate and pentose sugar
[B] Pentose sugar and organic base
[C] Phospahte and organic base
[D] Pentose, base and phosphate

49. Which form of DNA has the greatest diameter?

[A] B-DNA [B] Z-DNA
[C] C-DNA [D] A-DNA

50. Which form of DNA has the longest pitch?

[A] A-DNA [B] B-DNA
[C] C-DNA [D] Z-DNA

51. Which form of DNA is common *in vivo*?

[A] A-DNA [B] B-DNA
[C] C-DNA [D] Z-DNA

52. Who was the first to use the term transposon?

[A] Hedges and Jacob [B] Leder and Flavel
[C] Singer and Gilbert [D] None of the above

53. Viral genomes are packed into which of the following?

[A] Capsids [B] Nuclei
[C] Capsules [D] Nucleoids

54. Which of the following scientist discovered repetitive DNA in eukaryotes?

[A] Britten and Waring [B] Taylor and Temin

[C] Sanger and Berg [D] None of the above

55. Which of the following has the smallest genome size?

[A] Phase [B] SV 40

[C] MS2 [D] None of the above

56. C-Value represents DNA content of a nucleus that is in which of the following stage?

[A] Somatic cell G1 phase [B] Somatic cell telophase

[C] Somatic cell metaphase [D] Meiotic telophase II

57. Which of the following gene does not have introns?

[A] Actin gene [B] Histone gene

[C] Globin gene [D] None of the above

58. Which of the following mode of transposition is the most damaging to the genome?

[A] Replicative [B] Conservative

[C] Neoconservative [D] None of the above

59. *E. coli* DNA polymerase III is known for

[A] DNA replication

[B] DNA repair

[C] Primer digestion and gap fillig

[D] None of the above

60. Under optimum conditions, one complete replication of *E. coli* chromosome takes approximately how much time?

[A] 10 minute [B] 20 minute

[C] 40 minute [D] None of the above

61. In case of *E. coli*, what is the approximate rate of fork movement under optimum condition?

[A] 30, 000bp/minute [B] 50,000bp/minute

[C] 75,000bp/minute [D] 100,000bp/minute

62. Among the followings, which of the DNA polymerase is involved in DNA repair only?

[A] DNA polymerase I [B] DNA polymerase II

[C] DNA polymerase III [D] None of the above

63. For the discovery of DNA polymerase enzyme, Kornberg was awarded Nobel Prize in which of the following year?

[A] 1955 [B] 1959

[C] 1962 [D] None of the above

64. Pilot proteins have been studied in?

[A] M13 [B] MS2

[C] X174 [D] All of the above

65. In Prokaryotic DNA polymerase III, which of the following constitutes the catalytic core?

[A] α, ε [B] α, θ, ε
[C] α, β, ε [D] None of the above

66. The Classical book "Inborn Errors in Metabolism" was written by

[A] Beadle [B] Garrod
[C] Tatum [C] None of the above

67. One -gene-one enzyme hypothesis was proposed by

[A] Lederberg [B] Southern
[C] Woodward [D] Beadle and Tatum

68. Replica-plating technique was suggested by

[A] Woodward [B] Lederberg
[C] Southern [D] None of the above

69. Who is regarded as father of biochemical genetics?

[A] Garrod [B]Galton
[C] Beadle [D] Tatum

70. The term "Protein" was proposed by?

[A] Mulder [B] Dalton
[C] Berzelius [D] None of the above

71. The process of protein synthesis is generally known as

[A] Phosphodiester bond [B] Salt bridge
[C] Peptide bond [D] Ester bond

72. tRNA is found in which of the following part of the cell?

[A] ER [B] Cytoplasm
[C] Vacuole [D] None of the above

73. The process of transcription of rRNA is completed in which of the following?

[A] Nucleus [B] Cytoplasm
[C] Nucleolus [D] None of the above

74. The translation process occurs in association with which of the following?

[A] tRNA [B] ER
[C] Mitochondria [D] Ribosome and RNA
[E] None of the above

75. Among the followings, which one is the smallest?

[A] mRNA [B] hnRNA
[C] tRNA [D] None of the above

76. Which one of the following is known as antisense RNA?

[A] rRNA [B] tRNA
[C] mRNA [D] None of the above

77. In most of the organisms RNA is single stranded structure except
[A] Wound tumor virus [B] Rheo group of virus
[C] Both [D] None of the above

78. In eukaryotes how many types of RNA are found?
[A] hn-RNA [B] m-RNA
[C] t-RNA [D] r-RNA [E] All of the above

79. m-RNA is also known as
[A] Messenger RNA [B] Template RNA
[C] Both [D] None of the above

80. Amount of m-RNA in cell is
[A] 5-10% [B] 20-30%
[C] 40-60% [D] None of the above

81. "Prokaryotic m-RNA is metabolically less stable than eukaryotic m-RNA", this statement is
[A] True [B] False
[C] Not known [D] None of the above

82. Minimum length of m-RNA is
[A]100 codon [B] 200 codon
[C]300 codon [D] None of the above

83. Minimum sedimentation coefficient of m-RNA is
[A] 8 'S' [B] 10 'S'
[C] 15 'S' [D] None of the above

84. In eukaryotic cell m-RNA is synthesized from DNA by transcription inside nucleus and then such m-RNA comes in cytoplasm in form of
[A] Informosome [B] Carboform
[C] Both [D] None of the above

85. Informosome stage is
[A] m-RNA +Protein [B] r-RNA + protein
[C] t-RNA + Protein [D] None of the above

86. t-RNA is also known as
[A] Transfer RNA [B] Soluble RNA
[C] Adopter RNA [D] All of the above

87. Which of the following RNA is known as smallest RNA in the cell?
[A] m-RNA [B] t-RNA
[C] r-RNA [D] None of the above

88. What is the size of t-RNA?
[A] 80 nucleotides [B] 100 nucleotides
[C] 200 nucleotides [D] None of the above

89. Molecular weight of t-RNA is
[A] 2.5×10^4 to 3×10^4 [B] 4.5×10^4 to 6×10^4
[C] 6.5×10^4 to 9×10^4 [D] None of the above

90. Sedimentation coefficient of t-RNA is

[A] 3.8 'S' [B] 4.8 'S'

[C] 5.8 'S' [D] None of the above

91. What are the unusual bases of t-RNA?

[A] Methyl guanin [B] Dimethyl guanin

[C] Methyl cytocin [D] Dihudrouridine

[E] Inosine [F] Ribothymime

[G] Pseudouridine [H] All of the above

92. Which of the following ion provides stability to t-RNA?

[A] Mg^{2+} [B] Na^{+}

[C] K^{+} [D] None of the above

93. Clover leaf model of t-RNA was given by

[A] Robert Holly [B] S. H. Kim

[C] Wilson [D] None of the above

94. 3-D model of t-RNA was given by

[A] Robert Holly [B] S. H. Kim

[C] Wilson [D] None of the above

95. According to 3-D model of t-RNA, it is .... shaped thickness of 20Å.

[A] L [B] S

[C] M [D] None of the above

96. Which of the following is the most stable RNA?

[A] r-RNA [B] t-RNA

[C] m-RNA [D] None of the above

97. "r-RNA has both paired and unpaired condition of base". This statement is

[A] True [B] False

[C] Not known [D] None of the above

98. In eukaryote .... is known as precursor of m-RNA?

[A] r-RNA [B] t-RNA

[C] hn-RNA [D] None of the above

99. Removal of unwanted base of hn-RNA is known as

[A] Editing [B] Splicing

[C] Both [D] None of the above

100. Polyadenylation occurs at .... end of hn-RNA

[A] 3' [B] 5'

[C] Both [D] None of the above

101. Capping is addition of .... at 5' end of hn-RNA

[A] 7 methyl guanosine tri phosphate

[B] 5 methyl guanosine tri phosphate

[C] 2 methyl guanosine tri phosphate

[D] None of the above

102. Capping occurs at the .... end of the hn-RNA?

[A] 5' [B] 3'

[C] 2' [D] None of the above

103. The orientation of sugar of 7-methyl guanosine triphosphates remains inverted at an angle of .... with respect to hn RNA or m-RNA

[A] 180° [B] 170°

[C] 160° [D] None of the above

104. In methylation process, there is addition of methyl group at .... position of adenine of internal adenylate of poly A Tail.

[A] 6th [B] 7th

[C] 8th [D] None of the above

105. Cap and Tail is part of

[A] hn-RNA [B] m-RNA

[C] Both [D] None of the above

106. In prokaryotic m-RNA, cap and tail both are

[A] Present [B] Absent

[C] Not known [D] None of the above

107. Informosome stage of m-RNA is absent in ....

[A] Prokaryote [B] Eukaryote

[C] Both [D] None of the above

108. What is the role of cap in eukaryotic m-RNA?

[A] Cap help in formation of m-RNA-ribosomal complex during protein synthesis

[B] Cap help the m-RNA to recognized and bind with the ribosome, during protein synthesis

[C] Both

[D] None of the above

109. Repeating unit of Z-DNA is

[A] Dinucleotide [B] Mononucleotide

[C] Both [D] None of the above

110. Tilt/Dimer of Z-DNA is

[A] -60° [B] -30°

[C] Both [D] None of the above

111. Tilt/Monomer of Z-DNA is

[A] -40° [B] -30°

[C] -90° [D] None of the above

112. Angle of curvature between two consecutive base pair of B-DNA is

[A] +36° [B] -36°

[C] +20° [D] None of the above

113. Vertical distance between two consecutive base pairs of B-DNA is

[A] 3.4Å [B] 3.0Å

[C] 3.9Å [D] None of the above

114. Angle of curvature between two consecutive major and minor groves is

[A] 360° [B] 300°

[C] 280° [D] None of the above

115. Angle between base pair and main axis of B-DNA is

[A] 90° [B] 80°

[C] 70° [D] None of the above

116. Tilt per base pair of A-DNA is

[A] +19° [B] +20°

[C] +22° [D] None of the above

117. Pith of the helix of B-DNA is

[A] 34Å [B] 40Å

[C] 50Å [D] None of the above

118. Pith of the helix of A-DNA is

[A] 34Å [B] 25Å

[C] 35Å [D] None of the above

119. Base pair per turn of helix of B-DNA is

[A] 10 [B] 20

[C] 30 [D] None of the above

120. Base pair per turn of helix of A-DNA is

[A] 11 [B] 21

[C] 31 [D] None of the above

121. Base pair per turn of helix of Z-DNA is

[A] 12 [B] 22

[C] 32 [D] None of the above

122. Base pair per turn of helix of D-DNA is

[A] 8 [B] 9

[C] 10 [D] None of the above

123. Distance between two base pairs of D-DNA is

[A] 3.03Å [B] 4.03Å

[C] 5.03Å [D] None of the above

124. Right left handed DNA was reported by

[A] G.A. Rodley's group [B] V. Sashisekharan

[C] Both [D] None of the above

125. .... is the example of largest palindrome

[A] r-DNA [B] m-RNA

[C] t-RNA [D] None of the above

126. Usually DNA is double stranded in most of the organism except

[A] Φ X174 bacteriophase [B] M13 bacteriophase

[C] S13 coliphase [D] Parvovirus

[E] All of the above

127. Synthesis of new complementary polynucleotide chain over the old polynucleotide chain of mother DNA is known as

[A] Replication [B] Translation

[C] Transcription [D] None of the above

128. Topoisomerase .... is known as DNA gyrase

[A] I [B] II

[C] Both [D] None of the above

129. What are the properties of DNA replication?

[A] DNA replication is semi conservative

[B] DNA replication is semi discontinous

[C] DNA replication is uni or bidirectional

[D] All of the above

130. In which of the following DNA replication is continuous?

[A] Φ29 adenovirus [B] Φ30 adenovirus

[C] Φ31 adenovirus [D] None of the above

131. Class A RNA polymerase synthesizes

[A] r-RNA [B] m-RNA

[C] t-RNA [D] None of the above

132. Class B RNA polymerase synthesizes

[A] r-RNA [B] m-RNA

[C] t-RNA [D] None of the above

133. Class C RNA polymerase synthesizes

[A] r-RNA [B] m-RNA

[C] t-RNA [D] None of the above

134. .... factor helps in recognition of and initiating from correct site and correct strand of citrons for transcription.

[A] Sigma factor [B] Rho factor

[C] Delta factor [D] None of the above

135. What are the properties of codon?

[A] Codon is triplet [B] Codon is degenerate

[C] Codon is coma less [D] Codon is none overlapping

[E] Codon is no ambiguous [F] Codon is universal

[G] Codon is polar [H] All of the above

136. What is the name of codon UAA?

[A] Ochre [B] Amber

[C] Opal or Umber [D] None of the above

137. What is the name of codon UAG?

[A] Ochre [B] Amber

[C] Opal or Umber [D] None of the above

138. What is the name of codon UGA?
[A] Ochre [B] Amber
[C] Opal or Umber [D] None of the above

139. Which of the following is the start codons?
[A] AUG [B] GUG
[C] Both [D] None of the above

140. Which of the following is the most common start codons?
[A] AUG [B] GUG
[C] Both [D] None of the above

141. In eukaryote which of the following is the start codons?
[A] AUG [B] GUG
[C] Both [D] None of the above

142. If total number of codon in a m-RNA is 10, codon located before start codon is 2 and codon located after a stop codon is 2 then what is total number of amino acids in protein chain synthesized?
[A] 6 [B] 7
[C] 8 [D] None of the above

143. Codon is non overlapping except
[A] X174 Bacteriophage [B] X176 Bacteriophage
[C] X178 Bacteriophage [D] None of the above

144. If total number of base on m-RNA is 12 and number of amino acids in protein chain synthesized by such m-RNA is 3 then the codon on m-RNA is
[A] Overlapping [B] No overlapping
[C] Both [D] None of the above

145. If total number of base including stop codon on m-RNA is 27 and the number of amino acid in protein chain synthesized by such m-RNA is 9 then the codon is
[A] Overlapping [B] No-overlapping
[C] Both [D] None of the above

146. Which of the following cap is found in the lower eukaryote?
[A] Cap 1 [B] Cap 0
[C] Cap 2 [D] None of the above

147. Which of the following is the predominant cap type in higher eukaryotes?
[A] Cap 1 [B] Cap 0
[C] Cap 2 [D] None of the above

148. Long-lived mRNA is found in which of the following?
[A] Roots [B] Seeds
[C] Leaves [D] None of the above

22

PHYSIOLOGY OF CROP PLANTS

1. **The value of solar constant is .... cal cm^{-2} min^{-1}**

[A] 2 [B] 4

[C] 6 [D] None of the above

2. **PAR is ....% of total solar radiation entering the earth's atmosphere.**

[A] 44-50% [B] 55-65%

[C] 75-85% [D] None of the above

3. **What are the factors that influence the amount of solar radiation received in one day?**

[A] Angle of sun's rays

[B] Day length

[C] The amount of atmospheric particle

[D] All of the above

4. **Leaf resistance to CO_2 assimilation =**

[A] ra + rs + rm [B] ra + rs

[C] ra [D] None of the above

5. **Glucose is completely converted to CO_2 in approximately .... steps.**

[A] 30 [B] 40

[C] 50 [D] None of the above

6. **LAI term was coined by**

[A] Watson [B] Mann

[C] Naylor [D] None of the above

7. **What is the unit of Crop Growth Rate (CGR)?**

[A] g. m^{-2} (land area). Day^{-1} [B] g. m^{-1}(land area). Day^{-1}

[C] g. m^{-2} (crop area). Day^{-1} [D] None of the above

8. **What is the unit of Net Assimilation Rate (NAR)?**

[A] g. m^{-2} (leaf area). Day^{-1} [B] g. m^{-2} (land area). Day

[C] g. m^{2} (land area). Day^{-1} [D] None of the above

9. **CGR=**

[A] NAR x LAI [B] NAR/LAI

[C] NAR + LAI [D] None of the above

10. Leaf area index at which the canopy first reaches maximum CGR is known as

[A] Critical LAI [B] Optimal LAI

[C] Both [D] None of the above

11. LAI at which there is maximum CGR is known as

[A] Critical LAI [B] Optimal LAI

[C] Both [D] None of the above

12. Critical LAI mostly found in

[A] Cereals [B] Grasses

[C] Both [D] None of the above

13. Optimal LAI mostly found in

[A] Cereals [B] Grasses

[C] Legumes [D] None of the above

14. A legume canopy intercepts 95% of radiation at LAI of ....

[A] 5 [B] 6

[C] 7 [D] None of the above

15. In grasses the critical LAI is attained at an LAI of

[A] 9 [B] 10

[C] 11 [D] None of the above

16. The concept of leaf inclination was given by

[A] De Wit [B] Lang

[C] Leopold [D] None of the above

17. In erectophile canopy most of the leaves are .... .The angle formed by the leaf from the perpendicular of the main stem is more than ....

[A] Vertical, 60° [B] Horizontal, 70°

[C] Vertical, 30° [D] None of the above

18. Erectophile canopy are mostly found in

[A] Cereals [B] Legumes

[C] Vegetables [D] Fruit plants

19. In planophile canopy most of the leaves are .... The angle formed by the leaf from perpendicular of the main stem is lesser than

[A] Vertical, 45° [B] Horizontal, 35°

[C] Vertical, 30° [D] None of the above

20. Theoretically, a .... canopy would be more efficient if radiations were distributed more evenly over leaf surfaces.

[A] Planophile [B] Erectophile

[C] Both [D] None of the above

21. Planophile canopy is predominant in

[A] Cereals [B] Legumes

[C] Vegetables [D] Fruit plants

22. Photosynthetic response to radiation is .... and radiation efficiency is greatest at ...., the .... leaf is more efficient per unit of radiation intercepted.

[A] Curvilinear, low radiation level, vertical

[B] Curvilinear, high radiation level, vertical

[C] Curvilinear, low radiation level, horizontal

[D] None of the above

23. What are the strategies for maximizing solar radiation utilization and crop yields?

[A] Early planting for earlier leaf area development

[B] Planting at seeding rate

[C] Planting at a time that provide s total ground cover during the period of maximum solar radiations level

[D] All of the above

24. 'Partitioning" of assimilates and inorganic nutrients affects both productivity and survival of the plant'. This statement is

[A] True [B] False

[C] Not known [D] None of the above

25. What is the concentration of non reducing sugar in phloem sap?

[A] 10-25% [B] 30-45%

[C] 5-10% [D] None of the above

26. Specific mass transfer (SMT) =

[A] Weight of the assimilate moved per cross-sectional area of phloem per unit of time.

[B] Weight of the assimilate moved per cross-sectional area of xylem per unit of time

[C] Weight of the assimilate produced per cross-sectional area of phloem per unit of time

[D] None of the above

27. Specific mass transfer (SMT) rate for several species is

[A] 3-5 $g.cm^{-2}.hr^{-1}$ [B] 9-10 $g.cm^{-2}.hr^{-1}$

[C] 12-15$g.cm^{-2}.hr^{-1}$ [D] None of the above

28. Yield component=

[A] Grain yield + number of reproductive units per unit of ground area + the number of grains per reproductive unit+ the average weight of the grain

[B] Grain yield X number of reproductive units per unit of ground area X the number of grains per reproductive unit X the average weight of the grain

[C] Grain yield + number of reproductive units per unit of ground area - the number of grains per reproductive unit X the average weight of the grain

[D] None of the above

29. The movement of compounds from an area where they were once deposited to an area where they can be reutilized is referred to as

[A] Remobilization [B] Mobilization

[C] Utilization [D] None of the above

30. Photosynthates produced and deposited in grain come from

[A] Current leaf photosynthesis

[B] Current photosynthesis from parts other than leaf

[C] Remobilization of assimilates from storage parts

[D] All of the above

31. The relationship between the growth rates of individual parts of an organ or organism is referred to as

[A] Allometry [B] Allometer

[C] Leaf area ratio [D] None of the above

32. Plant growth can be expressed as

[A] Genotype X environment [B] Genotype + environment

[C] Genotype – environment [D] None of the above

33. Plant development includes

[A] Only growth of plant

[B] Only differentiation of plant

[C] Both growth and differentiation of plant

[D] None of the above

34. What are the internal factors which affect the growth and development of plants?

[A] Photosynthetic rate [B] Respiration

[C] Differentiation [D] Heterosis

[E] Epistasis [F] Type and location of meristems

[G] Partitioning of assimilates

[H] Resistance to climate, edaphic and biological stresses

[I] All of the above

35. *"A deficiency or absence of one necessary constituent, all others being present, renders the soil barren for crops for which that nutrient is needed"*. **This statement was made by**

[A] J.V. Liebig [B] F.F. Blackman's

[C] Macy [D] None of the above

36. Theory of "optimum and limiting factors" was given by

[A] F.F. Blackman [B] Mitscherlich

[C] Macy [D] None of the above

37. *"When a process is conditioned as to its rapidity by a number of separate factors, the rate of the process is limited by the pace of the slowest factors."* **This famous statement was made by**

[A] F.F. Blackman [B] Mitscherlich

[C] Macy [D] None of the above

38. Law of diminishing returns for plant growth was given by

[A] F.F. Blackman [B] Mitscherlich

[C] Macy [D] None of the above

39. ***"The increase in any crop produced by a unit increment of a deficient factor is proportional to the decrement of that factor from the maximum."*** **This famous statement was made by**

[A] F.F. Blackman [B] Mitscherlich

[C] Macy [D] None of the above

40. Which of the following is the pre requisites for differentiation processes in plant?

[A] Available assimilate in excess of most metabolic uses

[B] A favorable temperature

[C] A proper enzyme system to mediate the differentiation process

[D] All of the above

41. Biomass duration is analogous to

[A] LAD [B] LAI

[C] CGR [D] None of the above

42. .... expresses the leafiness during the period of crop growth.

[A] LAD [B] NAR

[C] CGR [D] None of the above

43. Which of the following technique is used for studying of root growth in plant?

[A] Trench profile method

[B] Framed monolith and Pinboard method

[C] Soil moisture depletion method

[D] Core sampling method

[E] Minirhizotron method

[F] Radioactive isotope method

[G] Allometry method

[H] All of the above

44. Which of the following stain is used in the staining of root cell?

[A] Congo red [B] Congo blue

[C] Congo pink [D] None of the above

45. What is the formula of Relative Growth Rate?

[A] $\frac{1}{W} \cdot \frac{dw}{dt}$ [B] $\frac{2}{W} \cdot \frac{dw}{dt}$

[C] $\frac{3}{W} \cdot \frac{dw}{dt}$ [W=weight, t=time]

[D] None of the above

46. What is the formula of Leaf Area Ratio?

[A] $\frac{LA}{W}$ [B] $\frac{WA}{L}$

[C] Both [D] None of the above

47. What is the formula of specific leaf area?

[A] $\frac{LA}{LW}$ [B] $\frac{Lw}{LA}$

[C] $\frac{\text{Total Weight}}{\text{Total Area}}$ [D] None of the above

48. What is the formula of net assimilation rate?

[A] $\frac{1}{A}\cdot\frac{dw}{dt}$ [B] $\frac{2}{A}\cdot\frac{dw}{dt}$

[C] $\frac{3}{A}\cdot\frac{dw}{dt}$ [D] None of the above

49. What is the unit of leaf area duration?

[A] W.T [B] T

[C] A.T [D] None of the above

50. What is the unit of biomass duration?

[A] W.T [B] T

[C] A.T [D] None of the above

51. Quiescent center is not present in apical meristems of

[A] Stem [B] Root

[C] Both [D] None of the above

52. Formation of lateral roots is the result of which of the following factor?

[A] Production of -inhibitor in root tip, which is related to apical dominance

[B] Production of growth-promoting substances in the shoot, which are transported to the roots

[C] A balance or interaction between growth-promoting and growth-inhibiting substances.

[D] All of the above

53. The average life span of root hair is .... at moderate temperature.

[A] 50 h [B] 20 h

[C] 100 h [D] None of the above

54. Adventitious roots are also referred to as

[A] Nodal root [B] Crown root

[C] Coronal root [D] All of the above

55. If we increase bulk density of soil then root growth will be

[A] Decreasing [B] Increasing

[C] Not affected [D] None of the above

56. *'Shoot-Root ratio can be increased by increasing the amount of nitrogen in the soil'* **This statement is**

[A] True [B] False

[C] Unpredictable [D] None of the above

57. In soybean, the oxygen requirement and nutrient uptake were greater during .... stages than during .... stages.

[A] Vegetative, reproductive [B] Reproductive, vegetative

[C] Flowering, seedling [D] None of the above

58. .... is considered as carbohydrate factories of the crop plants.

[A] Shoot [B] Leaf

[C] Root [D] None of the above

59. Phytomers develop

[A] Acropetally [B] Indeterminately

[C] Both [D] None of the above

60. The interval of time between the appearance of successive leaf primordia is termed as

[A] Plastochron [B] Phyllochron

[C] Both [D] None of the above

61. The interval of time between the appearance of the tip of successive leaves is termed as

[A] Phyllochron [B] Phyllomer

[C] Both [D] None of the above

62. The characteristic leaf number for wheat, oat and barley is

[A] 7-9 [B] 10-11

[C] 12-14 [D] None of the above

63. The characteristic leaf number for sorghum is

[A] 7-14 [B] 10-11

[C] 12-14 [D] None of the above

64. Removal of inflorescence in tobacco is known as

[A] Topping [B] Nipping

[C] Plugging [D] None of the above

23 WATER RELATIONS

1. **The movement of particle from a region of higher concentration to its lower concentration is known as**

[A] Osmosis [B] Diffusion

[C] Inhibition [D] None of the above

2. **The rate of diffusion increases if**

[A] Density of diffusing particles is lesser

[B] The temperature is increased

[C] Both

[D] None of the above

3. **Consumptive use coefficient (K) =**

[A] $\frac{\text{Actuale evapotranspiration}}{\text{Potential evapotranspiration}}$

[B] $\frac{\text{Potential evapotranspiration}}{\text{Actuale evapotranspiration}}$

[C] Potential evapotranspiration + actual evapotranspiration

[D] None of the above

4. **What are the properties of water?**

[A] It has high dielectric constant

[B] It has more specific heat

[C] It has more latent heat of vaporization

[D] Water is polar compound

[E] Water molecules remain connected together by hydrogen bond

[F] It has more surface tension

[G] Water is colourless

[H] All of the above

5. **Which of the following is correct relation?**

[A] $\alpha \frac{1}{\sqrt{\text{density}}}$ [B] $\alpha \sqrt{\text{Density}}$

[C] $\alpha \frac{1}{\sqrt{\text{density}}}$ [D] None of the Above

6. Imbibition is absorption of water/liquid by solid substance .... forming solution.

[A] Without [B] With
[C] Both [D] None of the above

7. During the process of imbibition, water is known as

[A] Imbibet [B] Imbibent
[C] Not known [D] None of the above

8. During the process of imbibition, solid substance is known as

[A] Imbibet [B] Imbibent
[C] Not known [D] None of the above

9. After imbibition water is held between and over the surface of hydrophilic particle of imbibent through ....

[A] Capillary action only
[B] Adsorption only
[C] Capillary action and adsorption respectively
[D] None of the above

10. During imbibition movement of water molecule up to the hydrophilic particle of imbibant occurs by

[A] Diffusion [B] Capillary action
[C] Both [D] None of the above

11. Adsorption of water on hydrophilic substance is called

[A] Imbibition [B] Diffusion
[C] Osmosis [D] Capillary action
[E] More than one

12. Imbibition involves

[A] Diffusion of water
[B] Movement of water molecule into imbibent by capillary action
[C] Movement of water into imbibent by diffusion and capillary action
[D] Adsorption of water
[E] None of the above

13. Imbibition involves

[A] Diffusion [B] Capillary action
[C] Adsorption [D] All of the above

14. What are the characters of imbibition?

[A] In course of imbibitions the volume of system increases or swelling occurs
[B] During the process of imbibition there is production of heat
[C] Development of pressure
[D] All of the above

15. Which of the following represents imbibition?

[A] Rubber kept in kerosene oil [B] Rubber kept in ethanol
[C] Wood kept in water [D] All of the above

16. Which of the following does not represent imbibition?

[A] Rubber kept in water
[B] Rubber kept in kerosene oil
[C] Wood kept in ethanol
[D] All of the above
[E] None of the above

17. If water enters (move) into a plant cell from outer side up to cell sap of vacuole then which of the following process will occur?

[A] Imbibitions [B] Osmosis
[C] Diffusion [D] Two of the above
[E] All of the above

18. In between isotonic solution net flow of water is not possible but exchange of water in equal amount is possible. This statement is

[A] True [B] False
[C] Not known [D] None of the above

19. If both plant and animal cells are placed in equally hypotonic solutions then which of the following will burst easily

[A] Animal cell [B] Plant cell
[C] Both will burst easily [D] None of the above

20. DPD term was coined by

[A] Mayer [B] Rennor
[C] Went [D] None of the above

21. Which of the following equation is correct?

[A] DPD= O.P-T.P [B] DPD= O.P+ T.P.
[C] DPD= O.P X T.P. [D] None of the above

22. DPD =

[A] $\frac{1}{\text{Suction pressure}}$
[B] Suction pressure
[C] Another name for suction pressure
[D] None of the above

23. If water enters osmotically into a cell (plant) from outside then

[A] Osmotic pressure decreases
[B] Turgor pressure increases
[C] Wall pressure increases
[D] DPD decreases
[E] All of the above

24. If water is osmotically removed from a cell then

[A] Osmotic pressure increases [B] Turgor pressure decreases

[C] Wall pressure decreases [D] DPD increases

[E] All of the above

25. If DPD of cell (Sol) A and B is equal then which of the following is possible?

[A] Net flow of water A to B

[B] Net flow of water from B to A

[C] Net flow of water is zero

[D] Exchange of water molecule in equal quantity between A and B is possible.

[E] Both C and D options are correct

26. If concentration of cell (soln) A and B is unequal or if cell A and B is heterotonic then which of the following is possible?

[A] Flow of water from low DPD to High DPD

[B] Flow of water from hypo to hyper concentration

[C] Both

[D] None of the above

27. If turgor pressure of two cells is equal but osmotic pressure is unequal then flow of water occurs from

[A] Lower O.P to Higher O.P [B] Higher O.P to lower O.P

[C] Both [D] None of the above

28. To make the value of O.P equal to T.P a cell should be placed in

[A] Hypotonic solution [B] Hypertonic solution

[C] Isotonic solution [D] None of the above

29. To make DPD equal to OP the cell should be placed in

[A] Hypotonic solution [B] Hypertonic solution

[C] Isotonic solution [D] None of the above

30. If a cell is placed in hypertonic solution then which of the following is possible?

[A] Exosmosis [B] Plasmolysis

[C] Flaccidity [D] Cytorrhysis

[E] All of the above

31. Osmotic movement of water from cell to outside resulting in net loss of water to cell is known as

[A] Exosmosis [B] Osmosis

[C] Endosmosis [D] None of the above

32. Cytological effect of exosmosis resulting in shrinkage of protoplast of cell is known as

[A] Plasmolysis [B] Flaccidity

[C] Exosmosis [D] None of the above

33. Earliest stage of plasmolysis in plant cell is

[A] Tonoplast plasmolysis [B] Incipient plasmolysis
[C] Cap plasmolysis [D] None of the above

34. Shrinkage of protoplast is followed by separation of protoplast and cell wall but connection between protoplast and cell wall remains maintained at many point, and this plasmolysis is known as

[A] Incipient plasmolysis [B] Evident plasmolysis
[C] Cytorrhysis [D] None of the above

35. What is the type of incipient plasmolysis on the basis of shape of protoplast?

[A] Cap plasmolysis [B] Concave plasmolysis
[C] Convex plasmolysis [D] All of the above

36. Complete shrinkage of protoplast and almost no connection of protoplast with the cell wall is known as

[A] Complete plasmolysis [B] Evident plasmolysis
[C] Both [D] None of the above

37. After plasmolysis which of the following is present in between protoplast and cell wall?

[A] Solvent [B] Solute
[C] Solution [D] None of the above

38. When a cell is placed in hypotonic solution then which of the following is possible?

[A] Endosmosis [B] Deplasmolysis
[C] Turgidity [D] All of the above

39. When a cell is placed in isotonic solution then its size

[A] Increases [B] Decreases
[C] Remains unchanged [D] None of the above

40. When a cell is placed in hypotonic solution then its size

[A] Increases [B] Decreases
[C] Remains unchanged [D] None of the above

41. When a cell is placed in hypertonic solution then its size

[A] Increases [B] Decreases
[C] Remains unchanged [D] None of the above

42. When a cell is placed in 5% glucose solution then its size remains unchanged, if such cell is placed in 5% NaCl solution, then its size will

[A] Increase [B] Decrease
[C] Unchanged [D] None of the above

43. When a cell is placed in 2% NaCl then its size remains unchanged, if such cell is placed in 2% glucose solution then its size will

[A] Increase [B] Decrease
[C] Unchanged [D] None of the above

44. Amount of water held as thin film around soil particle due to strong force of attraction is known as

[A] Hygroscopic water [B] Capillary water

[C] Gravitational water [D] None of the above

45. Holard is

[A] Capillary water+ hygroscopic water

[B] Capillary water+ hygroscopic water+ gravitational water

[C] Hygroscopic water+ gravitational water

[D] None of the above

46. Chesard is

[A] Capillary water+ hygroscopic water

[B] Capillary water

[C] Hygroscopic water+ gravitational water

[D] None of the above

47. Echard of field capacity is

[A] Hygroscopic water [B] Gravitational water

[C] Capillary water [D] None of the above

48. Amount of water held in crystal lattice structure of mineral in soil is called

[A] Water of crystallization [B] Chemically bounded water

[C] Bounded water [D] Combined water

[E] All of the above

49. Water available for plant is known as

[A] Chesard [B] Echard

[C] Both [D] None of the above

50. Instrument used to measure the water tension in soil is known as

[A] Tensiometer [B] Porometer

[C] Photometer [D] None of the above

24 NITROGEN METABOLISM AND BIOLOGICAL NITROGEN FIXATION

1. **In terms of total quantity required for production of a crop, N ranks .... among the 16 essential elements.**

   [A] 4th [B] 5th
   [C] 6th [D] None of the above

2. **Which of the following organism can utilize gaseous N_2?**

   [A] Bacteria [B] Actinomycetes
   [C] BG Alga [D] All of the above

3. **High amount of N in relation to C in the soil greatly .... the N_2 fixation.**

   [A] Decreases [B] Increases
   [C] Both [D] None of the above

4. **....................... mol of ATP are considered necessary to convert 1 mol N_2 to NH_4^+ and then to glutamic acid.**

   [A] 20 to 30 mol [B] 40 to 50 mol
   [C] 60-70 mol [D] None of the above

5. **Molecular weight of Fe protein in nitrogenase enzyme is**

   [A] 50,000-70,000 [B] 80,000-100,000
   [C] 20,000-30,000 [D] None of the above

6. **Which of the following genus of bacteria is most common in soils of temperate region?**

   [A] *Azotobacter* [B] *Azospirillum*
   [C] *Beijerinckia* [D] None of the above

7. **Which of the following is the well known host plant for *Frankia*?**

   [A] Alnus [B] Soybean
   [C] Clovers [D] None of the above

8. **Nitrogen fixation is mostly limited to**

   [A] Prokaryote [B] Eukaryote
   [C] Archebacteria [D] None of the above

9. **Which of the following is an example of obligate anaerobe nitrogen fixing bacteria?**

   [A] *Clostridium pasteurianum* [B] *Klebsiella*
   [C] *Rhodobacter* [D] None of the above

10. **Which of the following is an example of facultative anaerobe nitrogen fixing bacteria?**
[A] *Clostridium pasteurianum* [B] *Klebsiella*
[C] *Rhodobacter* [D] None of the above

11. **Which of the following is known as close relative of *E. coli*?**
[A] *Clostridium pasteurianum* [B] *Klebsiella*
[C] *Rhodobacter* [D] None of the above

12. **In nitrogenase enzyme, molybdoferredoxin is an .... tetramer.**
[A] Alpha2/beta2 [B] Alpha1/beta3
[C] Alpha0/beta4 [D] None of the above

13. **What is the turnover number of nitrogenase enzyme?**
[A] 50 moles/min per mole of Mo [B] 30 moles/min per mole of Mo
[C] 20 moles/min per mole of Mo [D] None of the above

14. **Chief form of nitrogen taken up by majority of plants from soil is**
[A] Molecular nitrogen [B] Amino acids
[C] Ammonia [D] Nitrates and nitrites

15. **Conversion of nitrate into ammonia is a**
[A] Reductive process [B] Amination process
[C] Oxidative process [D] None of the above

16. **Nitrate reductase enzyme was first isolated in highly purified form by**
[A] Evans and Nason [B] Fritz
[C] Haber-Bosch [D] None of the above

17. **What is the correct formula of biological nitrogen fixation catalyzed by nitrogenase?**
[A] $N_2 + 16\ ATP + 8e^- + 8H \rightarrow 2NH_3 + H_2 + 16ADP + 16Pi$
[B] $N_2 + 8\ ATP + 16e^- + 8H \rightarrow 2NH_3 + H_2 + 16ADP + 16Pi$
[C] $N_2 + 6\ ATP + 8e^- + 8H \rightarrow 2NH_3 + H_2 + 16ADP + 16Pi$
[D] None of the above

18. **What is the correct formula of reduction of nitrate to nitrite catalyzed by Nitrate Reductase?**
[A] $NO_3^- + NAD(P)H + H^+ \rightarrow NO_2^- + NAD(P)^+ + H_2O$
[B] $NO_3^- + NAD(P)H + H^+ \rightarrow NO_2^- + NAD(P)^+ + H_2O$
[C] $NO_3^- + NAD(P)H + H^+ \rightarrow NO_2^- + NAD(P)^+ + H_2O$
[D] None of the above

19. **The seeds of which of the following plant contain highest known protein content in any plant?**
[A] *Glycine max* [B] *Pisum sativum*
[C] *Phaseolus mungo* [D] None of the above

20. **Nitrate reduction takes place chiefly in**
[A] Green leaves [B] Roots
[C] Both [D] None of the above

21. The enzyme nitrate reductase is inducible type of enzyme which is found in cells, in

[A] Cytosol [B] Mitochondria
[C] Amyloplast [D] All of the above

22. What is the source of electron in reduction process of nitrate?

[A] NADH [B] Reduced ferredoxin
[C] NADPH [D] All of the above

23. Which of the following is not involved in reduction of nitrate to nitrite?

[A] Mo [B] FAD
[C] Cytochrome b 557 [D] None of the above

24. Conversion of nitrite to ammonia is mediated by transfer of

[A] Six electron [B] Four electron
[C] Two electron [D] None of the above

25. Immediate source of electron for reduction of nitrate to ammonia in chloroplast is

[A] NADH [B] NADPH
[C] Reduced ferredoxin [D] None of the above

26. Nitrogen fixation is the process of conversion of atmospheric nitrogen into

[A] Nitrites [B] Nitrates
[C] Ammonia [D] None of the above

27. The prosthetic group of nitrate reductase is

[A] Iron-sulphur cluster
[B] Iron-sulphur cluster and FAD
[C] Iron-sulphur cluster and special heme
[D] None of the above

28. Which of the following is associated with prosthetic group of nitrate reductase?

[A] Mo [B] Mn
[C] Fe [D] None of the above

29. Nitrate reductase is a single polypeptide that contains

[A] Two prosthetic groups [B] One prosthetic group
[C] Three prosthetic groups [D] None of the above

30. The enzyme nitrogenase is most sensitive to

[A] Oxygen [B] Nitrogen
[C] Helium [D] None of the above

31. Which of the following is the component of nitrogenase?

[A] Fe-protein [B] Fe-Mo protein
[C] Both [D] None of the above

32. The alpha and beta subunits of FeMo- protein of nitrogenase are encoded by

[A] *nifK* gene [B] *nifD* gene

[C] Both [D] None of the above

33. Biological nitrogen fixation requires

[A] Aerobic conditions [B] Anaerobic conditions

[C] Presence of hydrogen [D] None of the above

34. Cortical cells of host root infected by Rhizobia are

[A] Diploid [B] Haploid

[C] Tetraploid [D] All of the above

35. The amount of nitrogen fixed by lightening in the atmosphere is

[A] <10 Tg/year [B] 90-140 Tg/year

[C] 30-300 Tg/year [D] 80 Tg/year

[E] None of the above

36. The amount of nitrogen fixed by biological nitrogen fixation in terrestrial system is

[A] <10 Tg/year [B] 90-140 Tg/year

[C] 30-300 Tg/year [D] 80 Tg/year

[E] None of the above

37. The amount of nitrogen fixed by biological nitrogen fixation in marine system is

[A] <10 Tg/year [B] 90-140 Tg/year

[C] 30-300 Tg/year [D] 80 Tg/year

[E] None of the above

38. Dinitrogenase is ....KDa heterotetramer enzyme.

[A] 240-KDa [B] 200-KDa

[C] 150 KDa [D] None of the above

39. The midpoint reduction potential of the reaction catalyzed by Dinitrogenase reductase is about

[A] -400mV [B] -200 mV

[C] -100 mV [D] None of the above

40. Fe-protein of nitrogenase complex is encoded by

[A] *nif*H [B] *nif*D

[C] *nif*K [D] None of the above

41. Which of the following enzyme(s) is involved in assimilation of ammonia in plants?

[A] GDH [B] GOGAT

[C] GS [D] All of the above

42. Which of the following is not found in root nodules of legumes?

[A] Infected diploid cells [B] Infected tetraploid cells

[C] Uninfected diploid cells [D] None of the above

43. From root hairs, Rhizobia penetrate deep into cortex of host root through

[A] Cortical cell walls [B] Plasmodesmata
[C] Infection thread [D] All of the above

44. Which of the following is not required for biological nitrogen fixation?

[A] Mg^{++} [B] ATP
[C] Reduced ferredoxin [D] None of the above

45. The Rhizobia enter into roots of leguminous host plant through

[A] Root hairs [B] Epidermal cells
[C] Meristematic regions [D] None of the above

46. Which scientist first described rhizosphere?

[A] Lorenz Hiltner [B] Lorenz William
[C] T.R.G. Gray [D] None of the above

47. Which of the following stage of *Rhizobium* is nitrogen fixing form?

[A] Bacteroid stage [B] Symbiosome stage
[C] Nod factor stage [D] None of the above

48. What is the viable cell count of *Rhizobium* biofertilizers?

[A] 10^7 cell/g [B] 10^8 cell/g
[C] 10^9 cell/g [D] None of the above

49. What is the pH for *Rhizobium* fertilizer?

[A] 6.5-7.5 [B] 2.5-3.5
[C] 4.5-5.5 [D] None of the above

50. What is the EC number for nitrogenase enzyme?

[A] 1.18.6.1 [B] 1.17.6.1
[C] 1.16.5.1 [D] None of the above

51. .... is an organism that is able to grow without external source of fixed nitrogen.

[A] Diazotroph [B] Dixon
[C] Postgate [D] None of the above

52. *"All diazotroph contain iron-molybdenum nitrogenase system"*. This statement is

[A] True [B] False
[C] Not known [D] None of the above

53. The first Rhizobia was isolated from nodules by ....

[A] Beijerinck [B] Ladha
[C] Tate [D] None of the above

54. In root nodules the oxygen level is regulated by a special hemoglobin-leghemoglobin. The globin protein is encoded by .... but the heme cofactor is made by the ....

[A] Plant genes, symbiotic bacteria [B] Symbiotic bacteria, plant genes
[C] Plant genes, plant genes [D] None of the above

55. Nodules which have a large .... or .... region, will usually be active in N_2 fixation, and are said to be effective. If the nodule is .... or .... region is said to be senescing.

[A] Pink, red, white, greenish
[B] Pink, white, red, greenish
[C] Greenish, pink, red, white
[D] White, black, violet, pink

56. Successful infections of Rhizobia may be visible .... to .... days after inoculation and active N_2 fixation beginning .... to .... days thereafter.

[A] 5 to 6, 8 to 15 days [B] 10 to 16, 8 to 15 days
[C] 1 to 2, 8 to 15 days [D] None of the above

57. "Soil ecology is the study of the interactions among soil organisms and between biotic and abiotic aspects of the soil environment." This statement is

[A] True [B] False
[C] Not known [D] None of the above

58. Heterotrophic nutrition is found in

[A] Pistia [B] Opuntia
[C] Drosera [D] Vallisneria

59. *Agrobacterium* is closest relative of genus

[A] *Rhizobium* [B] *Bradyrhizobium*
[C] *Azorhizobium* [D] None of the above

60. What are the characteristics of genus *Rhizobium*?

[A] Grow fast on culture media
[B] Genes of nitrogen fixation and nodulation reside on symplasmid
[C] Both
[D] None of the above

61. What are the characteristics of genus *Bradyrhizobium*?

[A] Grow slow on culture media
[B] Genes of nitrogen fixation and nodulation reside on chromosome
[C] Both
[D] None of the above

62. What are the characteristics of genus *Azorhizobium*?

[A] Grow slow on culture media
[B] Genes of nitrogen fixation and nodulation reside on chromosome
[C] Capable of free-living nitrogen fixation.
[D] All of the above

63. Rhizobium is

[A] Gram-ve [B] Gram +ve
[C] Not known [D] None of the above

64. Actinomycete association with plant roots is called

[A] Actinorhizae [B] Mycorrhizae

[C] Gigaspora [D] None of the above

65. The concept "altruism in the rhizosphere" was suggested by

[A] I. Olivieri [B] S.A. Frank

[C] Both [D] None of the above

66. In most of the plants, activation of nif genes transcription is done by the nitrogen sensitive .... protein.

[A] NifA [B] NifB

[C] NifC [D] None of the above

25 SULFUR ASSIMILATION AND METABOLISM IN PLANTS

1. **Sulfate enters a plant primarily through the roots by way of .... uptake mechanism.**

[A] Active [B] Passive

[C] Both [D] None of the above

2. **Which of the following is the correct equation of reduction of sulfate to sulfide?**

[A] SO_4^{2-} + ATP + 8e- +8H+ → S_2^- + $4H_2O$ + AMP +PPi

[B] SO_4^{2-} + ATP + 6e- +8H+ → S_2^- + $2H_2O$ + AMP +PPi

[C] SO_4^{2-} + ATP + 5e- +8H+ → S_2^- + $3H_2O$ + AMP +PPi

[D] None of the above

3. **In the process of sulfur uptake, reduction and transport in plants, sulfate is stored in**

[A] Vacuole [B] Chloroplast

[C] Mitochondria [D] None of the above

4. **Reduction of sulfate and its assimilation takes place in**

[A] Plastids of root [B] Plastids of shoot

[C] Plastids of leaves [D] More than one

5. **What are the roles of sulfur in higher plants?**

[A] Sulfur is important constituent of amino acid cysteine and methionine

[B] Disulfide linkage in proteins helps to stabilize their structure and also play important regulatory role

[C] It play important role in electron transport through iron-sulphur cluster

[D] All of the above

6. **Sulphur containing amino acids which constitute proteins are**

[A] Cysteine and methionine [B] Methionine and histidine

[C] Cysteine and serine [D] None of the above

7. **Which of the following compound does not contain sulphur?**

[A] Coenzyme A and glutathione [B] Cysteine and thiamine

[C] Biotin and lipoic acid [D] None of the above

8. **Reduction of sulphate into cysteine entails transfer of**
   [A] 9 electrons [B] 10 electrons
   [C] 11 electrons [D] 12 electrons

9. **Higher plants absorb sulphur from the surrounding environment mainly in the form of**
   [A] Sulphide [B] Sulphite
   [C] Sulphate [D] None of the above

10. **Sulphur assimilation in higher plants occur mainly in**
   [A] Roots [B] Fruits
   [C] Leaves [D] None of the above

11. **Assimilation of sulphate into cysteine in higher plants is**
   [A] ATP-dependent oxidative process
   [B] ATP-dependent reductive process
   [C] Oxidative process
   [D] None of the above

12. **Find the missing enzymes in sulfur assimilation pathways in higher plants?**

Sulfate (So_4^{2-})
E1=? ↓
Adenosine-5'-Phosphosulfate (APS)
E2=? ↓
Sulfate
E3=? ↓
Sulfide
E4=? ↓
Cysteine

   [A] E1= ATP-Sulfurylase, E2= APS-Reductase, E3= Sulfite reductase, E4= O-Acetylserine thiolyase
   [B] E1= ATP-Reductase, E2= APS-Sulfurylase, E3= Sulfite reductase, E4= O-Acetylserine thiolyase
   [C] E1=ATP-Sulfurylase, E2=APS-Reductase, E3=O-Acetylserine thiolyase E4=Sulfite reductase.
   [D] None of the above

13. **Synthesis of methionine from cysteine occurs in**
   [A] Chloroplast in leaves [B] Chloroplast in shoots
   [C] Amyloplast in roots [D]None of the above

14. **Which of the following amino acid is considered as sulfur donor in many reactions?**
   [A] Cysteine [B] Methionine
   [C] Aspartate [D] None of the above

15. Disulfide bond can be formed by .... of thiol groups of two cysteine residues.

[A] Oxidation [B] Reduction
[C] Both [D] None of the above

16. Tertiary structure of proteins is stabilized by

[A] H-bond [B] Disulphide bonds
[C] Ionic bonds [D] All of the above

17. *"Glucosinolates contain both reduced and oxidized sulfur"*. **This statement is**

[A] True [B] False
[C] Not known [D] None of the above

18. Glutathione produced during sulfur assimilation is involved in

[A] Redox control [B] Storage of reduced sulfur
[C] Transport of reduced sulfur [D] Metabolism of herbicides
[E] Response to environmental stress
[F] All of the above

19. Which of the following compound is sometimes referred to as vitamin U?

[A] S-methylmethionine [B] Biotin
[C] SAM [D] None of the above

20. Sulfur containing phytoalexins such as .... are produced by members of the Brassicaceae in response to pathogen attack.

[A] Camalexin [B] Pamalexin
[C] Zamalexin [D] None of the above

21. In crop plants, particularly ...., are methionine deficient.

[A] Cereals [B] Legumes
[C] Oil seeds [D] None of the above

22. The interconversion of oxidized and reduced sulfur states on earth is known as

[A] Biogeochemical cycle of sulfur
[B] Reduction cycle of sulfur
[C] Oxidation cycle of sulfur
[D] None of the above

23. Some anaerobic bacteria use sulfate as a terminal electron acceptor analogous to the way that aerobic organisms use oxygen. This process is known as

[A] Dissimilation [B] Assimilation
[C] Utilization [D] None of the above

24. .... is the greatest contributor of atmospheric sulfur.

[A] Burning of fossil fuel [B] Atmospheric lightning
[C] Cultivation of soil [D] None of the above

25. Marine alga produced ...., a tertiary sulfur analog of the quaternary nitrogen compounds known as betaines.

[A] Dimethylsulfoniopropionate (DMSP)
[B] Oxidized product of DMSP i.e DMSO
[C] Both
[D] None of the above

26. What are the roles of DMSP?

[A] Osmoprotectant
[B] Cryoprotectants
[C] Repellant against planktonic herbivores
[D] All of the above

27. Sulfur uptake is mediated by a transporter that is powered by a .... across root cell plasma membranes.

[A] Proton gradient [B] Sodium electrogenic gradient
[C] Potassium gradient [D] None of the above

28. The reduction of sulfate to sulfide is an energy intensive process that consumes ....kj per mol.

[A] 732 [B] 632
[C] 532 [D] None of the above

29. Sulfate assimilation can be divided into three steps such as

[A] Activation
[B] Reduction to sulfide
[C] Incorporation of sulfide into cysteine
[D] All of the above

30. Sulfate activation is catalyzed by

[A] ATP sulfurylase [B] ATP sulfotransferase
[C] ATP kinase [D] None of the above

31. In the reaction given below, what is the free energy of adenylylsulfate (APS) formation?

$$SO_4^{2-} + MgATP = MgPP_i + 5'\text{-adenylylsulfae (APS)}$$

[A] +41.8 kJ mol^{-1} [B] +51.8 kj mol^{-1}
[C] +61.8 kj mol^{-1} [D] None of the above

26 CARBOHYDRATE METABOLISM IN PLANT

1. **Some organic compound follows the general formula of carbohydrate but these compounds are not carbohydrates. Examples of such types of compounds are**
   [A] Formaldehyde [B] Acetic acid
   [C] Lactic acid [D] All of the above
2. **Besides containing C, H and O, some carbohydrates also contain**
   [A] Nitrogen [B] Sulfur
   [C] Both [D] None of the above
3. **Carbohydrates are commonly referred as**
   [A] Polyhydroxyaldehydes [B] Polyhydroxyketones
   [C] Both [D] None of the above
4. **Which of the following statement is correct?**
   [A] Carbohydrates are hydrates of carbon
   [B] Carbohydrates corresponds to general formula $(CH_2O)_n$
   [C] Carbohydrates are polyhydroxyaldehydes and
   [D] All of the above
5. **Among the followings, which one is known as structural polysaccharide?**
   [A] Cellulose [B] Starch
   [C] Inulin [D] More than one
6. **Among the followings, which one is known as storage polysaccharide?**
   [A] Cellulose [B] Starch
   [C] Inulin [D] More than one
7. **What are the characteristics of monosaccharides?**
   [A] These cannot be hydrolysed.
   [B] These are the building units of complex carbohydrates.
   [C] These are simplest of carbohydrate and are known as sugars.
   [D] These are sweet-tasting, crystalline and soluble in water
   [E] All of the above

8. **In aldose sugars, aldehyde group or reducing centre always lies at carbon number**

   [A] 1 [B] 2
   [C] 3 [D] None of the above

9. **In ketose sugars, reducing centre always lies at carbon number**

   [A] 1 [B] 2
   [C] 3 [D] None of the above

10. **Which of the following character is correct regarding oligosaccharides?**

    [A] They are sweet in test [B] They are crystalline
    [C] They are soluble sugar [D] All of the above

11. **Which of the following is the example of disaccharides?**

    [A] Sucrose [B] Maltose
    [C] Lactose [D] All of the above

12. **Which of the following is the example of trisaccharides?**

    [A] Raffinose [B] Gentianose
    [C] Both [D] None of the above

13. **Which of the following is not a reducing sugar?**

    [A] Sucrose [B] Maltose
    [C] Ribose [D] Glucose

14. **Which of the following is not an oligosaccharide?**

    [A] Raffinose [B] Sucrose
    [C] Maltose [D] Inulin

15. **Six membered ring structure of aldohexoses is known as**

    [A] Furanose form [B] Cyclic form
    [C] Both [D] None of the above

16. **Which of the following statement is best suited for identification of glucose?**

    [A] It forms part of the disaccharide sucrose [B] It is a ketose
    [C] Carbon 2 is the anomeric carbon atom [C] None of the above

17. **The sugar residue of amylase is**

    [A] Fructose unit only [B] In -1, 4 linkages
    [C] Galactose units only [D] None of the above

18. **What are the characteristics of polysaccharides?**

    [A] Amorphous in nature [B] Tasteless
    [C]Insoluble in water [D] All of the above

19. **Among the followings, which one is structural polysaccharide?**

    [A] Cellulose [B] Hemi-cellulose
    [C] Pectic substances [D] Chitin
    [E] Gum [F] Mucilage
    [G] All of the above

20. Almost all the sugars in plants are

[A] D-Sugars [B] L-Sugars
[C] Both [D] None of the above

21. The highest number of asymmetric carbon atom is present in

[A] Arabinose [B] Glucose
[C] Fructose [D] None of the above

22. Arabinose is also known as

[A] L-Sugar [B] M-Sugar
[C] N-sugar [D] None of the above

23. D-Glucose is also known as

[A] Grape-sugar [B] Dextrose
[C] Both [D] None of the above

24. D-Fructose is also known as

[A] Levulose [B] Fruit sugar
[C] Both [D] None of the above

25. Maltose is a reducing sugar because one of the glucose units has free-OH at carbon number.......... ?

[A] 1 [B] 2
[C] 3 [D] None of the above

26. Sucrose is a

[A] Non-reducing sugar [B] Reducing sugar
[C] Both [D] None of the above

27. Raffinose consists of how many monosaccharide units?

[A] 3 [B] 4
[C] 5 [D] None of the above

28. Which of the following is not a component of pectic substances?

[A] Arabinan [B] Galacturonan
[C] Liginin [D] None of the above

29. Which of the following is the component of hemi-cellulose?

[A] Mannans and Glucomannans [B] Xylans
[C] Galactans and arabogalactans [D] All of the above

30. In polysaccharide inulin, the fructofurnase units are jointed together by

[A] α ($1 \rightarrow 2$) glycosidic bonds [B] β ($1 \rightarrow 2$) glycosidic bonds
[C] α ($2 \rightarrow 1$) glycosidic bonds [D] None of the above

31. The enzyme invertase hydrolyses sucrose into

[A] Glucose phosphate and fructose [B] Glucose and fructose
[C] Fructose [D] None of the above

32. In general, it is considered that plant metabolism is highly

[A] Flexible [B] Rigid
[C] Both [D] None of the above

33. One of the carbon atoms of a glucose molecule is [14C]-labelled. If $^{14}CO_2$ is released during the conversion of pyruvate to acetyl coenzyme-A, which carbon atom of glucose was radiolabelled?

[A] C3 but not C4 [B] C3 or C4
[C] C1 or C6 [D] C1 but not C6

34. Pick the correct statement.

[A] In O-linked glycosylation, sugars are attached to the protein via O-glycosidic bonds to the carboxyl groups of Asp and Glu

[B] In O-linked glycosylation, preformed oligosaccharides are attached to the relevant protein

[C] O-linked glycosylation, N-acetylgalactosamine is added via O-glycosidic bonds to the OH groups of Ser and Thr after which other sugars are added sequentially

[D] O-linked glycosylation is inhibited by the passage of the newly synthesized protein through the Golgi complex

35. Which of the following is a mucopolysaccharide?

[A] Pectin [B] Chitin
[C] Hemi-cellulose [D] None of the above

36. Which of the following has α (1 → 4) glycosidic linkage in its molecule?

[A] Maltose [B] Cellulose
[C] Sucrose [D] None of the above

37. Which of the following stage is known as reactive form of monosaccharide?

[A] De-phosphorylated [B] Sulphonated
[C] Phosphorylated [D] None of the above

38. Which of the following is known as optically active and dextrorotatory?

[A] Glucose [B] Fructose
[C] Both [D] None of the above

39. Diastase enzyme consists of

[A] R-enzyme [B] α-amylase and β-amylase
[C] Maltase [D] All of the above

40. Mutarotation is phenomenon where

[A] α and β-forms of the sugar change into one another to maintain the equilibrium

[B] α and α-forms of sugar change into one another to maintain the equilibrium

[C] β and β-forms of sugar change into one another to maintain the equilibrium

[D] All of the above

41. Among the followings, which one is known as most abundant organic polysaccharide on the earth?

[A] Cellulose [B] Chitin
[C] Starch [D] None of the above

27 SPECIAL MODES OF NUTRITION

1. **Parasitic angiosperms obtain their nutrition from other living plants by means of special roots called**
[A] Haustoria [B] Hauspie
[C] Both [D] None of the above
2. **What are the types of parasitic angiosperm?**
[A] Total Stem Parasite [B] Partial Stem Parasite
[C] Total Root Parasites [D] Partial Root Parasites
[E] All of the above
3. **Dodder or Cuscuta is very common example of**
[A] Total Stem Parasite [B] Partial Stem Parasite
[C] Total Root Parasites [D] Partial Root Parasites
4. ***Loranthus, Viscum* or mistletoe are the best examples of**
[A] Total Stem Parasite [B] Partial Stem Parasite
[C] Total Root Parasites [D] Partial Root Parasites
5. **In partial stem parasite, the Haustoria penetrate the host tissue and make connection only with its**
[A] Xylem [B] Phloem
[C] Both [D] None of the above
6. ***Orobanche, Striga, Balanophora* and *Rafflesia* are the best examples of**
[A] Total Stem Parasite [B] Partial Stem Parasite
[C] Total Root Parasites [D] Partial Root Parasites
7. **Which of the following is a saprophytic angiosperm?**
[A] *Santalum* [B] *Striga*
[C] *Dionaea* [D] *Monotropa*
8. **Which of the following flower is known as largest in diameter among the plant kingdom?**
[A] *Rafflesia* [B] *Albizzia*
[C] *Dalbergia* [D] None of the above
9. **Santalum album is very common example of ....**
[A] Total Stem Parasite [B] Partial Stem Parasite
[C] Total Root Parasites [D] Partial Root Parasites

10. The Pitcher plants are usually ....

[A] Small herbs [B] Shrubs

[C] Both [D] None of the above

11. In pitcher plants, ... are modified into pitchers.

[A] Leaves [B] Shoots

[C] Roots [D] None of the above

12. The pitcher contains .... at its base.

[A] Acidic fluid [B] Basic fluid

[C] Neutral fluid [D] None of the above

13. The insectivorous plants grow in swampy soils which are deficient in ....

[A] Nitrogen [B] Phosphorous

[C] Potassium [D] None of the above

14. In insectivorous plants, nectar gland is present at ....

[A] Mouth of the pitcher [B] Base of the pitcher

[C] Middle of the pitcher [D] None of the above

15. In the insectivorous plant, pitcher is formed by the modification of ....

[A] Lamina of the leaves [B] Midribs of the leaves

[C] Petiole of the leaves [D] None of the above

16. .... plant is also known as sundew?

[A] *Drosera* [B] *Neottia*

[C] *Monotropa* [D] None of the above

17. .... is also known as bladderwort.

[A] *Utricularia* [B] *Dionaea*

[C] *Striga* [D] None of the above

18. What is the feature of *Utricularia*?

[A] Is an aquatic plant [B] Free-floating

[C] Rootless [D] Small herbaceous

[E] All of the above

19. Which of the following is known as Venus Fly Trap?

[A] *Utricularia* [B] *Dionaea*

[C] *Striga* [D] None of the above

20. Rosette of leaves are found in

[A] Pinguicula [B] Dionaea

[C] Both [D] None of the above

21. Which plant is known as Butterwort?

[A] *Pinguicula* [B] *Orobanche*

[C] *Striga* [D] None of the above

22. Insectivorous plants digest the bodies of insects by secreting chiefly

[A] Proteolytic enzymes [B] Carbohydrates

[C] Both [D] None of the above

23. Mycorrhizae are fungus-root associations, first discovered by ....

[A] Albert Bernhard Frank [B] T.G.R. Gray

[C] S.T. Williams [D] None of the above

24. Ectomycorrhizal are found primarily in

[A] Temperate region [B] Tropical region

[C] Subtropical region [D] None of the above

25. Predominantly .... are involved in ectomycorrhizal symbiotic relationships.

[A] Basidiomycetes [B] Ascomycetes

[C] Zygomycetes [D] None of the above

28 RESPIRATION AND PHOTORESPIRATION

1. **Respiration is**
   [A] A catabolic process [B] An anabolic process
   [C] A chemical process [D] All of the above
2. **In plants, most common respiratory substrate is**
   [A] Organic acids [B] Carbohydrates
   [C] Proteins [D] None of the above
3. **The process of respiration in green plants takes place**
   [A] Only when stomata are open
   [B] Only when photosynthesis ceases
   [C] Only when photosynthesis is in progress
   [D] At all time
4. **Respiration differs from combustion because in combustion**
   [A] Light is produced
   [B] ATPs are not synthesized
   [C] Flame is produced
   [D] All of the above
5. **..................... is a highly coordinated multistep process; carried out by living organisms in their cells at body temperature, that involves the biological oxidation of organic substance (through electron transfer) resulting in their stepwise breakdown into simpler ones, with consequent but gradual and stepwise release of energy.**
   [A] Respiration [B] Photosynthesis
   [C] Combustion [D] None of the above
6. **ATP is the .... universal energy carrier of the cells.**
   [A] Mobile [B] Stagnate
   [C] Both [D] None of the above
7. **Glycolysis is the part of**
   [A] Anaerobic respiration [B] Aerobic respiration
   [C] Both [D] None of the above

8. Glycolysis is

[A] Obligate aerobe [B] Facultative aerobe
[C] Obligate anaerobe [D] Facultative anaerobe
[E] More than one

9. How many ATP is net gained during citric acid cycle of aerobic respiration of one glucose?

[A] 24 [B] 30
[C] 12 [D] More than one

10. What are the possible organic compounds produced in citric acid cycle on the basis of number of carbon atom?

[A] 5C [B] 4C
[C] 6C [D] 2C
[E] More than one

11. What are the proper sequences of organic compound produced in citric acid cycle on the basis of number of carbon atom?

[A] 6C-5C-4C [B] 4C-5C-6C -
[C] 5C-4C-6C [D] More than one
[E] None of the above

12. Name the organic compound produced in citric acid cycle having 5C atom

[A] -ketoglutaric acid [B] Isocitric acid
[C] Oxalosuccenic acid [D] None of the above

13. Name the organic compound produced in citric acid cycle having 6C atom

[A] Isocitric acid [B] Oxalosuccenic acid
[C] Cis-A conitate [D] All of the above

14. Name the organic compound dehydrogenated during Krebs cycle

[A] Isocitrate [B] α-ketoglutarate
[C] Succinate [D] Malate
[E] All of the above

15. Name the organic compound decarboxylated during Krebs cycle

[A] α-ketoglutarate [B] Isocitrate
[C] Both [D] None of the above

16. Name the organic compound which undergoes simultaneous oxidative decarboxylation during Krebs cycle

[A] α-ketoglutarate [B] Succinate
[C] Malate [D] None of the above

17. Name the hydrogen acceptor or carrier used during Krebs cycle

[A] NAD+ [B] FAD+
[C] Both [D] None of the above

18. Which organic compound donates its hydrogen to FAD+ during Krebs cycle?

[A] NAD+ [B] FAD+
[C] Both [D] None of the above

19. Which organic compound donates its hydrogen to FAD+ during Krebs cycle?

[A] Succinate [B] α-ketoglutarate
[C] Both [D] None of the above

20. Which organic compound donates its hydrogen to NAD+ during Krebs cycle?

[A] Malate [B] Isocitrate
[C] α-ketoglutarate [D] All of the above

21. Name the organic compound which gives ATP by substrate phosphorylation during Krebs cycle

[A] Succinyl -CoA [B] Isocitrate
[C] Both [D] None of the above

22. TCA cycle is also known as

[A] Krebs cycle [B] Citric acid cycle
[C] Both [D] None of the above

23. Who among the followings is not Noble laureate?

[A] Peter Mitchell [B] Hans A. Krebs
[C] Paul Boyer and John Walker [D] None of the above

24. Which of the following is not a form of anaerobic respiration?

[A] Lactic acid fermentation [B] Alcoholic fermentation
[C] Butyric acid fermentation [D] None of the above

25. In mitochondria, succinate dehydrogenase is located in

[A] Outer membrane [B] Inner membrane
[C] Matrix [D] Intermembrane space
[D] None of the above

26. Except succinate dehydrogenase, all other enzymes of TCA cycle are found in mitochondria in

[A] Outer membrane [B] Inner membrane
[C] Matrix [D] All of the above

27. Most of the reactions of glycolysis requires presence of

[A] Mg^{++} ions [B] Mn^{++} ions
[C] Ca^{++} ions [D] K^{+} ions

28. ATP molecule in glycolysis is synthesized by

[A] Substrate level phosphorylation
[B] Oxidative phosphorylation
[C] Photophosphorylation
[D] None of the above

29. Net gain of ATP molecule in oxidation of one molecule of glucose in glycolysis is

[A] 4 [B] 3

[C] 2 [D] 1

30. Krebs cycle is also called amphibolic cycle because

[A] It is involved in breakdown of Acetyl Co A

[B] Some intermediate of Krebs' cycle like α-ketoglutaric acid and Oxaloacetic acid are used to produce important biomolecules like glutamic acid and aspartic acid respectively.

[C] It is involved in the production of ATP

[D] All of the above

31. Name the intermediate compound of Krebs cycle which can produce glutamic acid and aspartic acid, respectively.

[A] α-ketoglutaric acid produces glutamic acid

[B] OAA produces Aspartic acid

[C] Both

[D] None of the above

32. Which of the following is the common oxidation pathway for carbohydrate, protein and fat metabolism?

[A] Glycolysis [B] Link reaction

[C] Krebs cycle [D] All of the above

33. Which of the following is the site of electron transport system?

[A] F1-F0 Complex [B] F0-F1 complex

[C] F1 complex [D] None of the above

34. Cytochrome a and a_3 are collectively known as

[A] Cytochrome oxidase [B] Cytochrome hydrogenase

[C] Both [D] None of the above

35. Cytochrome oxidase is

[A] Cyt a [B] Cyt a3

[C] Both [D] None of the above

36. Cytochrome oxidase is

[A] Cyt a [B] Cyt a_3

[C] Cyt b [D] None of the above

37. What is proton gradient?

[A] High concentration of H+ ion in perimitochondrial space and low concentration of H+ in matrix

[B] Low concentration of H+ ion in perimitochondrial space and low concentration of H+ in matrix

[C] High concentration of H+ ion in perimitochondrial space and High concentration of H+ in matrix

[D] None of the above

38. Energy librated due to which of the following movement of proton pair ($2H^+$) is responsible for synthesis of ATP?

[A] From matrix to PMS [B] From PMS to Matrix

[C] Both [D] None of the above

39. Which one of the following arrangement of Hydrogen carrier and cytochrome used during ETS is according to increasing order of potential?

[A] FMN/FAD-CoQ-Cytb-CytC1-CytC-Cyta-Cyta3

[B] FMN/FAD-Cytb-CoQ-CytC1-Cyta-Cyta3-CytC

[C] Both

[D] None of the above

40. Chemiosmotic theory deals with

[A] How proton gradient is established

[B] Formation of ATP by utilizing the energy librated due to diffusion of proton pair through proton channel

[C] Both

[D] None of the above

41. Final or terminal or ultimate electron acceptor in ETS is

[A] Oxygen [B] $Cyta_3$

[C] Cyt b [D] More than one

42. Final or terminal or ultimate electron acceptor in ETS is

[A] $Cyta_3$ [B] Cyt b

[C] Cyt c [D] None of the above

43. Final or terminal or ultimate electron acceptor in ETS is

[A] Cyt a_3 [B] Cyt b

[C] Cyt C [D] Cyt a

[E] More than one

44. ATP is produced in which of the following step of ETS?

[A] NAD-FM/FAD [B] Cytb-Cytc

[C] Cyta-$Cyta_3$ [D] All of the above

45. ATP is produced in which of the following step of ETS?

[A] Cytb-Cytc [B] Cyta-Cytc

[C]Cytc-Cytc1 [D]FAD-UQ

46. How many proton is pumped from matrix to PMS through inner membrane during single turn of ETS starting with one $NADH_2$?

[A] 3 pair proton [B] 2 pair proton

[C] 1 pair proton [D] None of the above

47. How many protons are pumped from matrix to PMS through inner membrane during single turn of ETS starting with one FADH?

[A] 3 pair proton [B] 2 pair proton

[C] 1 pair proton [D] None of the above

48. One molecule of $NADH_2$/$NADPH_2$ has sufficient energy to generate 3 ATP through ETS, the value of this energy present in 1 $NADH_2$/1 $NADPH_2$ (per mole) is

[A] 52 K cal [B] 15.2 K cal

[C] 24.5 Kcal [D] None of the above

49. One molecule of $FADH_2$ has sufficient energy to generate 2ATP through ETS. The value of this energy present in 1 $FADH_2$ is

[A] 40 K cal [B] 50 k cal

[C] 55 k cal [D] None of the above

50. Total ATP output (Net gain) after complete oxidation (or, aerobic oxidation) of 1 glucose molecule through ETS (oxidative phosphorylation) in prokaryote is

[A] 32 ATP [B] 34 ATP

[C] Both [D] None of the above

51. Total ATP output (Net gain) after complete oxidation (or, aerobic oxida-tion) of 1 glucose molecule through ETS (oxidative phosphorylation) in eukaryote is

[A] 32 ATP [B] 34 ATP

[C] Both [D] None of the above

29 MATCHING TYPE QUESTIONS

1. **Match the list I with the list II and select the correct answer using the codes given below the lists**

List I [Terms]	List II [Remarks]
A. ABA	1. Endogenous
B. Abiotic	2. High and low temperature
C. ABA biosynthesis	3. Sesquiterpene
D. Optical rotary dispersion and circular dichroism	4. Physical method of measurement of ABA
E. ABA: Water	5. Drought tolerance

[A] A: 1; B: 3; C: 4; D: 5; E: 2 [B] A: 1; B: 2; C: 3; D: 4; E: 5
[C] A: 4; B: 1; C: 2; D: 3; E: 5 [D] A: 4; B: 2; C: 3; D: 5; E: 1

2. **Match the list I with the list II and select the correct answer using the codes given below the lists**

List I [Terms]	List II [Remarks]
A. Abscisin II	1. Dormin
B. Abscission	2. Fall or detachment of leaves
C. Absolute counting	3. It is the measurement of radioactivity in which every radioactive disintegration occurring in a sample is accounted for
D. Absorption spectra	4. The curve showing the percentage absorption of light at each wavelength by a pigment
E. Addicott F.T.	5. First identified and characterized the hormone ABA

[A] A: 1; B: 3; C: 4; D: 5; E: 2 [B] A: 4; B: 2; C: 3; D: 5; E:
[C] A: 4; B: 1; C: 2; D: 3; E: 5 [D] A: 1; B: 2; C: 3; D: 4; E: 5

3. Match the list I with the list II and select the correct answer using the codes given below the lists

List I [Scientist]	List II [Contributions]
A. Alexander Resuss	1. Principle of electrophoresis
B. Always	2. Hygroscopic coefficient
C. Alwine	3. Northern blotting technique for RNA separation
D. Arnon	4. Discovered photophosphorylation
E. Arnon and Stout	5. Criteria of essential elements

[A] A: 1; B: 2; C: 3; D: 4; E: 5 [B] A: 1; B: 3; C: 4; D: 5; E: 2
[C] A: 4; B: 1; C: 2; D: 3; E: 5 [D] A: 4; B: 2; C: 3; D: 5; E: 1

4. Match the list I with the list II and select the correct answer using the codes given below the lists

List I [Scientist]	List II [Contributions]
A. Asana R.D.	1. Worked on drought tolerance
B. Atkins and Priestly	2. Osmotic theory of active absorption
C. Barbara McClintock	3. Jumping gene, Transposons
D. Bateson	4. Defined genetics
E. Beadle and Tatum	5. One gene-one enzyme hypothesis

[A] A: 1; B: 3; C: 4; D: 5; E: 2 [B] A: 4; B: 1; C: 2; D: 3; E: 5
[C] A: 1; B: 2; C: 3; D: 4; E: 5 [D] A: 4; B: 2; C: 3; D: 5; E: 1

5. Match the list I with the list II and select the correct answer using the codes given below the lists

List I [Scientist]	List II [Contributions]
A. Beijernick W.M.	1. First to isolate the root nodule bacterium
B. Belling	2. Demonstrated dark phase of photosynthesis
C. Benda	3. Copy choice theory
D. Blackman	4. Discovered Mitochondria
E. Blackman F.F.	5. Discovered chemical phase in photosynthesis

[A] A: 1; B: 3; C: 4; D: 5; E: 2 [B] A: 1; B: 2; C: 3; D: 4; E: 5
[C] A: 4; B: 1; C: 2; D: 3; E: 5 [D] A: 4; B: 2; C: 3; D: 5; E: 1

6. **Match the list I with the list II and select the correct answer using the codes given below the lists**

	List I [Scientist]		List II [Contributions]
A.	Blackman R.F.	1.	Law of inhibitory factor
B.	Bose J.C.	2.	Moisture equivalent
C.	Bray	3.	Vital theory for the ascent of sap
D.	Brayer	4.	Nutrient mobility concept
E.	Briggs and McLane	5.	Chlorine essentiality

[A] A: 1; B: 3; C: 4; D: 5; E: 2 [B] A: 1; B: 2; C: 3; D: 4; E: 5
[C] A: 4; B: 1; C: 2; D: 3; E: 5 [D] A: 4; B: 2; C: 3; D: 5; E: 1

7. **Match the list I with the list II and select the correct answer using the codes given below the lists**

	List I [Scientist]		List II [Contributions]
A.	Briggs and Santz	1.	Permanent Wilting Point
B.	Buchner	2.	Discovered first the enzymatic activity
C.	Buckinghum	3.	Capillary potential
D.	Bunning	4.	Biological clock
E.	Camillo Golgi	5.	Golgi complex

[A] A: 1; B: 3; C: 4; D: 5; E: 2 [B] A: 1; B: 2; C: 3; D: 4; E: 5
[C] A: 4; B: 1; C: 2; D: 3; E: 5 [D] A: 4; B: 2; C: 3; D: 5; E: 1

8. **Match the list I with the list II and select the correct answer using the codes given below the lists**

	List I [Scientist]		List II [Contributions]
A.	Chailakhyan	1.	Given that transpiration is a necessary evil
B.	Charls Darwin	2.	Origin of Species, Theory of Pangenesis
C.	Charles Darwin and Wallace	3.	Evolution through natural selection
D.	Christian de Duve	4.	Hormonal theory of flowering
E.	Curtis	5.	Discovered lysosomes

[A] A: 1; B: 3; C: 4; D: 5; E: 2 [B] A: 1; B: 2; C: 3; D: 4; E: 5
[C] A: 4; B: 1; C: 2; D: 3; E: 5 [D] A: 4; B: 2; C: 3; D: 5; E: 1

9. Match the list I with the list II and select the correct answer using the codes given below the lists

List I [Scientist]	List II [Contributions]
A. Dangeared	1. Concept of plant Ideotype
B. Decker	2. Coined the term photorespiration
C. Denny	3. Stated ethylene is highly effective in inducing fruit ripening
D. Dixon and Jolly	4. Discovered spherosomes
E. Donald C.M.	5. Transpiration pull

[A] A: 1; B: 3; C: 4; D: 5; E: 2 [B] A: 1; B: 2; C: 3; D: 4; E: 5
[C] A: 4; B: 1; C: 2; D: 3; E: 5 [D] A: 4; B: 2; C: 3; D: 5; E: 1

10. Match the list I with the list II and select the correct answer using the codes given below the lists

List I [Scientist]	List II [Contributions]
A. Fisher	1. Mesosomes
B. Fitz-James	2. Gene for gene hypothesis
C. Flor	3. Suggested that principle nourishment of plants was water
D. Francis Bacon	4. Lock and Key model
E. Frankel C. and Singer	5. RNA is a genetic material

[A] A: 1; B: 3; C: 4; D: 5; E: 2 [B] A: 1; B: 2; C: 3; D: 4; E: 5
[C] A: 4; B: 1; C: 2; D: 3; E: 5 [D] A: 4; B: 2; C: 3; D: 5; E: 1

11. Match the list I with the list II and select the correct answer using the codes given below the lists

List I [Scientist]	List II [Contributions]
A. Fraenkel	1. Cyclic phosphorylation
B. Fraenkel G.	2. Token Stimuli
C. Frits Went	3. Auxin
D. Funk	4. Vitamin
E. Galston A.W.	5. Worked on plant water relationship

[A] A: 1; B: 3; C: 4; D: 5; E: 2 [B] A: 1; B: 2; C: 3; D: 4; E: 5
[C] A: 4; B: 1; C: 2; D: 3; E: 5 [D] A: 4; B: 2; C: 3; D: 5; E: 1

12. Match the list I with the list II and select the correct answer using the codes given below the lists

List I [Scientist]	List II [Contributions]
A. Gammow G.	1. Triple codon
B. Gardner and Allard	2. Ribosomes
C. Godlewski	3. Photoperiod phenomena
D. Gregory F.G.	4. Relay Pump theory of ascent of sap
E. Haguenan	5. Phytochrome theory of flowering

[A] A: 1; B: 3; C: 4; D: 5; E: 2 [B] A: 1; B: 2; C: 3; D: 4; E: 5
[C] A: 4; B: 1; C: 2; D: 3; E: 5 [D] A: 4; B: 2; C: 3; D: 5; E: 1

13. Match the list I with the list II and select the correct answer using the codes given below the lists

List I [Scientist]	List II [Contributions]
A. Hilgard	1. Moisture holding capacity
B. Hill	2. Worked on isolated chloroplast
C. Hill and Bendell	3. Z-scheme of electron transport
D. Holley	4. First reported the base sequence of a t-RNA molecule
E. Ingenhouse	5. First to recognize the participation of chlorophyll and light in photosynthesis

[A] A: 1; B: 3; C: 4; D: 5; E: 2 [B] A: 1; B: 2; C: 3; D: 4; E: 5
[C] A: 4; B: 1; C: 2; D: 3; E: 5 [D] A: 4; B: 2; C: 3; D: 5; E: 1

14. Match the list I with the list II and select the correct answer using the codes given below the lists

List I [Scientist]	List II [Contributions]
A. Ingram	1. Concept of seed vigour
B. Isloy	2. Field capacity
C. Israelson & West	3. Operon concept
D. Jacob and Monod	4. One gene one polypeptide hypothesis
E. Kerry Mullis	5. PCR technique

[A] A: 1; B: 3; C: 4; D: 5; E: 2 [B] A: 1; B: 2; C: 3; D: 4; E: 5
[C] A: 4; B: 1; C: 2; D: 3; E: 5 [D] A: 4; B: 2; C: 3; D: 5; E: 1

15. Match the list I with the list II and select the correct answer using the codes given below the lists

List I [Scientist]		List II [Contributions]
A. Khorana	1.	Discovered anaerobic respiration
B. Khus G.S.	2.	Idea of super rice
C. Kogl	3.	IAA from Urine
D. Koshland	4.	Genetic code
E. Kostychev	5.	Induced fit hypothesis

[A] A: 1; B: 3; C: 4; D: 5; E: 2 [B] A: 1; B: 2; C: 3; D: 4; E: 5
[C] A: 4; B: 1; C: 2; D: 3; E: 5 [D] A: 4; B: 2; C: 3; D: 5; E: 1

16. Match the list I with the list II and select the correct answer using the codes given below the lists

List I [Scientist]		List II [Contributions]
A. Kuhne W.	1.	Introduced the term enzyme
B. Letham	2.	Cytokinine
C. Levitt	3.	Proton transport concept
D. Liebig	4.	Law of minimum
E. P. Maheswari	5.	Intracellular location of phytochrome in wheat and maize

[A] A: 1; B: 3; C: 4; D: 5; E: 2 [B] A: 1; B: 2; C: 3; D: 4; E: 5
[C] A: 4; B: 1; C: 2; D: 3; E: 5 [D] A: 4; B: 2; C: 3; D: 5; E: 1

17. Match the list I with the list II and select the correct answer using the codes given below the lists

List I [Scientist]		List II [Contributions]
A. Mayer	1.	DPD
B. Mc Collum and Davis	2.	Chemiosmotic hypothesis
C. Micheal Tswett	3.	Vitamin A
D. Miller and Skoog	4.	Initiated the idea of chromatography
E. Mitchell P.	5.	Kinetin from Yeast

[A] A: 1; B: 3; C: 4; D: 5; E: 2 [B] A: 1; B: 2; C: 3; D: 4; E: 5
[C] A: 4; B: 1; C: 2; D: 3; E: 5 [D] A: 4; B: 2; C: 3; D: 5; E: 1

18. Match the list I with the list II and select the correct answer using the codes given below the lists

List I [Scientist]		List II [Contributions]
A. Mollister	1.	Red light promoted and far-red light inhibited seed germination in lettuce
B. Moore and Mc Alear	2.	Discovered lomasomes
C. Muller and Borger	3.	Given Phytoalexin concept
D. Munch	4.	Mass Flow Mechanism
E. Nagaheru	5.	Triangle of brassicas

[A] A: 1; B: 3; C: 4; D: 5; E: 2 [B] A: 1; B: 2; C: 3; D: 4; E: 5
[C] A: 4; B: 1; C: 2; D: 3; E: 5 [D] A: 4; B: 2; C: 3; D: 5; E: 1

19. Match the list I with the list II and select the correct answer using the codes given below the lists

List I [Scientist]		List II [Contributions]
A. Nagarjan S.	1.	Idea of super wheat
B. Neal O.	2.	Give permeability classification into total seven classes
C. Nicholas J.D.	3.	Functional element
D. Ogren and Bowes	4.	Shown the first time the oxygenase function of Rubisco
E. Palade	5.	First time reported the presence of ribosome in cell

[A] A: 1; B: 3; C: 4; D: 5; E: 2 [B] A: 1; B: 2; C: 3; D: 4; E: 5
[C] A: 4; B: 1; C: 2; D: 3; E: 5 [D] A: 4; B: 2; C: 3; D: 5; E: 1

20. Match the list I with the list II and select the correct answer using the codes given below the lists

List I [Scientist]		List II [Contributions]
A. Palissy	1.	The ash content of plants
B. Paul field and Woods	2.	Explained the Lock and Key arrangement
C. Perner	3.	Named Spherosomes
D. Ping Y.L.	4.	Father of Hybrid Rice
E. Porter	5.	Discovered Endoplasmic Reticulum

[A] A: 1; B: 3; C: 4; D: 5; E: 2 [B] A: 1; B: 2; C: 3; D: 4; E: 5
[C] A: 4; B: 1; C: 2; D: 3; E: 5 [D] A: 4; B: 2; C: 3; D: 5; E: 1

21. Match the list I with the list II and select the correct answer using the codes given below the lists

	List I [Scientist]		List II [Contributions]
A.	Priestly	1.	Tensiometer
B.	Purkinje	2.	Protoplasm
C.	Renner O.	3.	Water potential concept
D.	Rhodin	4.	Microbodies
E.	Richard and Gardner	5.	Idea of gas exchange in photosynthesis

[A] A: 1; B: 3; C: 4; D: 5; E: 2 [B] A: 1; B: 2; C: 3; D: 4; E: 5

[C] A: 4; B: 1; C: 2; D: 3; E: 5 [D] A: 5; B: 2; C: 3; D: 4; E: 1

22. Match the list I with the list II and select the correct answer using the codes given below the lists

	List I [Scientist]		List II [Contributions]
A.	Rimpu	1.	Triticale
B.	Robert Brown	2.	Discovered cell
C.	Robert Hooke	3.	Discovered nucleus
D.	Saches	4.	Imbibitions theory of ascent of sap
E.	Salisbury	5.	Coined the term stomatal index

[A] A: 1; B: 3; C: 4; D: 5; E: 2 [B] A: 1; B: 2; C: 3; D: 4; E: 5

[C] A: 1; B: 3; C: 2; D: 4; E: 5 [D] A: 4; B: 2; C: 3; D: 5; E: 1

23. Match the list I with the list II and select the correct answer using the codes given below the lists

	List I [Scientist]		List II [Contributions]
A.	Sayre	1.	Cell theory
B.	Schleiden and Swann	2.	Joint assay with Darwin
C.	Schimper	3.	Plastid
D.	Schloesing and Muntz	4.	Demonstrated nitrification
E.	Slanntterback	5.	Discovered microtubules

[A] A: 2; B: 1; C: 3; D: 4; E: 5

[B] A: 1; B: 2; C: 3; D: 4; E: 5

[C] A: 4; B: 1; C: 2; D: 3; E: 5

[D] A: 4; B: 2; C: 3; D: 5; E: 1

24. Match the list I with the list II and select the correct answer using the codes given below the lists

List I [Scientist]	List II [Contributions]
A. Sorenson S.P.L.	1. pH
B. Southern	2. Given Southern Blotting for DNA separation
C. Stephan Hales	3. Root pressure theory of Ascent of Sap
D. Strasburger	4. Discovered chromosome
E. Strasburger and Herrtwig	5. Character transmit from generation to generation

[A] A: 1; B: 3; C: 4; D: 5; E: 2 [B] A: 1; B: 2; C: 3; D: 4; E: 5
[C] A: 4; B: 1; C: 2; D: 3; E: 5 [D] A: 4; B: 2; C: 3; D: 5; E: 1

25. Match the list I with the list II and select the correct answer using the codes given below the lists

List I [Scientist]	List II [Contributions]
A. Strong and Ray	1. Given the term skototropism
B. Strutvant	2. First genetic map in Drosophila
C. Summer JB	3. Isolate the enzyme "urease" as a cattaline protein
D. Sutton	4. Given "chromosome theory'
E. Sutton and Boveri	5. Established parallel behavior of chromosomes and genes (1903)

[A] A: 1; B: 3; C: 4; D: 5; E: 2 [B] A: 1; B: 2; C: 3; D: 4; E: 5
[C] A: 4; B: 1; C: 2; D: 3; E: 5 [D] A: 4; B: 2; C: 3; D: 5; E: 1

26. Match the list I with the list II and select the correct answer using the codes given below the lists

List I [Scientist]	List II [Contributions]
A. Swammerdam and Bonnet	1. Performation theory
B. Talbert	2. Peroxisome
C. Taylor	3. Ultimate Wilting Percentage
D. Thimann K.V.	4. Isolation and characterization of auxins
E. Thimann and Went	5. Discovered first time IAA

[A] A: 1; B: 3; C: 4; D: 5; E: 2 [B] A: 1; B: 2; C: 3; D: 4; E: 5
[C] A: 4; B: 1; C: 2; D: 3; E: 5 [D] A: 4; B: 2; C: 3; D: 5; E: 1

27. Match the list I with the list II and select the correct answer using the codes given below the lists

List I [Scientist]		List II [Contributions]
A. Towbin	1.	Western blotting for protein separation
B. Tucker and Nason	2.	Isolated "Hydroxlamine" Reductase" from Neurosporay
C. Van den Honest	3.	Carrier concept of ion transport
D. Van Helmont B.	4.	Given Centre's of Origin
E. Vavilov N.I.	5.	Water

[A] A: 1; B: 3; C: 4; D: 5; E: 2 [B] A: 1; B: 2; C: 3; D: 4; E: 5
[C] A: 4; B: 1; C: 2; D: 3; E: 5 [D] A: 1; B: 2; C: 3; D: 5; E: 4

28. Match the list I with the list II and select the correct answer using the codes given below the lists

List I [Scientist]		List II [Contributions]
A. Von Mohl	1.	Given that protoplasm is the physical basis of life
B. Wadleigh & Ayres	2.	Total soil moisture stress
C. Wahua	3.	Given the term chromosome
D. Waldeyer	4.	Nutrient supplementary index
E. Walther & Frau Hesse	5.	Gave the use of Agar-Agar as a solid medium

[A] A: 1; B: 3; C: 4; D: 5; E: 2 [B] A: 1; B: 2; C: 3; D: 4; E: 5
[C] A: 1; B: 2; C: 4; D: 3; E: 5 [D] A: 4; B: 2; C: 3; D: 5; E: 1

29. Match the list I with the list II and select the correct answer using the codes given below the lists

List I [Scientist]		List II [Contributions]
A. Watson and Crick	1.	Coined the term Dormin
B. Wareig	2.	Double helix model of DNA
C. Wright	3.	First discovered mutation
D. Yabuta and Hayashi	4.	Isolated gibberellins from fungus
E. Yoshida	5.	Ideal plant type in rice

[A] A: 2; B: 1; C: 3; D: 4; E: 5 [B] A: 1; B: 2; C: 3; D: 4; E: 5
[C] A: 4; B: 1; C: 2; D: 3; E: 5 [D] A: 4; B: 2; C: 3; D: 5; E: 1

30. Match the list I with the list II and select the correct answer using the codes given below the lists

	List I [Terms]		List II [Meaning]
A.	Biotic stress	1.	Joint assay with Darwin
B.	Bisexual	2.	A disorder of certain fruits, especially pears, in which the apical fleshy tissue dry out, turn black and become hard and woody
C.	Black end	3.	Collected Essays
D.	Black heart	4.	A diseased condition of the woody tissues in which the inner layers become darkened in cortex, impregnated with certain substances, lose their ability to serve as conducting and storage tissue, apparently caused by low winter temperature
E.	Blanching	5.	Heat treatment given to fruits and vegetables before processing

[A] A: 1; B: 3; C: 2; D: 4; E: 5 [B] A: 1; B: 2; C: 3; D: 4; E: 5
[C] A: 4; B: 1; C: 2; D: 3; E: 5 [D] A: 4; B: 2; C: 3; D: 5; E: 1

31. Match the list I with the list II and select the correct answer using the codes given below the lists

	List I [Terms]		List II [Meaning]
A.	Biological yield	1.	The sum total of biomass by a plant or of a crop excluding root in unit land area
B.	Biometrics	2.	The science, which deals with the application of statistical procedures to the study of biological problems
C.	Biometrical genetics	3.	The branch of genetics, which attempts to explain the inheritance of quantitative characters using statistical concepts and procedures
D.	Biosynthesis	4.	A technology based on the knowledge of life process which is used for mass production of useful substances/products for industry, medicine and agriculture
E.	Biotechnology	5.	The building up or synthesis of complex compound from simple substances in a living body

[A] A: 1; B: 3; C: 4; D: 5; E: 2 [B] A: 1; B: 2; C: 3; D: 4; E: 5
[C] A: 4; B: 1; C: 2; D: 3; E: 5 [D] A: 1; B: 2; C: 3; D: 5; E: 4

32. Match the list I with the list II and select the correct answer using the codes given below the lists

List I [Terms]		List II [Meaning]
A. Basic number	1.	The genetic chromosome number of a true diploid species
B. Biennial	2.	The kind of plant that produces vegetative growth during the first year of rowing seasons
C. Bifacial leaf	3.	A leaf having structurally distinct upper and lower surfaces
D. Bioassay	4.	To test for the presence or quality of substance by using an organisms response as an indicator
E. Biodiversity	5.	The variability exists among living organisms from all sources, including terrestrial, marine and other ecosystems

[A] A: 1; B: 3; C: 4; D: 5; E: 2 [B] A: 1; B: 2; C: 3; D: 4; E: 5
[C] A: 4; B: 1; C: 2; D: 3; E: 5 [D] A: 4; B: 2; C: 3; D: 5; E: 1

33. Match the list I with the list II and select the correct answer using the codes given below the lists

List I [Terms]		List II [Meaning]
A. Bactericide	1.	A substance capable of destroying bacteria
B. Baramasi	2.	Group of cultivation of mango whose flowering and cropping is spread throughout the year
C. Bark	3.	The external tissue of a woody stem or root, from the vascular cambium outward
D. Barometer	4.	Atmospheric pressure
E. Barren	5.	Unproductive, unable to support vegetation, incapable of producing offspring or seed

[A] A: 1; B: 3; C: 4; D: 5; E: 2 [B] A: 1; B: 2; C: 3; D: 4; E: 5
[C] A: 4; B: 1; C: 2; D: 3; E: 5 [D] A: 4; B: 2; C: 3; D: 5; E: 1

34. Match the list I with the list II and select the correct answer using the codes given below the lists

List I [Terms]		List II [Meaning]
A. Autonomic	1.	Parthenocarpy
B. Autotropic	2.	Capable of utilizing carbon dioxide and of obtaining energy for the reduction of carbon dioxide and other life processes from the oxidation of inorganic elements or from light

C.	Auxin	3.	A substance which regulates plant growth and controls cell elongations
D.	Avogadro's hypothesis	4.	Equal volume of all gas under the same conditions of temperature and pressure, contain the same number of molecules
E.	Bateson	5.	Materials for the study of variations

[A] A: 1; B: 3; C: 4; D: 5; E: 2 [B] A: 1; B: 2; C: 3; D: 4; E: 5
[C] A: 4; B: 1; C: 2; D: 3; E: 5 [D] A: 4; B: 2; C: 3; D: 5; E: 1

35. Match the list I with the list II and select the correct answer using the codes given below the lists

	List I [Terms]		List II [Meaning]
A.	Assimilation number	1.	Amount of CO_2 absorbed in grams per hour for each gram of chlorophyll during photosynthesis
B.	Atavism	2.	The appearance of a character in an organism of ancestor
C.	Atmometer	3.	An instrument which is used for measurement of evaporation rates
D.	Auricle	4.	Transfer of pollen grains from the anther to the stigma of the same flower
E.	Autogamy	5.	A small ear like projection from the base of a leaf or petal

[A] A: 1; B: 2; C: 3; D: 5; E: 4 [B] A: 1; B: 2; C: 3; D: 4; E: 5
[C] A: 4; B: 1; C: 2; D: 3; E: 5 [D] A: 4; B: 2; C: 3; D: 5; E: 1

36. Match the list I with the list II and select the correct answer using the codes given below the lists

	List I [Terms]		List II [Meaning]
A.	Artificial seed	1.	Somatic embryo
B.	Ascent of sap	2.	Does not require union of gametes
C.	Asexual reproduction	3.	Raise against gravity
D.	Asphyxiant	4.	Gas for killing pests through respiratory difficulties
E.	Assimilation	5.	Conversion of substances taken in from the outside into living tissue of plants or animals

[A] A: 1; B: 3; C: 4; D: 5; E: 2 [B] A: 1; B: 2; C: 3; D: 4; E: 5
[C] A: 1; B: 3; C: 2; D: 4; E: 5 [D] A: 4; B: 2; C: 3; D: 5; E: 1

37. Match the list I with the list II and select the correct answer using the codes given below the lists

List I [Terms]		List II [Meaning]
A. Apical dominance	1.	Growth of a bud at the apex of a stem or tube while growth of all other buds on the stem or tuber is inhibited
B. Apical meristem	2.	A mass of meristematic cells
C. Apomixis	3.	Asexual reproduction of seedling in the flower without fusion of gametes
D. Arid	4.	Litchi
E. Aril	5.	Water losses by evaporation and transpiration more than precipitation

[A] A: 1; B: 2; C: 3; D: 5; E: 4 [B] A: 1; B: 2; C: 3; D: 4; E: 5
[C] A: 4; B: 1; C: 2; D: 3; E: 5 [D] A: 4; B: 2; C: 3; D: 5; E: 1

38. Match the list I with the list II and select the correct answer using the codes given below the lists

List I [Terms]		List II [Meaning]
A. Antipodal	1.	Cells at the end of the embryo sac opposite that of egg apparatus
B. Antitranspirants	2.	Retard or inhibit transpiration
C. Anthocyanin	3.	Pigments in plant
D. Antibiotics	4.	Interfere with the life cycle of the microbes
E. Anti-oxidant	5.	Discoloration of food

[A] A: 1; B: 3; C: 4; D: 5; E: 2 [B] A: 1; B: 2; C: 3; D: 4; E: 5
[C] A: 4; B: 1; C: 2; D: 3; E: 5 [D] A: 4; B: 2; C: 3; D: 5; E: 1

39. Match the list I with the list II and select the correct answer using the codes given below the lists

List I [Terms]		List II [Meaning]
A. Anoxia	1.	Suppression or interference with the normal growth of plant pathogens by another microorganism, such as bacteria or fungi
B. Antagonism	2.	Dying from lack of oxygen
C. Anther	3.	Pollen producing organ of flower
D. Anthesis	4.	The opening of the flower bud
E. Antibiosis	5.	Adverse effect of feeding

[A] A: 1; B: 3; C: 4; D: 5; E: 2 [B] A: 1; B: 2; C: 3; D: 4; E: 5
[C] A: 2; B: 1; C: 3; D: 4; E: 5 [D] A: 5; B: 2; C: 3; D: 4; E: 1

40. Match the list I with the list II and select the correct answer using the codes given below the lists

List I [Terms]	List II [Meaning]
A. Androdioecious	1. Papaya
B. Androgenesis	2. Development in which the embryo contains only paternal chromosome
C. Anion	3. Completion of life cycle within one year
D. Annual	4. Negative
E. Annual rings	5. Visible in cross-section of stems woody plants

[A] A: 1; B: 3; C: 4; D: 5; E: 2 [B] A: 1; B: 2; C: 3; D: 4; E: 5
[C] A: 1; B: 2; C: 4; D: 3; E: 5 [D] A: 4; B: 2; C: 3; D: 5; E: 1

41. Match the list I with the list II and select the correct answer using the codes given below the lists

List I [Terms]	List II [Meaning]
A. Anabolism	1. Building of body tissues from simple raw materials supplied by the products of digestion of food
B. Anaerobe	2. Able to live in the absence of oxygen
C. Analogues	3. Organs of different plants or animals with like function but of unlike origin
D. Anaphase	4. Chromosome begin to pull to opposite poles of the cell
E. Anatropous ovule	5. Ovule is inverted with respect to its funiculus

[A] A: 1; B: 3; C: 4; D: 5; E: 2 [B] A: 1; B: 2; C: 3; D: 4; E: 5
[C] A: 4; B: 1; C: 2; D: 3; E: 5 [D] A: 4; B: 2; C: 3; D: 5; E: 1

42. Match the list I with the list II and select the correct answer using the codes given below the lists

List I [Terms]	List II [Meaning]
A. Amorph	1. A mutant allele that completely lost the function of normal allele
B. Amphidiploid	2. An allopolyploid that has two copies of each genome present in it
C. Amphoteric compound	3. Compounds that can act as an acid and as a base
D. Amylase	4. When starch is predominant, the leucoplast is called as amyloplast
E. Amyloplast	5. Enzymes that act on starch

[A] A: 1; B: 3; C: 4; D: 5; E: 2 [B] A: 1; B: 2; C: 3; D: 4; E: 5
[C] A: 4; B: 1; C: 2; D: 3; E: 5 [D] A: 1; B: 2; C: 3; D: 4; E: 5

43. Match the list I with the list II and select the correct answer using the codes given below the lists

List I [Terms]	List II [Meaning]
A. Allotetraploid	1. An organism with four genomes derived from hybridization of different species
B. Alternate bearing	2. Bearing heavy and lean crop in alternative years
C. Ambia bahar	3. The second blossom in some fruit trees appearing in December-January
D. Amensalism	4. Materials for the study of variations
E. Bateson	5. A symbiotic relationship between organisms in which one species is harmed or inhibited and the other species is unaffected

[A] A: 1; B: 3; C: 4; D: 5; E: 2 [B] A: 1; B: 2; C: 3; D: 4; E: 5
[C] A: 1; B: 2; C: 3; D: 5; E: 4 [D] A: 4; B: 2; C: 3; D: 5; E: 1

44. Match the list I with the list II and select the correct answer using the codes given below the lists

List I [Terms]	List II [Meaning]
A. Aleurone	1. The outermost layer of the endosperm in a seed
B. Aleuroneplast	2. When protein granules or crystals are predominant
C. Allele	3. One or several alternative forms of a gene occupying the same locus on a particular chromosome
D. Allelopathy	4. The adverse effect on the growth of plants or microorganisms caused by the action of chemical produced by other living or decaying plants
E. Allopolyploid	5. A polyploidy, which contains genetically different sets of chromosomes from two or more species

[A] A: 1; B: 3; C: 4; D: 5; E: 2 [B] A: 1; B: 2; C: 3; D: 4; E: 5
[C] A: 4; B: 1; C: 2; D: 3; E: 5 [D] A: 4; B: 2; C: 3; D: 5; E: 1

45. Match the list I with the list II and select the correct answer using the codes given below the lists

List I [Terms]	List II [Meaning]
A. Air layer	1. The shoot of a plant on which roots are made to develop by means of some stimulation or manipulation

B.	Aitionomic	2.	Ability to develop parthenocarpic fruits only in response to some stimulus extending to ovary
C.	Albinism	3.	Absence of pigment in skin, hair and eyes of an animal or absence of chlorophyll in plants is called albinism
D.	Albino	4.	White coloured
E.	Albumen	5.	Collective term for the nutritional tissue between the embryo and the seed coat, including perisperm and endosperm

[A] A: 1; B: 3; C: 4; D: 5; E: 2 [B] A: 1; B: 2; C: 3; D: 4; E: 5
[C] A: 4; B: 1; C: 2; D: 3; E: 5 [D] A: 4; B: 2; C: 3; D: 5; E: 1

46. Match the list I with the list II and select the correct answer using the codes given below the lists

	List I [Terms]		List II [Meaning]
A.	Aerobiosis	1.	Existing in the presence of oxygen as in the atmosphere
B.	Aeropyte	2.	A plant growing attached to an aerial portion of another plant
C.	Aerosol	3.	Type of insecticide
D.	After-ripening	4.	Completion of maturation process within seeds to enable germination
E.	Aggregate fruit	5.	The product of the fusion of several ovaries in the same flower as in strawberry and raspberry

[A] A: 1; B: 3; C: 4; D: 5; E: 2 [B] A: 1; B: 2; C: 3; D: 4; E: 5
[C] A: 4; B: 1; C: 2; D: 3; E: 5 [D] A: 4; B: 2; C: 3; D: 5; E: 1

47. Match the list I with the list II and select the correct answer using the codes given below the lists

	List I [Terms]		List II [Meaning]
A.	Adsorption	1.	The adhesion of molecules of gases or dissolved substance to the surface of solid particles
B.	Adventitious	2.	Out of place, arising from the secondary meristem
C.	Adventative embryology	3.	Embryos develop directly from vegetative cells of ovules
D.	Aerobic	4.	Active in the presence of free oxygen
E.	Aerobic respiration	5.	Involving molecular oxygen

[A] A: 1; B: 3; C: 4; D: 5; E: 2 [B] A: 1; B: 2; C: 3; D: 4; E: 5
[C] A: 4; B: 1; C: 2; D: 3; E: 5 [D] A: 4; B: 2; C: 3; D: 5; E: 1

48. Match the list I with the list II and select the correct answer using the codes given below the lists

List I [Terms]	List II [Meaning]
A. Adaptation	1. The adjustment of an organism or population to a new or altered environment
B. Adaptability	2. The ability of a genotype to produce a relatively narrow range of phenotypes in different environment
C. Adenine	3. A sticking together of unlike things on materials
D. Adhesion	4. A purine base
E. Adsorption	5. The adhesion of molecules of gases or dissolved substance to the surfaces of solid particles

[A] A: 1; B: 2; C: 4; D: 3; E: 5 [B] A: 1; B: 2; C: 3; D: 4; E: 5
[C] A: 4; B: 1; C: 2; D: 3; E: 5 [D] A: 4; B: 2; C: 3; D: 5; E: 1

49. Match the list I with the list II and select the correct answer using the codes given below the lists

List I [Terms]	List II [Meaning]
A. Acuminate	1. Drawn out into a long point; tapering, pointed
B. Active transport	2. The transport of solutes across cellular membranes by mechanisms that expand metabolic energy, which can lead to transport against an electrochemical potential gradient
C. Active absorption	3. Absorption of water through expenditure of metabolic energy and it occurs as a result of activities in the root
D. Action spectra	4. A graph relating the degree of physiological response caused by different wavelength of light
E. Acre	5. A unit of land measure equal to 43560 sq. ft

[A] A: 1; B: 3; C: 4; D: 5; E: 2 [B] A: 1; B: 2; C: 3; D: 4; E: 5
[C] A: 4; B: 1; C: 2; D: 3; E: 5 [D] A: 4; B: 2; C: 3; D: 5; E: 1

50. Match the list I with the list II and select the correct answer using the codes given below the lists

List I [Terms]	List II [Meaning]
A. Achene	1. Dry, one seeded, small, indehiscent fruit of one carpel, seed like in appearance, as the so called seed of the strawberry

B. Acclimatization	2.	Adjustment in tolerance shown by a species in the course of several generation in changed environment, adapting to a new set of conditions
C. Absorption spectrum	3.	A graph relating to the ability of a substance to absorb lights of various wavelength
D. Absorption	4.	The process of taking up water and holding it in pores or open spaces as sponges
E. Abscission layer	5.	A layer of thin walled cells formed at the base of the organ

[A] A: 1; B: 3; C: 4; D: 5; E: 2 [B] A: 1; B: 2; C: 3; D: 4; E: 5
[C] A: 4; B: 1; C: 2; D: 3; E: 5 [D] A: 4; B: 2; C: 3; D: 5; E: 1

51. Match the list I with the list II and select the correct answer using the codes given below the lists

List I [Terms]		List II [Meaning]
A. Blanching	1.	Heat treatment given to fruits and vegetables before processing
B. Bleaching	2.	Lightens the colours
C. Bonsai	3.	Japanese art of growing
D. Bordeaux mixture	4.	Copper fungicide
E. Bordered pit	5.	A pit in a tracheid or vessel member having a distinct rim of cell wall overarching the pit membrane

[A] A: 1; B: 3; C: 4; D: 5; E: 2 [B] A: 1; B: 2; C: 3; D: 4; E: 5
[C] A: 4; B: 1; C: 2; D: 3; E: 5 [D] A: 4; B: 2; C: 3; D: 5; E: 1

52. Match the list I with the list II and select the correct answer using the codes given below the lists

List I [Terms]		List II [Meaning]
A. Bound water	1.	The volume varies inversely with the pressure applied to the gas
B. Boyle's law	2.	Water held by the cell against freezing or the water that is adsorbed strongly by colloids
C. Bud	3.	Meristematic in nature
D. Bud-break	4.	Resting buds results growth
E. Budding	5.	A special type of grafting using a single bud as a scion.

[A] A: 2; B: 1; C: 3; D: 4; E: 5 [B] A: 1; B: 2; C: 3; D: 4; E: 5
[C] A: 4; B: 1; C: 2; D: 3; E: 5 [D] A: 4; B: 2; C: 3; D: 5; E: 1

53. Match the list I with the list II and select the correct answer using the codes given below the lists

List I [Terms]		List II [Meaning]
A. Bud maturation	1.	A somatic mutation or variation arising in bud, producing an abnormal branch
B. Budstick	2.	A shoot or twig used as a source of bulbs for budding
C. Buffer solution	3.	A solution which resists change in pH by the addition and loss of hydrogen and hydroxyl ions
D. Bulb	4.	A plant structure having concentric layers of flashy leaves attached to a disc shaped stem plate at the base, functions as storage for the food and for reproduction
E. Bulblets	5.	Small bulb produced around the base of the mother bulb

[A] A: 1; B: 3; C: 4; D: 5; E: 2

[B] A: 1; B: 2; C: 3; D: 4; E: 5

[C] A: 4; B: 1; C: 2; D: 3; E: 5

[D] A: 4; B: 2; C: 3; D: 5; E: 1

54. Match the list I with the list II and select the correct answer using the codes given below the lists

List I [Terms]		List II [Meaning]
A. Callus	1.	Protective thin walled parenchyma cell converging which form
B. Calyx	2.	The sepals of the flower as a group
C. Cambium	3.	This mucilaginous, cellular layer between the wood of a tree and its inner bark which is converted into xylem on the inner side and the phloem on the outer side
D. Candy	4.	Fruit or vegetable product, a preserve which has been heavily impregnated with sugar
E. Canning	5.	A method of preserving food in a fresh condition by sealing in airtight containers

[A] A: 1; B: 3; C: 4; D: 5; E: 2

[B] A: 1; B: 2; C: 3; D: 4; E: 5

[C] A: 4; B: 1; C: 2; D: 3; E: 5

[D] A: 4; B: 2; C: 3; D: 5; E: 1

55. Match the list I with the list II and select the correct answer using the codes given below the lists

List I [Terms]	List II [Meaning]
A. Capillary water	1. Water held in soil between soil particles, and is available to plants
B. Carbohydrate	2. Energy containing product of plants consisting of carbon, hydrogen and oxygen
C. Carbon fixation	3. A class of fat soluble compounds that includes carotenes as well as xanthophylls; most of them absorbed light and appear yellow, organic or red
D. Carotenoid	4. The enzymatic reaction in which CO_2 is attached to a receiver compound such as ribulose biphophate, thereby adding to the supply of organic carbon
E. Carrier	5. A substance by which some elements or group is transferred from one compound to another

[A] A: 1; B: 3; C: 4; D: 5; E: 2
[B] A: 1; B: 2; C: 3; D: 4; E: 5
[C] A: 1; B: 2; C: 4; D: 3; E: 5
[D] A: 4; B: 2; C: 3; D: 5; E: 1

56. Match the list I with the list II and select the correct answer using the codes given below the lists

List I [Terms]	List II [Meaning]
A. Caryopsis	1. Superior, one celled, one seeded, indehiscent fruit with a thin dry membranous pericarp, inseparably united with seed as in barley, wheat
B. Catabolism	2. The breaking down in the body of chemical compounds into simpler ones, usually accompanied by the production of heat
C. Catabolic repression	3. Glucose mediated reduction in the rates of transcription of operons that specify enzymes involved in catabolic pathways
D. Catalyst	4. An ion with a positive electric charge
E. Cation	5. A substance that accelerates a chemical reaction

[A] A: 1; B: 3; C: 4; D: 5; E: 2
[B] A: 1; B: 2; C: 3; D: 4; E: 5
[C] A: 4; B: 1; C: 2; D: 3; E: 5
[D] A: 1; B: 2; C: 3; D: 5; E: 4

57. Match the list I with the list II and select the correct answer using the codes given below the lists

List I [Terms]	List II [Meaning]
A. Causal agent	1. The organism or condition directly responsible for causing a disease
B. Cell	2. Unit
C. Cell Wall	3. Outermost covering
D. Central leader	4. System of training fruit tree in which it is trained to from a trunk extending from the soil to the top of the tree
E. Charles law	5. Volume of the gas varies directly as the absolute temperature, if the pressure is kept constant

[A] A: 1; B: 3; C: 4; D: 5; E: 2 [B] A: 1; B: 2; C: 3; D: 4; E: 5
[C] A: 4; B: 1; C: 2; D: 3; E: 5 [D] A: 4; B: 2; C: 3; D: 5; E: 1

58. Match the list I with the list II and select the correct answer using the codes given below the lists

List I [Terms]	List II [Meaning]
A. Chasmogamy	1. Pollination
B. Chaubatia paste	2. Fungicide
C. Chelates	3. Metal containing organic compounds especially useful in supplying minerals to plants
D. Chemotropism	4. Related to the concentration of chemical substances
E. Chilling injury	5. Impaired viability caused by exposure to low temperature

[A] A: 1; B: 3; C: 4; D: 5; E: 2 [B] A: 1; B: 2; C: 3; D: 4; E: 5
[C] A: 4; B: 1; C: 2; D: 3; E: 5 [D] A: 4; B: 2; C: 3; D: 5; E: 1

59. Match the list I with the list II and select the correct answer using the codes given below the lists

List I [Terms]	List II [Meaning]
A. Chimera	1. Hybrid
B. Chip budding	2. A single bud with a large piece of bud
C. Chi-square	3. Test
D. Chlorophyll	4. Pigment
E. Chloroplast	5. Organelle

[A] A: 2; B: 1; C: 3; D: 4; E: 5 [B] A: 1; B: 2; C: 3; D: 4; E: 5
[C] A: 4; B: 1; C: 2; D: 3; E: 5 [D] A: 4; B: 2; C: 3; D: 5; E: 1

60. Match the list I with the list II and select the correct answer using the codes given below the lists

	List I [Terms]		List II [Meaning]
A.	Chlorosis	1.	Yellowing
B.	Chromosome	2.	Thread like structure
C.	Cistron	3.	A unit function in DNA
D.	Citric Acid Cycle	4.	Metabolic pathway
E.	Climatric peak	5.	The maximum point of respiration rate of mature fruit

[A] A: 1; B: 3; C: 4; D: 5; E: 2 [B] A: 1; B: 2; C: 3; D: 4; E: 5
[C] A: 4; B: 1; C: 2; D: 3; E: 5 [D] A: 4; B: 2; C: 3; D: 5; E: 1

61. Match the list I with the list II and select the correct answer using the codes given below the lists

	List I [Terms]		List II [Meaning]
A.	Clone	1.	A group of individuals of common ancestor
B.	Coenzyme	2.	Non protein and low molecular weight
C.	Cohesion theory	3.	Ascent of sap
D.	Coleoptile	4.	Pineapple
E.	Collective fruit	5.	Tubular structure

[A] A: 1; B: 3; C: 4; D: 5; E: 2 [B] A: 1; B: 2; C: 3; D: 4; E: 5
[C] A: 1; B: 2; C: 3; D: 5; E: 4 [D] A: 4; B: 2; C: 3; D: 5; E: 1

62. Match the list I with the list II and select the correct answer using the codes given below the lists

	List I [Terms]		List II [Meaning]
A.	Colloid	1.	A state of a substance where the units are very large molecule or molecule complex
B.	Colloidal	2.	Starch
C.	Companion crops	3.	Two non-competitive crops grown in the same area at the same time, of which, one is of short-term and the other is of long duration
D.	Compatible	4.	The ability of sex cells to units and form a fertilized egg that can grow to maturity
E.	Compatibility	5.	Varieties that will set fruit when cross-pollinated or make a successful graft union when inter-grafted

[A] A: 1; B: 3; C: 4; D: 5; E: 2 [B] A: 1; B: 2; C: 3; D: 4; E: 5
[C] A: 4; B: 1; C: 2; D: 3; E: 5 [D] A: 1; B: 2; C: 3; D: 5; E: 4

63. Match the list I with the list II and select the correct answer using the codes given below the lists

	List I [Terms]		List II [Meaning]
A.	Compensation point	1.	It is the light intensity, at which the photosynthetic intake of carbon dioxide is equal to the respiratory output of carbon dioxide
B.	Compost	2.	C:N ratio 10:1
C.	Conduction	3.	Act of moving water through the xylem in plant organs
D.	Conjugated protein	4.	Kills a plant primarily
E.	Contact herbicide	5.	Holoenzyme

[A] A: 1; B: 3; C: 4; D: 5; E: 2 [B] A: 1; B: 2; C: 3; D: 4; E: 5
[C] A: 4; B: 1; C: 2; D: 3; E: 5 [D] A: 1; B: 2; C: 3; D: 5; E: 4

64. Match the list I with the list II and select the correct answer using the codes given below the lists

	List I [Terms]		List II [Meaning]
A.	Cardon	1.	A disorder
B.	Corky core	2.	A system of training grape vine where the trunk is not developed straight but is trained in a horizontal direction along the lower wire of the trellis at a height of 60 cm
C.	Corm	3.	Structure similar to the bulbs in appearance
D.	Cornmeal	4.	Small corms produced around the base of the mother corm
E.	Corolla	5.	The petals of the flower as a group

[A] A: 1; B: 3; C: 4; D: 5; E: 2 [B] A: 1; B: 2; C: 3; D: 4; E: 5
[C] A: 2; B: 1; C: 3; D: 4; E: 5 [D] A: 4; B: 2; C: 3; D: 5; E: 1

65. Match the list I with the list II and select the correct answer using the codes given below the lists

	List I [Terms]		List II [Meaning]
A.	Colour	1.	Premature falling of flower
B.	Crazy top	2.	Citrus crop and alkaline soil
C.	Crinkle	3.	A definite sequence
D.	Crop rotation	4.	A disorder
E.	Cross pollination	5.	Materials for the study of variations

[A] A: 1; B: 3; C: 4; D: 5; E: 2 [B] A: 1; B: 2; C: 3; D: 4; E: 5
[C] A: 4; B: 1; C: 2; D: 3; E: 5 [D] A: 1; B: 2; C: 4; D: 3; E: 5

66. Match the list I with the list II and select the correct answer using the codes given below the lists

	List I [Terms]		List II [Meaning]
A.	Grapes	1.	The failure of the blossoms to set, resulting in a premature drop
B.	Crazy top	2.	A disorder of citrus trees associated with an alkaline soil reaction and resulting in peculiar type of growth
C.	Crinkle	3.	A disorder
D.	Crop rotation	4.	Growing of several different crops in a definite sequence, avoid continuous cropping with the same crop
E.	Cross pollination	5.	Materials for the study of variations

[A] A: 1; B: 3; C: 4; D: 5; E: 2 [B] A: 1; B: 2; C: 3; D: 4; E: 5
[C] A: 4; B: 1; C: 2; D: 3; E: 5 [D] A: 4; B: 2; C: 3; D: 5; E: 1

67. Match the list I with the list II and select the correct answer using the codes given below the lists

	List I [Terms]		List II [Meaning]
A.	Crown	1.	Wax
B.	Cuticle	2.	Perennial top
C.	Cutin	3.	Long chain of hydroxyl fatty acids
D.	Cuttage	4.	Propagation
E.	Cutting	5.	Dethatched part of plant

[A] A: 2; B: 1; C: 3; D: 4; E: 5 [B] A: 1; B: 2; C: 3; D: 4; E: 5
[C] A: 4; B: 1; C: 2; D: 3; E: 5 [D] A: 4; B: 2; C: 3; D: 5; E: 1

68. Match the list I with the list II and select the correct answer using the codes given below the lists

	List I [Terms]		List II [Meaning]
A.	Cytochrome	1.	Cell division
B.	Cytokinin	2.	Mitochondria
C.	Cytoplasm	3.	Protoplasm excluding the nucleus
D.	Day neutral plant	4.	Flowering
E.	Dead seeds	5.	Seeds which, at the end of test period, are neither hard nor fresh and have not produced seedlings, are classified as dead seeds

[A] A: 2; B: 1; C: 3; D: 4; E: 5 [B] A: 1; B: 2; C: 3; D: 4; E: 5
[C] A: 4; B: 1; C: 2; D: 3; E: 5 [D] A: 4; B: 2; C: 3; D: 5; E: 1

69. Match the list I with the list II and select the correct answer using the codes given below the lists

List I [Terms]		List II [Meaning]
A. Deciduous	1.	A chemical, or method of treatment, that causes only the leaves of a plant to fall or abscise
B. Defoliants	2.	Joint assay with Darwin
C. Denaturation	3.	Originally properties are greatly changed
D. Dendrograph	4.	Trunk diameter of tree
E. Dichogamy	5.	Maturity

[A] A: 2; B: 1; C: 3; D: 4; E: 5 [B] A: 1; B: 2; C: 3; D: 4; E: 5
[C] A: 4; B: 1; C: 2; D: 3; E: 5 [D] A: 4; B: 2; C: 3; D: 5; E: 1

70. Match the list I with the list II and select the correct answer using the codes given below the lists

List I [Terms]		List II [Meaning]
A. Dicliny	1.	Male and female organs separate and present in different flowers
B. Dicot	2.	Apple
C. Differentiation	3.	The process of change in composition, structure and functions of cells and tissues during growth
D. Diffusion	4.	Random migration
E. Dimorphism	5.	Presenting two forms as long and short constituents by enzymes

[A] A: 1; B: 3; C: 4; D: 5; E: 2 [B] A: 1; B: 2; C: 3; D: 4; E: 5
[C] A: 4; B: 1; C: 2; D: 3; E: 5 [D] A: 4; B: 2; C: 3; D: 5; E: 1

71. Match the list I with the list II and select the correct answer using the codes given below the lists

List I [Terms]		List II [Meaning]
A. Dioecious	1.	Unisexual
B. Diploidization	2.	Polyploidy show diploid like behavior
C. Diplospory	3.	A form of apomixes in which the embryo sac develops from the megaspore
D. Disbudding	4.	Removal of flowers or shoot buds
E. Dormancy	5.	Resting condition

[A] A: 1; B: 3; C: 4; D: 5; E: 2 [B] A: 1; B: 2; C: 3; D: 4; E: 5
[C] A: 4; B: 1; C: 2; D: 3; E: 5 [D] A: 4; B: 2; C: 3; D: 5; E: 1

72. Match the list I with the list II and select the correct answer using the codes given below the lists

List I [Terms]	List II [Meaning]
A. Dormant	1. Reduced physiological state
B. Double fertilization	2. Endosperm
C. Double reduction	3. The phenomena in which sister chromatids or their segments pass into the same gametes
D. Drought	4. A period of dryness
E. Drought avoidance	5. Ability of plants to maintain a favorable internal water balance under moisture stress

[A] A: 1; B: 3; C: 4; D: 5; E: 2 [B] A: 4; B: 1; C: 2; D: 3; E: 5
[C] A: 1; B: 2; C: 3; D: 4; E: 5 [D] A: 4; B: 2; C: 3; D: 5; E: 1

73. Match the list I with the list II and select the correct answer using the codes given below the lists

List I [Terms]	List II [Meaning]
A. Drought hardening	1. Drought tolerance
B. Drought resistance	2. The ability of a plant to show a minimum loss in yield under water deficit situation in yield and quality of the product
C. Drought tolerance	3. Ability of plant to withstand low tissue water content
D. Drupe	4. A stone fruit in which the seed is encased in a hard shell inside fleshy tissue
E. Dwarfed	5. Restricted plant size with our loss of health and vigour

[A] A: 1; B: 3; C: 4; D: 5; E: 2 [B] A: 1; B: 2; C: 3; D: 4; E: 5
[C] A: 4; B: 1; C: 2; D: 3; E: 5 [D] A: 1; B: 2; C: 3; D: 4; E: 5

74. Match the list I with the list II and select the correct answer using the codes given below the lists

List I [Terms]	List II [Meaning]
A. Early wood	1. Large cells and thin walled
B. Ecology	2. Study of the interrelationship of organisms
C. Ecotype	3. Specific environment
D. Edaphic	4. Soil
E. Elaioplast	5. Oil is predominant

[A] A: 1; B: 3; C: 4; D: 5; E: 2 [B] A: 4; B: 1; C: 2; D: 3; E: 5
[C] A: 1; B: 2; C: 3; D: 4; E: 5 [D] A: 4; B: 2; C: 3; D: 5; E: 1

75. Match the list I with the list II and select the correct answer using the codes given below the lists

List I [Terms]		List II [Meaning]
A. Electrophoresis	1.	Movement of colloids
B. Elliptical	2.	Narrows, tapering
C. Emasculation	3.	Artificial removal
D. Embryo	4.	Miniature of plant
E. Embryo culture	5.	Growing seedling from excised embryos

[A] A: 1; B: 3; C: 4; D: 5; E: 2 [B] A: 4; B: 1; C: 2; D: 3; E: 5
[C] A: 1; B: 2; C: 3; D: 4; E: 5 [D] A: 4; B: 2; C: 3; D: 5; E: 1

76. Match the list I with the list II and select the correct answer using the codes given below the lists

List I [Terms]		List II [Meaning]
A. Embryo sac	1.	The cell in the ovule in which the embryo is formed
B. Emulsion	2.	One liquid is dispersed throughout another
C. Endocarp	3.	The inner layer of the wall of a fructified ovary
D. Endodermis	4.	The layer of the living cells
E. Endosmosis	5.	If external water is at the higher chemical potential

[A] A: 1; B: 3; C: 4; D: 5; E: 2 [B] A: 4; B: 1; C: 2; D: 3; E: 5
[C] A: 1; B: 2; C: 3; D: 4; E: 5 [D] A: 4; B: 2; C: 3; D: 5; E: 1

77. Match the list I with the list II and select the correct answer using the codes given below the lists

List I [Terms]		List II [Meaning]
A. Endosperm	1.	The nutritive material stored within the cells
B. Environment	2.	Sum total of external condition, which influences the expression of genes of an organism
C. Enzyme	3.	Accelerates specific changes
D. Enzyme engineering	4.	Enhancing the activity or changing the substrate specificity of enzymes by changing their amino acid sequences
E. Enzyme technology	5.	Immobilization of enzymes in different matrices with a view to enhance their efficiency, reusability or hygiene

[A] A: 1; B: 3; C: 4; D: 5; E: 2 [B] A: 1; B: 2; C: 3; D: 4; E: 5
[C] A: 4; B: 1; C: 2; D: 3; E: 5 [D] A: 4; B: 2; C: 3; D: 5; E: 1

78. Match the list I with the list II and select the correct answer using the codes given below the lists

List I [Terms]	List II [Meaning]
A. Epicotyl	1. The upper portion of the embryo axis or seedling above the cotyledons and below the first true leaves
B. Epidermis	2. A superficial layer of cells
C. Epinasty	3. Causes leaf to curl downward
D. Episome	4. Fertility factor (F) in *E. coli*
E. Epistasis	5. The interaction before different genes affecting the same characters

[A] A: 1; B: 3; C: 4; D: 5; E: 2 [B] A: 1; B: 2; C: 3; D: 4; E: 5
[C] A: 4; B: 1; C: 2; D: 3; E: 5 [D] A: 4; B: 2; C: 3; D: 5; E: 1

79. Match the list I with the list II and select the correct answer using the codes given below the lists

List I [Terms]	List II [Meaning]
A. Equilibrium moisture content	1. The moisture content of the product when it is in equilibrium with the surrounding atmosphere is called the equilibrium moisture content or hygroscopic equilibrium
B. Erosion	2. Wind and moving water
C. Espalier	3. Training of fruit tree planted against the wall in which the main branches are trained in a plane parallel to the walls in a geometric design
D. Etiolation	4. Increased stem elongation
E. Eugenics	5. It deals with the application of the principle of heredity for the betterment of human race

[A] A: 1; B: 3; C: 4; D: 5; E: 2 [B] A: 1; B: 2; C: 3; D: 4; E: 5
[C] A: 4; B: 1; C: 2; D: 3; E: 5 [D] A: 4; B: 2; C: 3; D: 5; E: 1

80. Match the list I with the list II and select the correct answer using the codes given below the lists

List I [Terms]	List II [Significance]
A. Euphenics	1. An individual having a chromosome number that is an exact multiple of the basic chromosome number
B. Euploid	2. Control of the hereditary disease

C.	Excising	3.	Removing or extracting as an embryo from a seed or ovule
D.	Exocarp	4.	The outer layer of the wall of fructified ovary
E.	Exocortis	5.	Shelling of the bark of tree

[A] A: 2; B: 1; C: 3; D: 4; E: 5 [B] A: 1; B: 2; C: 3; D: 4; E: 5
[C] A: 4; B: 1; C: 2; D: 3; E: 5 [D] A: 4; B: 2; C: 3; D: 5; E: 1

81. Match the list I with the list II and select the correct answer using the codes given below the lists

	List I [Terms]		List II [Significance]
A.	Fascicular cambium	1.	The fusion of the two gametes to form a new cells
B.	Fecundation	2.	Vascular cambium that arises within a vascular bundle
C.	Fecundity	3.	The ability of the flower to produce seeds that will germinate
D.	Feedback inhibition	4.	Accumulated end product of a biochemical pathway stop synthesis of that product
E.	Feeder roots	5.	Fine roots an root branches with a large absorbing area

[A] A: 2; B: 1; C: 3; D: 4; E: 5 [B] A: 1; B: 2; C: 3; D: 4; E: 5
[C] A: 4; B: 1; C: 2; D: 3; E: 5 [D] A: 4; B: 2; C: 3; D: 5; E: 1

82. Match the list I with the list II and select the correct answer using the codes given below the lists

	List I [Terms]		List II [Significance]
A.	Amylose	1.	The stage of mitosis during which the two chromatids of each replicated chromosome are separated and move toward opposite poles
B.	Anaphase	2.	200-2000 glucose units
C.	Anaphase A	3.	Early anaphase, during which the sister chromatids separate and begin to move toward opposite poles
D.	Anaphase B	4.	The sister chromosomes are pushed to their respective poles
E.	Anaphase-promoting complex	5.	Controls the proteasomal destruction of cyclin proteins

[A] A: 2; B: 1; C: 3; D: 4; E: 5 [B] A: 1; B: 2; C: 3; D: 4; E: 5
[C] A: 4; B: 1; C: 2; D: 3; E: 5 [D] A: 4; B: 2; C: 3; D: 5; E: 1

83. Match the list I with the list II and select the correct answer using the codes given below the lists

List I [Terms]	List II [Significance]
A. Aminotransferases	1. Transamination
B. Amphipathic	2. Dual in nature
C. Amplitude	3. Biological rhythm
D. Amylopectin	4. Branched structure
E. Amyloplast	5. Root cap

[A] A: 1; B: 3; C: 4; D: 5; E: 2 [B] A: 1; B: 2; C: 3; D: 4; E: 5
[C] A: 4; B: 1; C: 2; D: 3; E: 5 [D] A: 4; B: 2; C: 3; D: 5; E: 1

84. Match the list I with the list II and select the correct answer using the codes given below the lists

List I [Terms]	List II [Meaning]
A. Auxin	1. A compound with biological activities
B. AEIs	2. Auxin efflux inhibitors
C. Auxin influx inhibitor	3. Synthetic substance
D. AuxRDs	4. Auxin response domains
E. AuxREs	5. Auxin response elements

[A] A: 1; B: 3; C: 4; D: 5; E: 2 [B] A: 1; B: 2; C: 3; D: 4; E: 5
[C] A: 4; B: 1; C: 2; D: 3; E: 5 [D] A: 4; B: 2; C: 3; D: 5; E: 1

85. Match the list I with the list II and select the correct answer using the codes given below the lists

List I [Terms]	List II [Significance]
A. BRs	1. Brassinosteroids
B. CAAT box	2. Initiation of transcriptional factors
C. Cadastral genes	3. Genes that act as spatial regulators of floral organ identity gene
D. CBLs	4. Calcineurin B-like proteins
E. Callose	5. Synthesized in plasmamembrane

[A] A: 1; B: 3; C: 4; D: 5; E: 2 [B] A: 1; B: 2; C: 3; D: 4; E: 5
[C] A: 4; B: 1; C: 2; D: 3; E: 5 [D] A: 4; B: 2; C: 3; D: 5; E: 1

86. Match the list I with the list II and select the correct answer using the codes given below the lists

List I [Terms]	List II [Significance]
A. Cambium	1. Layers of meristematic cells
B. CAM plant	2. Succulent

List I [Terms]		List II [Significance]
C. Campestrol	3.	The sterol progenitor of brassinosteroids, ultimately derived from cycloartenol
D. Capillarity	4.	Occurring in the stroma of the chloroplast
E. Carbon fixation reaction	5.	Due to cohesion, adhesion and surface tension

[A] A: 1; B: 2; C: 3; D: 5; E: 4 [B] A: 1; B: 2; C: 3; D: 4; E: 5
[C] A: 4; B: 1; C: 2; D: 3; E: 5 [D] A: 4; B: 2; C: 3; D: 5; E: 1

87. Match the list I with the list II and select the correct answer using the codes given below the lists

List I [Terms]		List II [Significance]
A. Veins	1.	Intricate network of leaf vascular bundles
B. Velocity	2.	Sieve element and photosynthetic assimilates
C. Vernalization	3.	Cold treatment
D. VLFRs	4.	*Glomus* spp.
E. VAM	5.	Very low-fluence responses

[A] A: 1; B: 2; C: 3; D: 5; E: 4 [B] A: 1; B: 2; C: 3; D: 4; E: 5
[C] A: 4; B: 1; C: 2; D: 3; E: 5 [D] A: 4; B: 2; C: 3; D: 5; E: 1

88. Match the list I with the list II and select the correct answer using the codes given below the lists

List I [Terms]		List II [Significance]
A. Waxes	1.	Hydrophobic lipids
B. Wetlands	2.	Land that is saturated with water on a regular basis
C. Whorl	3.	Organs around the flanks of meristem
D. Wilting	4.	An abnormal growth of shoots caused by infection with various pathogens
E. Witches's broom	5.	Loss of rigidity

[A] A: 1; B: 2; C: 3; D: 5; E: 4 [B] A: 1; B: 2; C: 3; D: 4; E: 5
[C] A: 4; B: 1; C: 2; D: 3; E: 5 [D] A: 4; B: 2; C: 3; D: 5; E: 1

89. Match the list I with the list II and select the correct answer using the codes given below the lists

	List I [Terms]		List II [Significance]
A.	Vascular cambium	1.	Consisting of fusiform and ray stem cells
B.	Vascular tissue	2.	Chloroplast and Mitochondria
C.	Vegetative segregation	3.	Specialized for transportation
D.	Vesicles	4.	Spherical, membrane-enclosed compartment
E.	Volatiles	5.	Ambient temperature

[A] A: 1; B: 3; C: 4; D: 5; E: 2 [B] A: 1; B: 2; C: 3; D: 4; E: 5
[C] A: 1; B: 3; C: 2; D: 4; E: 5 [D] A: 4; B: 2; C: 3; D: 5; E: 1

90. Match the list I with the list II and select the correct answer using the codes given below the lists

	List I [Terms]		List II [Significance]
A.	Toc	1.	Group of clock mutants in *Arabidopsis*
B.	Tonoplast	2.	A vascular membrane
C.	Torpedo stage embryo	3.	Third stage of embryogenesis
D.	Torus	4.	Gymnosperm
E.	Totipotency	5.	In differentiated cells

[A] A: 1; B: 3; C: 4; D: 5; E: 2 [B] A: 1; B: 2; C: 3; D: 4; E: 5
[C] A: 4; B: 1; C: 2; D: 3; E: 5 [D] A: 4; B: 2; C: 3; D: 5; E: 1

91. Match the list I with the list II and select the correct answer using the codes given below the lists

	List I [Terms]		List II [Significance]
A.	Tannins	1.	Phenolic polymers
B.	Taproot	2.	Single root axis from which lateral roots develop
C.	TATA box	3.	25 to 35 bp upstream
D.	T-DNA	4.	Ti plasmid
E.	Telomere	5.	Repetitive DNA

[A] A: 1; B: 3; C: 4; D: 5; E: 2 [B] A: 1; B: 2; C: 3; D: 4; E: 5
[C] A: 4; B: 1; C: 2; D: 3; E: 5 [D] A: 4; B: 2; C: 3; D: 5; E: 1

92. Match the list I with the list II and select the correct answer using the codes given below the lists

List I [Terms]	List II [Significance]
A. Tension	1. Refers to the partial pressure of the gas
B. Teratomas	2. A large class of plant compounds formed from five-carbon isoprene units, many of which are secondary metabolites with anti-herbivore activity
C. Terpenes	3. Tumors that contain partially developed structure
D. Terpenoids	4. Isoprenoids
E. Tertiary cross-regulation	5. Involves the outputs of the two distinct signaling pathways exerting influences on one another

[A] A: 1; B: 3; C: 4; D: 5; E: 2 [B] A: 1; B: 2; C: 3; D: 4; E: 5
[C] A: 1; B: 3; C: 2; D: 4; E: 5 [D] A: 4; B: 2; C: 3; D: 5; E: 1

93. Match the list I with the list II and select the correct answer using the codes given below the lists

List I [Terms]	List II [Significance]
A. TIBA	1. Inhibitor of polar auxin transport
B. Ti plasmid	2. Family of F box soluble protein
C. TIR1/AFB proteins	3. Virulent strain
D. tir3 mutant	4. Reduced polar auxin transport
E. tmr locus	5. Rooty mutation

[A] A: 1; B: 3; C: 4; D: 5; E: 2 [B] A: 1; B: 2; C: 3; D: 4; E: 5
[C] A: 1; B: 3; C: 2; D: 4; E: 5 [D] A: 4; B: 2; C: 3; D: 5; E: 1

94. Match the list I with the list II and select the correct answer using the codes given below the lists

List I [Terms]	List II [Significance]
A. Toxic zone	1. The range of the nutrient concentrations in excess of the adequate zone and where growth or yield decline
B. Toxicity effects	2. Face where vesicles and tubules are produced
C. trans Golgi	3. Caused by the high concentration of ions
D. trans-zeatine	4. Chemically similar to kinetin
E. trans-acting factors	5. Bind to cis-acting sequence

[A] A: 1; B: 3; C: 4; D: 5; E: 2 [B] A: 1; B: 2; C: 3; D: 4; E: 5
[C] A: 1; B: 3; C: 2; D: 4; E: 5 [D] A: 4; B: 2; C: 3; D: 5; E: 1

95. Match the list I with the list II and select the correct answer using the codes given below the lists

	List I [Terms]		List II [Significance]
A.	Stress tolerance/ Stress resistance	1.	A plant's ability
B.	Stroma	2.	Carbon fixation and reduction reaction
C.	Stroma reaction	3.	Fluid component
D.	Stromules	4.	Protrusions of the inner and outer membranes of the chloroplast
E.	Suberin	5.	A wax-like lipid polymer similar to cutin

[A] A: 1; B: 3; C: 4; D: 5; E: 2 [B] A: 1; B: 2; C: 3; D: 4; E: 5
[C] A: 1; B: 3; C: 2; D: 4; E: 5 [D] A: 4; B: 2; C: 3; D: 5; E: 1

96. Match the list I with the list II and select the correct answer using the codes given below the lists

	List I [Terms]		List II [Significance]
A.	Subfunctionalization	1.	The series by which evolution acts on duplicate genes, causing one copy to be either lost or changed in function, while the other retains its original function
B.	Subjective day	2.	When a organism is positioned in total darkness, the phase of the rhythum that coincides with the light period of a preceding light/dark cycle
C.	Subjective night	3.	When a organism is located in total darkness, the phase of the rhythum that coincides with the dark period of a preceding light/dark cycle
D.	Subsidiary cells	4.	Specialized epidermal cells
E.	Substomatal cavity	5.	The air space within the leaf

[A] A: 1; B: 3; C: 4; D: 5; E: 2 [B] A: 1; B: 2; C: 3; D: 4; E: 5
[C] A: 4; B: 1; C: 2; D: 3; E: 5 [D] A: 4; B: 2; C: 3; D: 5; E: 1

97. Match the list I with the list II and select the correct answer using the codes given below the lists

	List I [Terms]		List II [Significance]
A.	Q-cycle	1.	A mechanism for oxidation of plastohydroquinone in chloroplasts and ubihydroquinone in mitochondria
B.	Radical	2.	First organ to emerge on germination

C.	QC	3.	Cells donot divide
D.	Pulvinus	4.	Turgor-driven organ
E.	Pyrethroids	5.	Monoterpene

[A] A: 1; B: 3; C: 4; D: 5; E: 2 [B] A: 1; B: 2; C: 3; D: 4; E: 5
[C] A: 4; B: 1; C: 2; D: 3; E: 5 [D] A: 4; B: 2; C: 3; D: 5; E: 1

98. Match the list I with the list II and select the correct answer using the codes given below the lists

	List I [Terms]		List II [Significance]
A.	Quantum yield	1.	The ration of the yield of a particular product of a photochemical process to the total number of quanta absorbed
B.	Quenching	2.	The process by which energy stored in light excited chlorophyll is rapidly dissipated mainly by excitation transfer or photochemistry
C.	Quercetin	3.	An endogenous flavonoid compound that may inhibit and regulate auxin transport by blocking auxin efflux
D.	Rejuvenation	4.	The reversion of adult shoots to juvenile shoots
E.	Quantum efficiency	5.	Photosynthetic yield per photon flux of absorbed light

[A] A: 1; B: 3; C: 4; D: 5; E: 2 [B] A: 1; B: 2; C: 3; D: 4; E: 5
[C] A: 4; B: 1; C: 2; D: 3; E: 5 [D] A: 4; B: 2; C: 3; D: 5; E: 1

99. Match the list I with the list II and select the correct answer using the codes given below the lists

	List I [Terms]		List II [Significance]
A.	R genes	1.	Resistance genes that function in plant defense against fungi, bacteria and nematode in some cases by encoding protein receptors that bind to specific pathogens molecules, elicitors
B.	Rabs	2.	For selective fusion and fission of vesicles and tubules within the endo-membrane system
C.	Radial axis	3.	Studies in the Theory of Descent
D.	*RAN1*	4.	General Morphology of Organism
E.	Ray stem cells	5.	Materials for the study of variations

[A] A: 1; B: 3; C: 4; D: 5; E: 2 [B] A: 1; B: 2; C: 3; D: 4; E: 5
[C] A: 4; B: 1; C: 2; D: 3; E: 5 [D] A: 4; B: 2; C: 3; D: 5; E: 1

100. Match the list I with the list II and select the correct answer using the codes given below the lists

	List I [Terms]		List II [Significance]
A.	Prokaryotic pathways	1.	Elaborate semi-crystalline lattices of membrane tubules that develop in plastids that have not been exposed to light
B.	Prolamellar bodies	2.	In the chloroplast, the series of reactions for the synthesis of glycerolipids
C.	Promeristem	3.	That part of the shoot or root apical meristem containing the stem cells and their immediate derivatives which have not yet begun to differentiate
D.	Prometaphase	4.	Early metaphase stage in which the preprophase band disassembles and near microtubules polymerizes to form the mitotic spindle
E.	Promoter	5.	The region of the gene that binds RNA polymerase

[A] A: 2; B: 1; C: 3; D: 4; E: 5 [B] A: 1; B: 2; C: 3; D: 4; E: 5
[C] A: 4; B: 1; C: 2; D: 3; E: 5 [D] A: 4; B: 2; C: 3; D: 5; E: 1

101. Match the list I with the list II and select the correct answer using the codes given below the lists

	List I [Terms]		List II [Significance]
A.	Castasterone	1.	Peroxisomes
B.	Catalase	2.	Immediate precursor of brassinolide
C.	Cation exchange	3.	The substitution of mineral cations adsorbed to the surface of soil particles by supplementary cations
D.	Cavitation	4.	Sudden fall down of tension in a column of water resulting from the indefinite expansion of a tiny gas bubble
E.	CCCP	5.	Ionophore

[A] A: 2; B: 1; C: 3; D: 4; E: 5 [B] A: 1; B: 2; C: 3; D: 4; E: 5
[C] A: 4; B: 1; C: 2; D: 3; E: 5 [D] A: 4; B: 2; C: 3; D: 5; E: 1

102. Match the list I with the list II and select the correct answer using the codes given below the lists

List I [Terms]	List II [Significance]
A. Channels	1. Cell cycle
B. Checkpoint	2. Transmembrane protein
C. Chelates	3. Can form a complex with divalent cations
D. Chelator	4. Carbon compound that can form a non covalent complex
E. CF0-CF1 ATPase	5. Multi-protein complex

[A] A: 2; B: 1; C: 3; D: 4; E: 5 [B] A: 1; B: 2; C: 3; D: 4; E: 5
[C] A: 4; B: 1; C: 2; D: 3; E: 5 [D] A: 4; B: 2; C: 3; D: 5; E: 1

103. Match the list I with the list II and select the correct answer using the codes given below the lists

List I [Terms]	List II [Significance]
A. Carpels	1. Active or passive
B. Carrier-mediated transport	2. Develops into the fruit
C. Carriers	3. Membrane-transport protein
D. Cascade	4. A series of interaction that involves an amplification of the initial action
E. Casparian strip	5. Band in the cell walls of endodermis

[A] A: 2; B: 1; C: 3; D: 4; E: 5 [B] A: 1; B: 2; C: 3; D: 4; E: 5
[C] A: 4; B: 1; C: 2; D: 3; E: 5 [D] A: 4; B: 2; C: 3; D: 5; E: 1

104. Match the list I with the list II and select the correct answer using the codes given below the lists

List I [Terms]	List II [Significance]
A. Cell plate	1. Cell division
B. Cell wall	2. Rigid cell surface structure
C. Cell wall matrix	3. Structural protein
D. Cellobiose	4. Cellulose
E. Cellular dehydration	5. Reduction in the cell volume

[A] A: 1; B: 3; C: 4; D: 5; E: 2 [B] A: 1; B: 2; C: 3; D: 4; E: 5
[C] A: 4; B: 1; C: 2; D: 3; E: 5 [D] A: 4; B: 2; C: 3; D: 5; E: 1

105. Match the list I with the list II and select the correct answer using the codes given below the lists

List I [Terms]	List II [Significance]
A. Cellulose microfibril	1. Indeterminate length and variable width
B. Cellulose synthase	2. 1-4 linked β-D-glucose
C. Central zone	3. Quiescent center of root meristem
D. Centromere	4. Mitotic chromosome
E. Cereal grains	5. Seeds of grasses

[A] A: 1; B: 3; C: 4; D: 5; E: 2 [B] A: 1; B: 2; C: 3; D: 4; E: 5
[C] A: 4; B: 1; C: 2; D: 3; E: 5 [D] A: 4; B: 2; C: 3; D: 5; E: 1

106. Match the list I with the list II and select the correct answer using the codes given below the lists

List I [Terms]	List II [Significance]
A. Chilling	1. Leaf discoloration
B. Chilling injury	2. Can be used to release certain seeds from dormancy
C. Chlorophyte	3. Unicellular
D. Chloroplast	4. Avenue of photosynthesis
E. Chloroplast genome	5. 5 to 10 % of the cellular DNA

[A] A: 2; B: 1; C: 3; D: 4; E: 5 [B] A: 1; B: 2; C: 3; D: 4; E: 5
[C] A: 4; B: 1; C: 2; D: 3; E: 5 [D] A: 4; B: 2; C: 3; D: 5; E: 1

107. Match the list I with the list II and select the correct answer using the codes given below the lists

List I [Terms]	List II [Significance]
A. Chronic photoinhibition	1. Quantum efficiency and the maximum rate of photosynthesis are decreased
B. Chronobiology	2. Biological clock
C. Cinnamic acid	3. Phenol compounds
D. Circadian resonance	4. Match between a plant's circadian rhythms and the light-dark cycle of the environment
E. Circadian rhythum	5. Biological activity

[A] A: 1; B: 3; C: 4; D: 5; E: 2 [B] A: 1; B: 2; C: 3; D: 4; E: 5
[C] A: 4; B: 1; C: 2; D: 3; E: 5 [D] A: 4; B: 2; C: 3; D: 5; E: 1

108. Match the list I with the list II and select the correct answer using the codes given below the lists

	List I [Terms]		List II [Significance]
A.	Chromatin	1.	Nucleosomes wrapped with DNA
B.	30 nm chromatin fiber	2.	The DNA -protein complex found in the interphase nucleus
C.	Chromatin remodeling	3.	Epigenetic factors
D.	Chromophore	4.	A light-absorbing pigment molecules that is usually bound to a protein
E.	Chromoplasts	5.	High concentration of carotenoids

[A] A: 2; B: 1; C: 3; D: 4; E: 5 [B] A: 1; B: 2; C: 3; D: 4; E: 5
[C] A: 4; B: 1; C: 2; D: 3; E: 5 [D] A: 4; B: 2; C: 3; D: 5; E: 1

109. Match the list I with the list II and select the correct answer using the codes given below the lists

	List I [Terms]		List II [Significance]
A.	d8 mutant	1.	Maize
B.	DELLA domain	2.	17 amino acids
C.	Denitrification	3.	Anaerobic microbs
D.	Depolarization	4.	Activation of anion channels
E.	DICER-LIKE 1	5.	Plant nuclear protein

[A] A: 1; B: 3; C: 4; D: 5; E: 2 [B] A: 1; B: 2; C: 3; D: 4; E: 5
[C] A: 4; B: 1; C: 2; D: 3; E: 5 [D] A: 4; B: 2; C: 3; D: 5; E: 1

110. Match the list I with the list II and select the correct answer using the codes given below the lists

	List I [Terms]		List II [Significance]
A.	Diffuse growth	1.	Expansion occurs more or less uniformly over the entire surface
B.	Diffusion	2.	The movement of substance due to random thermal agitation from region of high free energy to region of low free energy
C.	Diffusion coefficient	3.	Proportionality constant
D.	Diffusion potential	4.	Result of the differential permeability of solutes with opposite charge
E.	Diffusion resistance	5.	Restriction applied by the boundary layer and the stomata

[A] A: 1; B: 3; C: 4; D: 5; E: 2 [B] A: 1; B: 2; C: 3; D: 4; E: 5
[C] A: 4; B: 1; C: 2; D: 3; E: 5 [D] A: 4; B: 2; C: 3; D: 5; E: 1

111. Match the list I with the list II and select the correct answer using the codes given below the lists

	List I [Terms]		List II [Significance]
A.	Desmotubule	1.	A narrow tubules of the ER that passes through plasmodesmata and connects the ER in adjacent cells
B.	Determinate growth	2.	Inability to grow beyond the maturity
C.	Determination	3.	The process by which a cell acquires a fixed identity or fate
D.	Dextranase	4.	Catalyzes
E.	Diacylglycerol	5.	A molecules consisting of the three-carbon glycerol molecules to which two fatty acids are covalently attached by ester linkage

[A] A: 1; B: 3; C: 4; D: 5; E: 2 [B] A: 1; B: 2; C: 3; D: 4; E: 5
[C] A: 4; B: 1; C: 2; D: 3; E: 5 [D] A: 4; B: 2; C: 3; D: 5; E: 1

112. Match the list I with the list II and select the correct answer using the codes given below the lists

	List I [Terms]		List II [Significance]
A.	Curvature test	1.	Bioassay for the auxin
B.	Cutan	2.	A lipid polymer made up of long-chain hydrocarbons that is a constituent of the cuticle
C.	Cuticle	3.	Coats the outer cell walls of the epidermis and restricts the passage water and gases into and out of the plant
D.	Cutin	4.	A rigid three-dimensional polymer of hydroxyl-bearing fatty acids that are attached to each other by ester linkage
E.	Cyanide	5.	An inhibitor of complex IV and other iron-containing enzymes

[A] A: 1; B: 3; C: 4; D: 5; E: 2 [B] A: 1; B: 2; C: 3; D: 4; E: 5
[C] A: 4; B: 1; C: 2; D: 3; E: 5 [D] A: 4; B: 2; C: 3; D: 5; E: 1

113. Match the list I with the list II and select the correct answer using the codes given below the lists

	List I [Terms]		List II [Significance]
A.	DCMU	1.	New synthesis/ or assembly from simple molecular species
B.	de novo synthesis	2.	Herbicide
C.	de-differentiation	3.	Process by which cells lose their differentiated characteristics

D.	de-embryonate	4.	Removal of the embryo from a cereal seed
E.	Deciduous trees	5.	Trees that fall leaves seasonally

[A] A: 2; B: 1; C: 3; D: 4; E: 5 [B] A: 1; B: 2; C: 3; D: 4; E: 5
[C] A: 4; B: 1; C: 2; D: 3; E: 5 [D] A: 4; B: 2; C: 3; D: 5; E: 1

114. Match the list I with the list II and select the correct answer using the codes given below the lists

	List I [Terms]		List II [Significance]
A.	Domains	1.	Regions within the gene that are similar to regions found in other genes
B.	Dormancy	2.	A living condition in which growth does not occur under conditions that are normally favorable to growth
C.	Drought	3.	A period of insufficient water availability for the plant that results in plant water deficit
D.	Drought avoidance	4.	Capacity of a plant to grow and complete its life cycle during the wet season, before the onset of drought
E.	Drought rhizogenesis	5.	During extended drought stress, the roots of the some plant species may initiate many additional lateral roots but inhibit their growth until the stress is relieved

[A] A: 1; B: 3; C: 4; D: 5; E: 2 [B] A: 1; B: 2; C: 3; D: 4; E: 5
[C] A: 4; B: 1; C: 2; D: 3; E: 5 [D] A: 4; B: 2; C: 3; D: 5; E: 1

115. Match the list I with the list II and select the correct answer using the codes given below the lists

	List I [Terms]		List II [Significance]
A.	E3 ligase	1.	A protein that attaches ubiquitin tags to proteins targeted for destruction by the 26S proteasome
B.	Early endosome	2.	100 nm
C.	Early genes	3.	Primary response
D.	Early response gene	4.	A gene that does not require protein synthesis for its expression and is therefore expressed rapidly in response to a stimulus
E.	Ectotrophic Mycorrhiza fungi	5.	A dense sheath of fungal mycelium around the roots and extending into the surrounding soil; some hyphae may penetrate between, but not into, the root cortical cells

[A] A: 1; B: 3; C: 4; D: 5; E: 2 [B] A: 1; B: 2; C: 3; D: 4; E: 5
[C] A: 4; B: 1; C: 2; D: 3; E: 5 [D] A: 4; B: 2; C: 3; D: 5; E: 1

116. Match the list I with the list II and select the correct answer using the codes given below the lists

	List I [Terms]		List II [Significance]
A.	Electrogenic transport	1.	Active ion transport involving the net movement of charge across a membrane
B.	Electron spin resonance(ESR)	2.	A magnetic resonance technique that detects unpaired electron in molecules
C.	Electronegative	3.	Having the capacity to attract electrons and thus producing a slightly negative electric charge
D.	Endosomes	4.	Early in endocytosis, vesicles that have lost their clathrin coats and moves away from the plasma membrane into the cell interior
E.	Endosymbiosis	5.	Evolutionary origin of the chloroplast and mitochondria through formation of a symbiotic relationship

[A] A: 1; B: 3; C: 4; D: 5; E: 2 [B] A: 1; B: 2; C: 3; D: 4; E: 5
[C] A: 4; B: 1; C: 2; D: 3; E: 5 [D] A: 4; B: 2; C: 3; D: 5; E: 1

117. Match the list I with the list II and select the correct answer using the codes given below the lists

	List I [Terms]		List II [Significance]
A.	Flavin Adenine Dinucleotide (FAD)	1.	A riboflavin-containing cofactor that undergoes a reversible two electron reduction to produce $FADH_2$
B.	Flavin hypothesis	2.	Riboflavin activated by blue light could participate the *in vivo* photo-destruction of an auxin
C.	FMN	3.	Group of ultraviolet light-absorbing protective flavonoids that may also r pollinating insect to flowers
D.	Flavones	4.	Riboflavin-containing cofactor
E.	Flavonids	5.	A large group of plant phenolics with the basic carbon structure of two aromatic rings connected by three carbons.

[A] A: 1; B: 3; C: 4; D: 5; E: 2 [B] A: 1; B: 2; C: 3; D: 4; E: 5
[C] A: 1; B: 2; C: 4; D: 3; E: 5 [D] A: 4; B: 2; C: 3; D: 5; E: 1

118. Match the list I with the list II and select the correct answer using the codes given below the lists

	List I [Terms]		List II [Significance]
A.	Epigneic modifications	1.	Chemical modifications to DNA and histones that cause heritable changes in gene activity without altering the underlying DNA sequence
B.	Epimers	2.	A pair of isomers that differ from each other only in their configuration at a single asymmetric center
C.	Epinasty	3.	For a particular solute, a state or condition in which there is no gradient of chemical potential, and therefore no net passive transport of the solute
D.	Equilibrium	4.	A downward curvature of leaves due to asymmetric growth of the petiole
E.	ER exit sites (ERES)	5.	Specialized sites characterized by the coat protein COPII from which delivery of vesicles to the Golgi occurs

[A] A: 1; B: 3; C: 4; D: 5; E: 2 [B] A: 1; B: 2; C: 3; D: 4; E: 5
[C] A: 1; B: 2; C: 4; D: 3; E: 5 [D] A: 4; B: 2; C: 3; D: 5; E: 1

119. Match the list I with the list II and select the correct answer using the codes given below the lists

	List I [Terms]		List II [Significance]
A.	Etioplast	1.	Photosynthetically inactive form of chloroplast found in etiolated seedlings
B.	Etiolation	2.	The form and growth of seedlings grown in darkness
C.	EMS	3.	Mutagen
D.	etr1	4.	Blocks the response to ethylene
E.	Euchromatin	5.	Transcriptionally active form of chromatin

[A] A: 1; B: 3; C: 4; D: 5; E: 2 [B] A: 1; B: 2; C: 3; D: 4; E: 5
[C] A: 4; B: 1; C: 2; D: 3; E: 5 [D] A: 4; B: 2; C: 3; D: 5; E: 1

120. Match the list I with the list II and select the correct answer using the codes given below the lists

	List I [Terms]		List II [Significance]
A.	Essential oils	1.	Pepper mint
B.	Ethephon	2.	Ethrel
C.	Ethylene	3.	Methionine
D.	Ethylene response element	4.	Key regulatory sequence found in genes regulated by ethylene
E.	Ethylene triple response	5.	Response of the most of the dicot

[A] A: 1; B: 3; C: 4; D: 5; E: 2 [B] A: 1; B: 2; C: 3; D: 4; E: 5
[C] A: 4; B: 1; C: 2; D: 3; E: 5 [D] A: 4; B: 2; C: 3; D: 5; E: 1

121. Match the list I with the list II and select the correct answer using the codes given below the lists

	List I [Terms]		List II [Significance]
A.	Cytochalasin B	1.	A drug that destroys actin filaments
B.	Cytochrome b6f complex	2.	A large multi-subunit protein containing two b-type hemes, one c-type heme and Rieske iron-sulfur protein
C.	Cytochrome c	3.	A peripheral, mobile component of the mitochondrial electron transport chain that oxidizes complex III and reduces complex IV
D.	Cytochrome f	4.	A membrane of the multi-protein cytochrome b6f complex that plays a role in electron transport between Photosystem I and II
E.	Cytochrome P450 monooxygenases	5.	Located on the endoplasmic reticulum

[A] A: 1; B: 3; C: 4; D: 5; E: 2 [B] A: 1; B: 2; C: 3; D: 4; E: 5
[C] A: 4; B: 1; C: 2; D: 3; E: 5 [D] A: 4; B: 2; C: 3; D: 5; E: 1

122. Match the list I with the list II and select the correct answer using the codes given below the lists

	List I [Terms]		List II [Contributions]
A.	Cytohistological zones	1.	Regions of the shoot apical meristem showing difference in morphological appearance and rate of mitosis
B.	Cytokinesis	2.	Separation of daughter cells by formation of new cell wall
C.	Cytokinin oxidase	3.	Transfers the Isopentenyl group to AMP
D.	Cytokinin synthase	4.	Removing the isoprene moiety
E.	Cytokinins	5.	May exist as free and bound forms

[A] A: 1; B: 3; C: 4; D: 5; E: 2 [B] A: 1; B: 2; C: 3; D: 4; E: 5
[C] A: 1; B: 2; C: 4; D: 3; E: 5 [D] A: 4; B: 2; C: 3; D: 5; E: 1

123. Match the list I with the list II and select the correct answer using the codes given below the lists

	List I [Terms]		List II [Significance]
A.	CONSTANS (CO)	1.	Gene for flowering of *Arabidopsis* in long days
B.	Constitutive	2.	Regularly present or expressed
C.	Constitutive defenses	3.	Ethylene triple response
D.	Constitutive ethylene response mutants	4.	Plant defenses that are always immediately available or operational; that is, defenses that are not induced
E.	Contact angle	5.	Quantitative measure

[A] A: 1; B: 3; C: 4; D: 5; E: 2 [B] A: 1; B: 2; C: 3; D: 4; E: 5
[C] A: 1; B: 2; C: 4; D: 3; E: 5 [D] A: 4; B: 2; C: 3; D: 5; E: 1

124. Match the list I with the list II and select the correct answer using the codes given below the lists

	List I [Terms]		List II [Significance]
A.	Cortex	1.	The outer region of the root delimited on the outside by the epidermal and on the inside by the endodermis
B.	Cortical cytoplasm	2.	The outer region or layer of cytoplasm adjacent to the plasma membrane
C.	Cortical-endodermal stem cells	3.	A ring of stem cells that surround the quiescent centre and generate the cortical and endodermal layers in the root
D.	Co-suppression	4.	Decreased expression of the gene when extra copies are introduced
E.	Co-transport	5.	Simultaneous transport of two solutes by the same carrier

[A] A: 1; B: 3; C: 4; D: 5; E: 2 [B] A: 1; B: 2; C: 3; D: 4; E: 5
[C] A: 4; B: 1; C: 2; D: 3; E: 5 [D] A: 4; B: 2; C: 3; D: 5; E: 1

125. Match the list I with the list II and select the correct answer using the codes given below the lists

	List I [Terms]		List II [Significance]
A.	Embryo dormancy	1.	Seed dormancy that is caused directly by the embryo
B.	Embryo sac	2.	Large oval cell
C.	Embryogenesis	3.	Development of zygote into a multi-cellular embryo

D.	Embryonic axis	4.	Hypothetical central line
E.	*emf* mutant	5.	Mutant produces flowers soon after germination

[A] A: 1; B: 3; C: 4; D: 5; E: 2 [B] A: 1; B: 2; C: 3; D: 4; E: 5
[C] A: 1; B: 2; C: 4; D: 3; E: 5 [D] A: 4; B: 2; C: 3; D: 5; E: 1

126. Match the list I with the list II and select the correct answer using the codes given below the lists

	List I [Terms]		List II [Significance]
A.	Endoreduplication	1.	Polyploidization
B.	Endosomes	2.	Vesicles that have lost their clathrin coats
C.	Endosymbiosis	3.	Evolutionary origin of mitochondria and chloroplast
D.	Endosymbiotic origin	4.	Semi-autonomous concept
E.	Endosymbiotic theory	5.	Widely accepted proposal related with mitochondria and chloroplast

[A] A: 1; B: 3; C: 4; D: 5; E: 2 [B] A: 1; B: 2; C: 3; D: 4; E: 5
[C] A: 4; B: 1; C: 2; D: 3; E: 5 [D] A: 4; B: 2; C: 3; D: 5; E: 1

127. Match the list I with the list II and select the correct answer using the codes given below the lists

	List I [Terms]		List II [Significance]
A.	Gall	1.	Tumor like plant tissue
B.	Galactan	2.	Polysaccharide
C.	*GAI* gene	3.	Della protein
D.	GAF domain	4.	Part of the N-terminal half of phytochrome with bilin-layer activity
E.	GABA shunt	5.	Supplement the citric acid cycle

[A] A: 1; B: 3; C: 4; D: 5; E: 2 [B] A: 1; B: 2; C: 3; D: 4; E: 5
[C] A: 4; B: 1; C: 2; D; 3; E: 5 [D] A: 4; B: 2; C: 3; D: 5; E: 1

128. Match the list I with the list II and select the correct answer using the codes given below the lists

	List I [Terms]		List II [Significance]
A.	GC box	1.	A sequence of nucleotides involves in the initiation of transcription in eukaryotes
B.	GDHs	2.	Glutamate dehydrogenase

C.	Gel	3.	Dispersed network of long polymers typically with elastic properties intermediate between that of a liquid and that of a solid
D.	Gene activation	4.	Proteins that, either alone or in concert with other proteins, increase gene expression
E.	Gene fusion	5.	An artificial construct that links a promoter for one gene with the coding sequence of another gene

[A] A: 1; B: 3; C: 4; D: 5; E: 2 [B] A: 1; B: 2; C: 3; D: 4; E: 5
[C] A: 4; B: 1; C: 2; D: 3; E: 5 [D] A: 4; B: 2; C: 3; D: 5; E: 1

129. Match the list I with the list II and select the correct answer using the codes given below the lists

	List I [Terms]		List II [Significance]
A.	Glyceroglycolipids	1.	Sugar form the polar head group
B.	Glycerolipids	2.	Polar lipids
C.	Glycerophospholipids	3.	The phosphate containing polar head group is attached to position 3 of the glycerol
D.	Glycine oxidation	4.	Glycine is converted into serine
E.	Glycolysis	5.	Pyruvates are produced

[A] A: 1; B: 3; C: 4; D: 5; E: 2 [B] A: 1; B: 2; C: 3; D: 4; E: 5
[C] A: 4; B: 1; C: 2; D: 3; E: 5 [D] A: 4; B: 2; C: 3; D: 5; E: 1

130. Match the list I with the list II and select the correct answer using the codes given below the lists

	List I [Terms]		List II [Significance]
A.	Floral reversion	1.	Conversion of floral meristem into a vegetative meristem
B.	Floral stimulus	2.	Photoperiodism
C.	Florigen	3.	Flowering hormone
D.	Flower	4.	Reproductive shoot
E.	FFLOWERING LOCUS C (FLC)	5.	Repress flowering

[A] A: 1; B: 3; C: 4; D: 5; E: 2 [B] A: 1; B: 2; C: 3; D: 4; E: 5
[C] A: 4; B: 1; C: 2; D: 3; E: 5 [D] A: 4; B: 2; C: 3; D: 5; E: 1

131. Match the list I with the list II and select the correct answer using the codes given below the lists

	List I [Terms]		List II [Significance]
A.	FLOWERING LOCUS T (FT)	1.	The gene coding for the protein that acts as florigen in Arabidopsis and other species
B.	Fluence	2.	The number of photons absorbed per unit area
C.	Fluence rate	3.	Universally accepted
D.	Fluid-mosaic model	4.	Watts per square meter
E.	Fluorescence	5.	Chlorophyll pigments

[A] A: 1; B: 3; C: 4; D: 5; E: 2 [B] A: 1; B: 2; C: 3; D: 4; E: 5
[C] A: 1; B: 2; C: 4; D: 3; E: 5 [D] A: 4; B: 2; C: 3; D: 5; E: 1

132. Match the list I with the list II and select the correct answer using the codes given below the lists

	List I [Terms]		List II [Significance]
A.	Flux density	1.	Moles per square meter per second
B.	F0F1-ATP synthase	2.	A multi protein complex associated with the inner mitochondrial membrane
C.	Foliar application	3.	Longitudinally oriented
D.	Formative division	4.	Spray
E.	Furanocoumarins	5.	Group of coumarine with attached furan rings whose toxicity may result from exposure to light

[A] A: 1; B: 3; C: 4; D: 5; E: 2 [B] A: 1; B: 2; C: 3; D: 4; E: 5
[C] A: 1; B: 2; C: 4; D: 3; E: 5 [D] A: 4; B: 2; C: 3; D: 5; E: 1

133. Match the list I with the list II and select the correct answer using the codes given below the lists

	List I [Terms]		List II [Significance]
A.	Induced thermo-tolerance	1.	Tolerance to lethal, high temperature accomplished by periodic brief exposure to sub-lethal heat stress
B.	Inducible	2.	The capacity for increased synthesis of a particular protein or proteins in response to a particular external signal, such as a hormone

List I [Terms]		List II [Significance]
C. Induction period	3.	Tubular extension
D. Infection thread	4.	Calvin-Benson cycle
E. InsP3	5.	Second messengers

[A] A: 1; B: 3; C: 4; D: 5; E: 2 [B] A: 1; B: 2; C: 3; D: 4; E: 5
[C] A: 1; B: 2; C: 4; D: 3; E: 5 [D] A: 4; B: 2; C: 3; D: 5; E: 1

134. Match the list I with the list II and select the correct answer using the codes given below the lists

List I [Terms]		List II [Significance]
A. HRGPs	1.	Hydroxyproline-rich proteins
B. Hyperpolarization	2.	An increase in the normally negative inside, electrical potential (mV) that exists across the plasma membrane of the cell
C. Hypersensitive response	3.	Plant defense response
D. Hyphae	4.	Tubular filaments of fungi
E. Hypocotyl	5.	Seedling stem below the cotyledons and above the root

[A] A: 1; B: 3; C: 4; D: 5; E: 2 [B] A: 1; B: 2; C: 3; D: 4; E: 5
[C] A: 4; B: 1; C: 2; D: 3; E: 5 [D] A: 4; B: 2; C: 3; D: 5; E: 1

135. Match the list I with the list II and select the correct answer using the codes given below the lists

List I [Terms]		List II [Significance]
A. Import	1.	The movement of photosynthetic products in sieve elements into sink organs
B. *In vitro*	2.	Cultivation in test tube or lab condition
C. *In vivo*	3.	Cultivation in nature
D. Intermediate growth	4.	Capacity for both vegetative growth and flowering over an extended period of time
E. Indole	5.	Auxin

[A] A: 1; B: 3; C: 4; D: 5; E: 2 [B] A: 1; B: 2; C: 3; D: 4; E: 5
[C] A: 4; B: 1; C: 2; D: 3; E: 5 [D] A: 4; B: 2; C: 3; D: 5; E: 1

136. Match the list I with the list II and select the correct answer using the codes given below the lists

	List I [Terms]		List II [Significance]
A.	Wallace	1.	Joint assay with Darwin
B.	Huxley	2.	Collected Essays
C.	Weismann	3.	Studies in the Theory of Descent
D.	Haeckel	4.	General Morphology of Organism
E.	Bateson	5.	Materials for the study of variations

[A] A: 1; B: 3; C: 4; D: 5; E: 2
[B] A: 1; B: 2; C: 3; D: 4; E: 5
[C] A: 4; B: 1; C: 2; D: 3; E: 5
[D] A: 4; B: 2; C: 3; D: 5; E: 1

137. Match the list I with the list II and select the correct answer using the codes given below the lists

	List I [Terms]		List II [Significance]
A.	Heat shock cognate proteins	1.	Constitutively expressed
B.	HSPs	2.	Most act as molecular chaperons
C.	Heat Shock Responses	3.	Nonlethal heat episode
D.	Heliotropism	4.	Movements of leaves toward or away from the sun
E.	Hemibiotrophic	5.	Plant pathogens and damage of tissue

[A] A: 1; B: 3; C: 4; D: 5; E: 2
[B] A: 1; B: 2; C: 3; D: 4; E: 5
[C] A: 4; B: 1; C: 2; D: 3; E: 5
[D] A: 4; B: 2; C: 3; D: 5; E: 1

138. Match the list I with the list II and select the correct answer using the codes given below the lists

	List I [Terms]		List II [Significance]
A.	Hemicellulose	1.	Polysaccharides
B.	Herb	2.	A plant with no enduring above-ground parts, as distinct from trees and shrubs
C.	Herbivores	3.	Plant feeding
D.	Histogenesis	4.	The differentiation of cells to produce various tissue
E.	Histogenic layers	5.	Tunica layers and Corpus

[A] A: 1; B: 3; C: 4; D: 5; E: 2
[B] A: 1; B: 2; C: 3; D: 4; E: 5
[C] A: 4; B: 1; C: 2; D: 3; E: 5
[D] A: 4; B: 2; C: 3; D: 5; E: 1

139. Match the list I with the list II and select the correct answer using the codes given below the lists

List I [Terms]		List II [Significance]
A. hy Mutants	1.	Cannot synthesized phytochrome
B. HSF	2.	Specific transcription factor
C. hpt protein	3.	How plants measure night length
D. Hoursglass hypothesis	4.	In *Arabidopsis*
E. H^+-PPase	5.	Electrogenic pump

[A] A: 1; B: 3; C: 4; D: 5; E: 2 [B] A: 1; B: 2; C: 3; D: 4; E: 5
[C] A: 4; B: 1; C: 2; D: 3; E: 5 [D] A: 1; B: 2; C: 4; D: 3; E: 5

140. Match the list I with the list II and select the correct answer using the codes given below the lists

List I [Terms]		List II [Significance]
A. Homeotic genes	1.	Caused body segments
B. Homeotic mutations	2.	Regulated location of body parts in flies
C. HIR	3.	High Irradiance Response
D. Heterotrophic cells	4.	Rely upon reduced forms of carbon for the synthesis of structural elements and supply of metabolic energy
E. Heterosis	5.	The increased vigour in offspring

[A] A: 2; B: 1; C: 3; D: 4; E: 5 [B] A: 1; B: 2; C: 3; D: 4; E: 5
[C] A: 4; B: 1; C: 2; D: 3; E: 5 [D] A: 4; B: 2; C: 3; D: 5; E: 1

141. Match the list I with the list II and select the correct answer using the codes given below the lists

List I [Terms]		List II [Significance]
A. Histone code	1.	Favour the formation of hetero-chromatine or euchromatin
B. Histone	2.	Nucleosomes
C. Hoagland solution	3.	Apoprotein
D. Holoprotein	4.	Formulated by Dennis R. Hoagland
E. Homeobox	5.	60-amino acid domain in transcription factors

[A] A: 1; B: 3; C: 4; D: 5; E: 2 [B] A: 1; B: 2; C: 3; D: 4; E: 5
[C] A: 4; B: 1; C: 2; D: 3; E: 5 [D] A: 2; B: 1; C: 3; D: 4; E: 5

142. Match the list I with the list II and select the correct answer using the codes given below the lists

List I [Terms]	List II [Significance]
A. Glyoxysome	1. Oil reach storage tissues of the seed
B. *GNOM* gene	2. Glutamate synthases
C. GOGAT	3. For development of roots and cotyledons
D. Goldman diffusion potential	4. Calculated diffusion potential
E. Goldman equation	5. Predicts the diffusion potential across a membrane

[A] A: 1; B: 3; C: 4; D: 5; E: 2 [B] A: 1; B: 2; C: 3; D: 4; E: 5
[C] A: 1; B: 3; C: 2; D: 4; E: 5 [D] A: 4; B: 2; C: 3; D: 5; E: 1

143. Match the list I with the list II and select the correct answer using the codes given below the lists

List I [Terms]	List II [Significance]
A. Gossypol	1. Sesquiterpine dimer
B. Grana lamellae	2. Stacked thylakoid membrane
C. Granum	3. Chloroplast
D. GRAS domain	4. C-terminal domain
E. GRAS proteins	5. GAI, RGA and SCR

[A] A: 1; B: 3; C: 4; D: 5; E: 2 [B] A: 1; B: 2; C: 3; D: 4; E: 5
[C] A: 4; B: 1; C: 2; D: 3; E: 5 [D] A: 4; B: 2; C: 3; D: 5; E: 1

144. Match the list I with the list II and select the correct answer using the codes given below the lists

List I [Terms]	List II [Significance]
A. Gravitropic response	1. The growth initiated by the root cap's perception of gravity and the signal
B. Gravitropism	2. Growth in response to gravity
C. Gravity potential	3. Component of water potential
D. Green Revolution	4. High-yielding dwarf varieties
E. Green-leaf volatiles	5. In response to mechanical damage

[A] A: 1; B: 3; C: 4; D: 5; E: 2 [B] A: 1; B: 2; C: 3; D: 4; E: 5
[C] A: 4; B: 1; C: 2; D: 3; E: 5 [D] A: 4; B: 2; C: 3; D: 5; E: 1

145. Match the list I with the list II and select the correct answer using the codes given below the lists

	List I [Terms]		List II [Significance]
A.	Greenhouse effect	1.	The Warming of Earth's climate
B.	Ground meristem	2.	In the plant meristems, cells that will gave rise to the cortical and pith tissues and, in the root and hypocotyls, will produce the endodermis
C.	Growth respiration	3.	Contrast to maintenance respiration
D.	GSA gene	4.	Blue light sensing system
E.	GT43	5.	Glycosyl transferase

[A] A: 1; B: 3; C: 4; D: 5; E: 2 [B] A: 1; B: 2; C: 3; D: 4; E: 5
[C] A: 4; B: 1; C: 2; D: 3; E: 5 [D] A: 4; B: 2; C: 3; D: 5; E: 1

146. Match the list I with the list II and select the correct answer using the codes given below the lists

	List I [Terms]		List II [Significance]
A.	Guard cell protoplast	1.	Removal of cell wall of guard cells
B.	Guard cells	2.	Specialized epidermal cells
C.	GURKE (GK)	3.	Used as reporter
D.	GUS	4.	Used in pattern formation
E.	Guttation	5.	Exudation of liquid from the leaves due to root pressure

[A] A: 1; B: 2; C: 4; D: 3; E: 5 [B] A: 1; B: 2; C: 3; D: 4; E: 5
[C] A: 4; B: 1; C: 2; D: 3; E: 5 [D] A: 4; B: 2; C: 3; D: 5; E: 1

147. Match the list I with the list II and select the correct answer using the codes given below the lists

	List I [Terms]		List II [Significance]
A.	Gymnosperm	1.	Naked cones
B.	H lignin	2.	Made from p-Coumaryl alcohol
C.	Halophytes	3.	Native to saline soil
D.	Hartig net	4.	CO homolog
E.	*Heading-date 1(Hd1)*	5.	Fungal network of hyphae

[A] A: 1; B: 2; C: 3; D: 5; E: 4 [B] A: 1; B: 2; C: 3; D: 4; E: 5
[C] A: 4; B: 1; C: 2; D: 3; E: 5 [D] A: 4; B: 2; C: 3; D: 5; E: 1

148. Match the list I with the list II and select the correct answer using the codes given below the lists

List I [Terms]		List II [Significance]
A. Gibberellic acid	1.	Identical to gibbereline GA3
B. Gibberellin glycosides	2.	Inactive or stored forms of gibberellins
C. Gibbs free energy	3.	Energy that is available to do work
D. Girdling	4.	Removal of ring of bark from a woody stem that severs the vascular system
E. Globular stage embryo	5.	First stage of embryogenesis

[A] A: 1; B: 3; C: 4; D: 5; E: 2 [B] A: 1; B: 2; C: 3; D: 4; E: 5
[C] A: 4; B: 1; C: 2; D: 3; E: 5 [D] A: 4; B: 2; C: 3; D: 5; E: 1

149. Match the list I with the list II and select the correct answer using the codes given below the lists

List I [Terms]		List II [Significance]
A. Glucan	1.	A polysaccharide made from glucose units
B. Glucomannan	2.	Both glucose and mannose units
C. Gluconeogenesis	3.	Reversal of glycolysis
D. Glucosinolates	4.	Defense compounds
E. Glutamate dehydrogenase	5.	Catalyzes a reversible reaction that synthesizes or deaminates glutamate as part of the nitrogen assimilation process

[A] A: 1; B: 3; C: 4; D: 5; E: 2 [B] A: 1; B: 2; C: 3; D: 4; E: 5
[C] A: 4; B: 1; C: 2; D: 3; E: 5 [D] A: 4; B: 2; C: 3; D: 5; E: 1

150. Match the list I with the list II and select the correct answer using the codes given below the lists

List I [Terms]		List II [Significance]
A. Glutamate synthase	1.	Transfer the amide group
B. Glutathione	2.	3 amino acid peptide
C. Glycan	3.	Polymer made of sugar units
D. Glyoxylate	4.	Plants that are not able to resist salts to the same degree as halophytes
E. Glycophytes	5.	A two-carbon acid aldehyde that is an intermediate of the glyoxylate cycle

[A] A: 1; B: 2; C: 3; D: 5; E: 4 [B] A: 1; B: 2; C: 3; D: 4; E: 5
[C] A: 4; B: 1; C: 2; D: 3; E: 5 [D] A: 4; B: 2; C: 3; D: 5; E: 1

151. Match the list I with the list II and select the correct answer using the codes given below the lists

List I [Terms]	List II [Significance]
A. Fusiform stem cells	1. Elongated
B. G1 Phase	2. Preceding the synthesis of DNA
C. G2 Phase	3. The phase of the cell cycle following the synthesis of DNA
D. G-lignin	4. Made from coniferyl alcohol
E. G-protein	5. GTP binding protein involved in the signal transduction

[A] A: 1; B: 3; C: 4; D: 5; E: 2 [B] A: 1; B: 2; C: 3; D: 4; E: 5
[C] A: 4; B: 1; C: 2; D: 3; E: 5 [D] A: 4; B: 2; C: 3; D: 5; E: 1

152. Match the list I with the list II and select the correct answer using the codes given below the lists

List I [Terms]	List II [Significance]
A. GC	1. Gas chromatography
B. Gate	2. Structural domain of the channel protein
C. Gating	3. Function of the biological clock
D. G-box	4. Specific sequence of the DNA nucleotides
E. Gene gun	5. Biolistics

[A] A: 1; B: 3; C: 4; D: 5; E: 2 [B] A: 1; B: 2; C: 3; D: 4; E: 5
[C] A: 4; B: 1; C: 2; D: 3; E: 5 [D] A: 4; B: 2; C: 3; D: 5; E: 1

153. Match the list I with the list II and select the correct answer using the codes given below the lists

List I [Terms]	List II [Significance]
A. General transcription factor	1. For proper positioning to RNA polymerase
B. Genetic tumors	2. Spontaneous
C. Genome	3. Refers to all the genes in a haploid complement of eukaryotic chromosomes, in an organelle, a microbe, or the DNA or RNA content of a virus
D. GGPP	4. Spore, seed or bud
E. Germination	5. A precursor

[A] A: 1; B: 2; C: 3; D: 5; E: 4 [B] A: 1; B: 2; C: 3; D: 4; E: 5
[C] A: 4; B: 1; C: 2; D: 3; E: 5 [D] A: 4; B: 2; C: 3; D: 5; E: 1

154. Match the list I with the list II and select the correct answer using the codes given below the lists

	List I [Terms]		List II [Significance]
A.	Floral homeotic genes	1.	Determine the positions and identity of the floral part
B.	Floral meristem	2.	Collected Essays
C.	Floral organ identity genes	3.	Studies in the Theory of Descent
D.	Floral organs	4.	General Morphology of Organism
E.	Floral primordial	5.	Materials for the study of variations

[A] A: 1; B: 3; C: 4; D: 5; E: 2 [B] A: 1; B: 2; C: 3; D: 4; E: 5
[C] A: 4; B: 1; C: 2; D: 3; E: 5 [D] A: 4; B: 2; C: 3; D: 5; E: 1

155. Match the list I with the list II and select the correct answer using the codes given below the lists

	List I [Terms]		List II [Significance]
A.	Floral evocation	1.	Commit the apical meristem to produce flower
B.	Flooding-tolerant	2.	Withstand the anoxia temporarily
C.	Flooding-sensitive	3.	24 hrs of flooding damage the plant
D.	Flippases	4.	Photosystem I
E.	Flavoprotein-ferredoxin-NADP reductase	5.	Symmetrical lipid composition

[A] A: 1; B: 2; C: 3; D: 5; E: 4 [B] A: 1; B: 2; C: 3; D: 4; E: 5
[C] A: 4; B: 1; C: 2; D: 3; E: 5 [D] A: 4; B: 2; C: 3; D: 5; E: 1

156. Match the list I with the list II and select the correct answer using the codes given below the lists

	List I [Terms]		List II [Significance]
A.	Exogenous	1.	From outside
B.	Expansins	2.	Class of wall-loosening proteins
C.	Export	3.	Movement away from source
D.	Extracellular space	4.	Polymerized protofilament
E.	F-actin	5.	Apoplast

[A] A: 1; B: 2; C: 3; D: 5; E: 4 [B] A: 1; B: 2; C: 3; D: 4; E: 5
[C] A: 4; B: 1; C: 2; D: 3; E: 5 [D] A: 4; B: 2; C: 3; D: 5; E: 1

157. Match the list I with the list II and select the correct answer using the codes given below the lists

List I [Terms]		List II [Significance]
A. Fe-S centers	1.	Prosthetic group
B. Fermentation	2.	Anaerobic condition
C. Fd	3.	Water soluble Iron-sulfur proteins
D. FNR	4.	Metabolism of pyruvate in the absence of oxygen
E. Fermentative metabolism	5.	Membrane associated flavoprotein

[A] A: 1; B: 2; C: 3; D: 5; E: 4 [B] A: 1; B: 2; C: 3; D: 4; E: 5
[C] A: 4; B: 1; C: 2; D: 3; E: 5 [D] A: 4; B: 2; C: 3; D: 5; E: 1

158. Match the list I with the list II and select the correct answer using the codes given below the lists

List I [Terms]		List II [Significance]
A. ELISA	1.	Assay
B. Elicitors	2.	Cell wall fragments
C. elf3 mutant	3.	Flowering time mutant
D. Electoneutral transport	4.	Active ion transport that involves no net movement of charge across a membrane
E. Electronegativity	5.	Attract electron

[A] A: 1; B: 3; C: 4; D: 5; E: 2 [B] A: 1; B: 2; C: 3; D: 4; E: 5
[C] A: 4; B: 1; C: 2; D: 3; E: 5 [D] A: 4; B: 2; C: 3; D: 5; E: 1

159. Match the list I with the list II and select the correct answer using the codes given below the lists

List I [Terms]		List II [Significance]
A. Endocytosis	1.	The formation of the small vesicles from the plasma membrane, which detach and move into the cytosol, where they fuse with elements of the endomembrane system
B. Endodermis	2.	Casparian strip
C. Endogenous	3.	Molecular pacemaker
D. Endogenous oscillator	4.	Interior of the living system
E. Energy transfer	5.	Chlorophyll to chlorophyll

[A] A: 1; B: 3; C: 4; D: 5; E: 2 [B] A: 1; B: 2; C: 3; D: 4; E: 5
[C] A: 1; B: 2; C: 4; D: 3; E: 5 [D] A: 4; B: 2; C: 3; D: 5; E: 1

160. Match the list I with the list II and select the correct answer using the codes given below the lists

	List I [Terms]		List II [Significance]
A.	Chemical potential	1.	Free energy
B.	CHLH protein	2.	Group of ABA receptor proteins localized to plastids
C.	Chlorophyllase	3.	Enzymes that remove the phytol from chlorophyll as part of the chlorophyll breakdown process
D.	Cholodny-Went model	4.	Related with bending of the plant axis by lateral transport of auxin in response to a stimulus
E.	Bateson	5.	Materials for the study of variations

[A] A: 1; B: 3; C: 4; D: 5; E: 2 [B] A: 1; B: 2; C: 3; D: 4; E: 5
[C] A: 4; B: 1; C: 2; D: 3; E: 5 [D] A: 4; B: 2; C: 3; D: 5; E: 1

161. Match the list I with the list II and select the correct answer using the codes given below the lists

	List I [Terms]		List II [Significance]
A.	Polar glycerolipids	1.	Main structural lipids in membrane
B.	Polar molecules	2.	Differences in the electronegativity
C.	Polar transport	3.	Distinct ends
D.	Polarity	4.	Directional movement
E.	Pollen	5.	Haploid male nuclei

[A] A: 1; B: 3; C: 4; D: 5; E: 2 [B] A: 1; B: 2; C: 3; D: 4; E: 5
[C] A: 1; B: 2; C: 4; D: 3; E: 5 [D] A: 4; B: 2; C: 3; D: 5; E: 1

162. Match the list I with the list II and select the correct answer using the codes given below the lists

	List I [Terms]		List II [Significance]
A.	Pollination	1.	Transfer of pollen from anther to stigma
B.	Polymer-trapping model	2.	Specific accumulation of sugars in the sieve elements of symplastically loading species
C.	Polyploidy	3.	1500 to 15,000 isoprene units
D.	Polyterpenes	4.	The condition of being polyploidy
E.	Position dependent	5.	Depends on the position of cell within the embryo

[A] A: 1; B: 3; C: 4; D: 5; E: 2 [B] A: 1; B: 2; C: 3; D: 4; E: 5
[C] A: 1; B: 2; C: 4; D: 3; E: 5 [D] A: 4; B: 2; C: 3; D: 5; E: 1

163. Match the list I with the list II and select the correct answer using the codes given below the lists

	List I [Terms]		List II [Significance]
A.	Primary cell wall	1.	>1 μm
B.	Primary cross regulation	2.	Regulation of transduction component
C.	Primary growth	3.	Meristem that produce stem which bears the inflorescence
D.	Primary inflorescence meristem	4.	It is the result of cell proliferation in meristem region
E.	Primary meristem	5.	Formed during embryogenesis

[A] A: 1; B: 3; C: 4; D: 5; E: 2 [B] A: 1; B: 2; C: 3; D: 4; E: 5
[C] A: 1; B: 2; C: 4; D: 3; E: 5 [D] A: 4; B: 2; C: 3; D: 5; E: 1

164. Match the list I with the list II and select the correct answer using the codes given below the lists

	List I [Terms]		List II [Significance]
A.	Positive regulator	1.	Presence causes a change in function
B.	PPFD	2.	Photosynthetic photon flux density
C.	Pr	3.	Red-light absorbing
D.	Preprophase	4.	The stage just before prophase
E.	Preprophase band	5.	A circular array

[A] A: 1; B: 3; C: 4; D: 5; E: 2 [B] A: 1; B: 2; C: 3; D: 4; E: 5
[C] A: 4; B: 1; C: 2; D: 3; E: 5 [D] A: 4; B: 2; C: 3; D: 5; E: 1

165. Match the list I with the list II and select the correct answer using the codes given below the lists

	List I [Terms]		List II [Significance]
A.	Pressure –flow model	1.	Phloem translocation
B.	Pressure potential	2.	Hydrostatic pressure
C.	PVC	3.	Late endosome
D.	Primary active transport	4.	ATP hydrolysis
E.	Primary plasmodesmata	5.	Tubular extensions

[A] A: 1; B: 3; C: 4; D: 5; E: 2 [B] A: 1; B: 2; C: 3; D: 4; E: 5
[C] A: 4; B: 1; C: 2; D: 3; E: 5 [D] A: 4; B: 2; C: 3; D: 5; E: 1

166. Match the list I with the list II and select the correct answer using the codes given below the lists

List I [Terms]		List II [Significance]
A. Photon	1.	Discrete physical unit
B. Photon irradiance	2.	Moles per square meter per second
C. Photoperiod	3.	Floral stimulus
D. Photoperiodic induction	4.	Control sexual and vegetative control of plant
E. Photoperiodism	5.	Response against length and timing of day and dark

[A] A: 1; B: 3; C: 4; D: 5; E: 2 [B] A: 1; B: 2; C: 3; D: 4; E: 5
[C] A: 1; B: 2; C: 4; D: 3; E: 5 [D] A: 4; B: 2; C: 3; D: 5; E: 1

167. Match the list I with the list II and select the correct answer using the codes given below the lists

List I [Terms]		List II [Significance]
A. Photophosphorylation	1.	Production of ATP
B. Photoprotection	2.	Photo oxidation
C. Photorespiration	3.	Phytochrome
D. Photoreversibility	4.	Glycolate
E. Photosynthate	5.	Carbon Product for assimilation

[A] A: 1; B: 3; C: 4; D: 5; E: 2 [B] A: 1; B: 2; C: 3; D: 4; E: 5
[C] A: 1; B: 2; C: 4; D: 3; E: 5 [D] A: 4; B: 2; C: 3; D: 5; E: 1

168. Match the list I with the list II and select the correct answer using the codes given below the lists

List I [Terms]		List II [Significance]
A. Photophosphorylation	1.	The production of ATP by using inorganic phosphate forming ADP
B. Photoprotection	2.	Carotenoid
C. Photorespiration	3.	Photorespiratory carbon oxidation
D. PCO cycle	4.	25 to 30% CO_2 loss
E. Photoreversibility	5.	Phytochrome

[A] A: 1; B: 3; C: 4; D: 5; E: 2 [B] A: 1; B: 2; C: 3; D: 4; E: 5
[C] A: 1; B: 2; C: 4; D: 3; E: 5 [D] A: 4; B: 2; C: 3; D: 5; E: 1

169. Match the list I with the list II and select the correct answer using the codes given below the lists

	List I [Terms]		List II [Significance]
A.	Phragmoplast	1.	Late anaphase or early telophase
B.	Phy	2.	Chromophore
C.	PHY	3.	Apoprotein
D.	PHY domain	4.	Necessary for establishment
E.	*PHY* gene	5.	*PHYD*

[A] A: 1; B: 3; C: 4; D: 5; E: 2 [B] A: 1; B: 2; C: 3; D: 4; E: 5
[C] A: 4; B: 1; C: 2; D: 3; E: 5 [D] A: 4; B: 2; C: 3; D: 5; E: 1

170. Match the list I with the list II and select the correct answer using the codes given below the lists

	List I [Terms]		List II [Significance]
A.	Phyllotaxy	1.	Arrangement of leaves
B.	Phyto-oncogenes	2.	Tumor
C.	Phytoalexins	3.	Antimicrobial activity
D.	Phytochelatins	4.	Glutathione
E.	Phytochrome	5.	Photoreceptor protein

[A] A: 1; B: 3; C: 4; D: 5; E: 2 [B] A: 1; B: 2; C: 3; D: 4; E: 5
[C] A: 4; B: 1; C: 2; D: 3; E: 5 [D] A: 4; B: 2; C: 3; D: 5; E: 1

171. Match the list I with the list II and select the correct answer using the codes given below the lists

	List I [Terms]		List II [Significance]
A.	PPFD	1.	Quanta (mol $m^{-2}s^{-1}$)
B.	Photosynthesis	2.	Physio-chemical process
C.	Quantasome	3.	Oxidizes plastocyanine
D.	Photosystem 1	4.	Functional unit in the chloroplast
E.	Photosystem 2	5.	Reduces plastoquinone

[A] A: 1; B: 3; C: 4; D: 5; E: 2 [B] A: 1; B: 2; C: 3; D: 4; E: 5
[C] A: 4; B: 1; C: 2; D: 3; E: 5 [D] A: 1; B: 2; C: 4; D: 3; E: 5

172. Match the list I with the list II and select the correct answer using the codes given below the lists

	List I [Terms]		List II [Significance]
A.	Period	1.	Peaks or troughs
B.	Peripheral proteins	2.	Membrane surface
C.	Peripheral zone	3.	Doughnut shaped region in meristem
D.	Permanent wilting point	4.	Plant cannot regain turgor
E.	Peroxisome	5.	Catalase

[A] A: 1; B: 3; C: 4; D: 5; E: 2 [B] A: 1; B: 2; C: 3; D: 4; E: 5
[C] A: 4; B: 1; C: 2; D: 3; E: 5 [D] A: 4; B: 2; C: 3; D: 5; E: 1

173. Match the list I with the list II and select the correct answer using the codes given below the lists

	List I [Terms]		List II [Significance]
A.	Petals	1.	Brightly colored
B.	Petiole	2.	Leaf stalk
C.	Pfr	3.	Maximum and minimum position
D.	Phase	4.	Far-red light
E.	Phase change	5.	Fates of the meristematic cells become altered in ways that cause them to produce new types of structure

[A] A: 1; B: 3; C: 4; D: 5; E: 2 [B] A: 1; B: 2; C: 3; D: 4; E: 5
[C] A: 4; B: 1; C: 2; D: 3; E: 5 [D] A: 1; B: 2; C: 4; D: 3; E: 5

174. Match the list I with the list II and select the correct answer using the codes given below the lists

	List I [Terms]		List II [Significance]
A.	Phenotypic plasticity	1.	Do not involve genetic changes
B.	PAL	2.	Cinnamic acid
C.	Phenylpropanoids	3.	A chlorophyll in which the central atom is occupied by two hydrogen atom
D.	Pheophytin	4.	Phenolic derivatives of Cinnamic acid
E.	Phloem	5.	Transportation

[A] A: 1; B: 3; C: 4; D: 5; E: 2 [B] A: 1; B: 2; C: 3; D: 4; E: 5
[C] A: 4; B: 1; C: 2; D: 3; E: 5 [D] A: 1; B: 2; C: 4; D: 3; E: 5

175. Match the list I with the list II and select the correct answer using the codes given below the lists

	List I [Terms]		List II [Significance]
A.	Phloem loading	1.	From mesophyll chloroplast to the sieve elements
B.	Phloem unloading	2.	From sieve element to the sink cells
C.	Phosphate	3.	Required for root development
D.	PA	4.	Phosphatidic acid
E.	PAPS	5.	First stable intermediate in the bacterial and fungal reduction of sulfate to sulfite

[A] A: 1; B: 3; C: 4; D: 5; E: 2 [B] A: 1; B: 2; C: 3; D: 4; E: 5
[C] A: 4; B: 1; C: 2; D: 3; E: 5 [D] A: 4; B: 2; C: 3; D: 5; E: 1

176. Match the list I with the list II and select the correct answer using the codes given below the lists

List I [Terms]		List II [Significance]
A. Photomorphogenesis	1.	Plant development
B. Photolyase	2.	Blue-light activated enzymes
C. Photoinhibition	3.	Rapid chemical reaction
D. Photochemistry	4.	Excess light inhibit the photosynthesis
E. Photoassimilation	5.	Nutrient assimilation to the photosynthetic electron transport

[A] A: 1; B: 3; C: 4; D: 5; E: 2 [B] A: 1; B: 2; C: 3; D: 4; E: 5
[C] A: 1; B: 2; C: 4; D: 3; E: 5 [D] A: 4; B: 2; C: 3; D: 5; E: 1

177. Match the list I with the list II and select the correct answer using the codes given below the lists

List I [Terms]		List II [Significance]
A. P protein	1.	Phloem protein
B. P-ATPase	2.	Ion pump
C. P-protein bodies	3.	Photosystem II
D. P680	4.	Spindle shaped
E. P700	5.	Photosystem 1

[A] A: 1; B: 3; C: 4; D: 5; E: 2 [B] A: 1; B: 2; C: 3; D: 4; E: 5
[C] A: 4; B: 1; C: 2; D: 3; E: 5 [D] A: 1; B: 2; C: 4; D: 3; E: 5

178. Match the list I with the list II and select the correct answer using the codes given below the lists

List I [Terms]		List II [Significance]
A. Oxygenic organism	1.	Photosynthetic organism
B. Oxygenases	2.	Add oxygen
C. Oxygenation	3.	Oxygen dependent reaction
D. Paraquat	4.	Active radiation for photosynthesis
E. PAR	5.	Herbicide

[A] A: 1; B: 2; C: 3; D: 5; E: 4 [B] A: 1; B: 2; C: 3; D: 4; E: 5
[C] A: 4; B: 1; C: 2; D: 3; E: 5 [D] A: 4; B: 2; C: 3; D: 5; E: 1

179. Match the list I with the list II and select the correct answer using the codes given below the lists

List I [Terms]	List II [Significance]
A. P870	1. Bacteriochlorophyll
B. Paclobutrazol	2. Inhibitor of gibberellins biosynthesis
C. Palisade cells	3. Below the leaf epidermis
D. Paleopolyploids	4. Sign of ancient genome
E. Patch-clamp	5. Electrophysiological method

[A] A: 1; B: 3; C: 4; D: 5; E: 2 [B] A: 1; B: 2; C: 3; D: 4; E: 5
[C] A: 4; B: 1; C: 2; D: 3; E: 5 [D] A: 4; B: 2; C: 3; D: 5; E: 1

180. Match the list I with the list II and select the correct answer using the codes given below the lists

List I [Terms]	List II [Significance]
A. Periderm	1. Cork cambium
B. Pericycle	2. Vascular cylinder
C. Periclinal	3. End walls of vessel element
D. Perforation plate	4. New cell walls form parallel to the tissue surface
E. Perfect flower	5. Hermaphroditic flowers

[A] A: 1; B: 3; C: 4; D: 5; E: 2 [B] A: 1; B: 2; C: 3; D: 4; E: 5
[C] A: 1; B: 2; C: 4; D: 3; E: 5 [D] A: 4; B: 2; C: 3; D: 5; E: 1

181. Match the list I with the list II and select the correct answer using the codes given below the lists

List I [Terms]	List II [Significance]
A. Pectins	1. Heterogenous group of complex cell wall polysaccharides
B. PPP	2. Cytosolic and plastidic pathway
C. PEP carboxylase	3. Differential distribution
D. Partioning	4. Cytosolic enzyme
E. Parthenocarpy	5. Banana and Pineapple

[A] A: 1; B: 3; C: 4; D: 5; E: 2 [B] A: 1; B: 2; C: 3; D: 4; E: 5
[C] A: 4; B: 1; C: 2; D: 3; E: 5 [D] A: 1; B: 2; C: 4; D: 3; E: 5

182. Match the list I with the list II and select the correct answer using the codes given below the lists

	List I [Terms]		List II [Significance]
A.	Nutrient film growth system	1.	Hydroponic culture
B.	Nutrient solution	2.	Having only inorganic salts
C.	Nyctinasty	3.	Linked via OH group
D.	O-linked oligosaccharides	4.	Sleep movement
E.	Octant embryo	5.	Radial symmetry

[A] A: 1; B: 3; C: 4; D: 5; E: 2 [B] A: 1; B: 2; C: 3; D: 4; E: 5
[C] A: 1; B: 2; C: 4; D: 3; E: 5 [D] A: 4; B: 2; C: 3; D: 5; E: 1

183. Match the list I with the list II and select the correct answer using the codes given below the lists

	List I [Terms]		List II [Significance]
A.	Opines	1.	Crown gall tumors
B.	Oleosin	2.	Specific protein
C.	Oleosomes	3.	Triacylglycerols
D.	Oligogalacturonans	4.	Pectin fragment
E.	Oligosaccharins	5.	Fragments of the plant cell wall degradation

[A] A: 1; B: 3; C: 4; D: 5; E: 2 [B] A: 1; B: 2; C: 3; D: 4; E: 5
[C] A: 4; B: 1; C: 2; D: 3; E: 5 [D] A: 4; B: 2; C: 3; D: 5; E: 1

184. Match the list I with the list II and select the correct answer using the codes given below the lists

	List I [Terms]		List II [Significance]
A.	Optimal temperature response	1.	Highest rate of photosynthesis
B.	Ordinary companion cell	2.	Few plasmodesmata
C.	Organ identity genes	3.	Floral development
D.	Organic fertilizers	4.	Embryogenesis
E.	Organogenesis	5.	Natural source

[A] A: 1; B: 2; C: 3; D: 5; E: 4 [B] A: 1; B: 2; C: 3; D: 4; E: 5
[C] A: 4; B: 1; C: 2; D: 3; E: 5 [D] A: 4; B: 2; C: 3; D: 5; E: 1

185. Match the list I with the list II and select the correct answer using the codes given below the lists

	List I [Terms]		List II [Significance]
A.	*orp* mutant	1.	Maize
B.	Orthostichy	2.	A set of leaves inserted in the stem directly above or below one another
C.	Osmolarity	3.	Moles per liter
D.	Outwardly rectifying	4.	Ion channels
E.	Osmosis	5.	Special type of diffusion

[A] A: 1; B: 3; C: 4; D: 5; E: 2
[B] A: 1; B: 2; C: 3; D: 4; E: 5
[C] A: 4; B: 1; C: 2; D: 3; E: 5
[D] A: 4; B: 2; C: 3; D: 5; E: 1

186. Match the list I with the list II and select the correct answer using the codes given below the lists

	List I [Terms]		List II [Significance]
A.	Nernst potential	1.	Electrical potential
B.	Night break	2.	Dark period ineffective
C.	NIT gene	3.	Nitrate to nitrite
D.	Nitrate reductase	4.	Nitrilase enzyme
E.	Nitric oxide	5.	Gas that acts as a second messenger

[A] A: 1; B: 3; C: 4; D: 5; E: 2
[B] A: 1; B: 2; C: 3; D: 4; E: 5
[C] A: 1; B: 2; C: 4; D: 3; E: 5
[D] A: 4; B: 2; C: 3; D: 5; E: 1

187. Match the list I with the list II and select the correct answer using the codes given below the lists

	List I [Terms]		List II [Significance]
A.	Nitrite reductase	1.	Nitrite to ammonium ion
B.	Nitrogen fixation	2.	Natural and industrial process
C.	Nitrogenase enzyme complex	3.	Two-component protein
D.	Nod factors	4.	Lipochitin oligosaccharide signal molecules
E.	Nodal root	5.	Form after the emergence of primary roots

[A] A: 1; B: 3; C: 4; D: 5; E: 2
[B] A: 1; B: 2; C: 3; D: 4; E: 5
[C] A: 4; B: 1; C: 2; D: 3; E: 5
[D] A: 4; B: 2; C: 3; D: 5; E: 1

188. Match the list I with the list II and select the correct answer using the codes given below the lists

List I [Terms]	List II [Significance]
A. Nodule primordium	1. Rhizobia infected root
B. Nodules	2. Specialized organ
C. Nodulin (Nod) genes	3. Nodule formation
D. Nonclimacteric fruits	4. Citrus
E. Non-Mendelian	5. Pattern of inheritance

[A] A: 1; B: 3; C: 4; D: 5; E: 2 [B] A: 1; B: 2; C: 3; D: 4; E: 5
[C] A: 4; B: 1; C: 2; D: 3; E: 5 [D] A: 4; B: 2; C: 3; D: 5; E: 1

189. Match the list I with the list II and select the correct answer using the codes given below the lists

List I [Terms]	List II [Significance]
A. Nuclear envelope	1. Double membrane
B. Nuclear genome	2. Entire complement of DNA found in the nucleus
C. Nuclear localization signal	3. Specific amino acid sequence
D. NPC	4. 120 nm wide
E. Nuclear pores	5. Varies from 8Å to 215Å

[A] A: 1; B: 3; C: 4; D: 5; E: 2 [B] A: 1; B: 2; C: 3; D: 4; E: 5
[C] A: 4; B: 1; C: 2; D: 3; E: 5 [D] A: 4; B: 2; C: 3; D: 5; E: 1

190. Match the list I with the list II and select the correct answer using the codes given below the lists

List I [Terms]	List II [Significance]
A. Nucleoids	1. Prokaryotic genome
B. Nucleolar organizer region	2. Nucleolus
C. Nucleolus	3. Densely granular region
D. Nucleoporins	4. Nuclear pore
E. Nucleosome	5. Eight histone protein

[A] A: 1; B: 3; C: 4; D: 5; E: 2 [B] A: 1; B: 2; C: 3; D: 4; E: 5
[C] A: 4; B: 1; C: 2; D: 3; E: 5 [D] A: 4; B: 2; C: 3; D: 5; E: 1

191. Match the list I with the list II and select the correct answer using the codes given below the lists

	List I [Terms]		List II [Significance]
A.	Nonphotochemical quenching	1.	Heat
B.	Nonprotein amino acids	2.	GABA
C.	Nonreducing end	3.	C-1 is β-linked to the C-4
D.	Nonreducing sugar	4.	RNAs
E.	Northern-blot analysis	5.	Sucrose

[A] A: 1; B: 2; C: 3; D: 5; E: 4 [B] A: 1; B: 2; C: 3; D: 4; E: 5
[C] A: 4; B: 1; C: 2; D: 3; E: 5 [D] A: 4; B: 2; C: 3; D: 5; E: 1

192. Match the list I with the list II and select the correct answer using the codes given below the lists

	List I [Terms]		List II [Significance]
A.	*MONOPTEROS (MP)*	1.	Embryonic patterning
B.	Monosomies	2.	Only one chromosome
C.	Monoterpenes	3.	10 carbons
D.	Morphogenesis	4.	Developmental process
E.	Morphogens	5.	Substances that play key role in providing positional cues in certain types of position-dependent development

[A] A: 1; B: 3; C: 4; D: 5; E: 2 [B] A: 1; B: 2; C: 3; D: 4; E: 5
[C] A: 4; B: 1; C: 2; D: 3; E: 5 [D] A: 4; B: 2; C: 3; D: 5; E: 1

193. Match the list I with the list II and select the correct answer using the codes given below the lists

	List I [Terms]		List II [Significance]
A.	Mycelium	1.	Mass of hyphae
B.	Mycorrhizae	2.	Mututalistic
C.	InsP6	3.	Broad bean
D.	Myosins	4.	Recessive mutant allele
E.	*na* mutant	5.	Motor protein

[A] A: 1; B: 2; C: 3; D: 5; E: 4 [B] A: 1; B: 2; C: 3; D: 4; E: 5
[C] A: 4; B: 1; C: 2; D: 3; E: 5 [D] A: 4; B: 2; C: 3; D: 5; E: 1

194. Match the list I with the list II and select the correct answer using the codes given below the lists

List I [Terms]		List II [Significance]
A. NCED	1.	ABA biosynthesis
B. ncRNAs	2.	RNA interference
C. Necrosis	3.	Cell death
D. Nectar	4.	Contains amino acids
E. Nectar guides	5.	Symmetric pattern of strips, spots, or concentric circles

[A] A: 1; B: 3; C: 4; D: 5; E: 2 [B] A: 1; B: 2; C: 3; D: 4; E: 5
[C] A: 4; B: 1; C: 2; D: 3; E: 5 [D] A: 4; B: 2; C: 3; D: 5; E: 1

195. Match the list I with the list II and select the correct answer using the codes given below the lists

List I [Terms]		List II [Significance]
A. Methylation	1.	Modification of cytosine residues in DNA
B. MEP pathway	2.	Chloroplast and other plastids
C. Mevalonic acid pathway	3.	Terpene biosynthesis
D. Michaelis-Menten constant	4.	Binding affinity
E. Microaerobic condition	5.	*Azotobacter*

[A] A: 1; B: 3; C: 4; D: 5; E: 2 [B] A: 1; B: 2; C: 3; D: 4; E: 5
[C] A: 4; B: 1; C: 2; D: 3; E: 5 [D] A: 4; B: 2; C: 3; D: 5; E: 1

196. Match the list I with the list II and select the correct answer using the codes given below the lists

List I [Terms]		List II [Significance]
A. Microarray	1.	DNA sequence
B. MAMPs	2.	Microbe-associated general molecular patterns
C. Microbodies	3.	Spherical organelles
D. Microfilament	4.	Heterochromatic
E. Microsatellites	5.	Cell cytoskeleton

[A] A: 1; B: 2; C: 3; D: 5; E: 4 [B] A: 1; B: 2; C: 3; D: 4; E: 5
[C] A: 4; B: 1; C: 2; D: 3; E: 5 [D] A: 4; B: 2; C: 3; D: 5; E: 1

197. Match the list I with the list II and select the correct answer using the codes given below the lists

	List I [Terms]		List II [Significance]
A.	Microtubule	1.	Tubulin
B.	Middle lamella	2.	Strasburger
C.	Middle region	3.	Studies in the Theory of Descent
D.	Mineral nutrients	4.	Formation of soil
E.	Mineralization	5.	Essential for completion of life

[A] A: 1; B: 2; C: 3; D: 5; E: 4 [B] A: 1; B: 2; C: 3; D: 4; E: 5
[C] A: 4; B: 1; C: 2; D: 3; E: 5 [D] A: 4; B: 2; C: 3; D: 5; E: 1

198. Match the list I with the list II and select the correct answer using the codes given below the lists

	List I [Terms]		List II [Significance]
A.	MADS box genes	1.	Conserved sequence
B.	MADS domain	2.	Dimerization region
C.	Magnesium dechelatase	3.	Remove magnesium
D.	Maintenance respiration	4.	Needed to support the function and turnover of existing tissue
E.	Malic enzyme	5.	Malate to pyruvate

[A] A: 1; B: 3; C: 4; D: 5; E: 2 [B] A: 1; B: 2; C: 3; D: 4; E: 5
[C] A: 4; B: 1; C: 2; D: 3; E: 5 [D] A: 4; B: 2; C: 3; D: 5; E: 1

199. Match the list I with the list II and select the correct answer using the codes given below the lists

	List I [Terms]		List II [Significance]
A.	MS	1.	Use fragmentation pattern
B.	Mass transfer ratio	2.	Sieve elements per unit time
C.	Maternal inheritance	3.	A non Mendelian pattern
D.	Matrixules	4.	Protrusions of the outer and inner membrane in mitochondria
E.	Matric potential	5.	Related with hydrostatic pressure

[A] A: 1; B: 3; C: 4; D: 5; E: 2 [B] A: 1; B: 2; C: 3; D: 4; E: 5
[C] A: 4; B: 1; C: 2; D: 3; E: 5 [D] A: 4; B: 2; C: 3; D: 5; E: 1

200. Match the list I with the list II and select the correct answer using the codes given below the lists

	List I [Terms]		List II [Significance]
A.	Membrane permeability	1.	Movement of a substance
B.	Meristem identity genes	2.	For initial induction of the meristem
C.	Meristematic zone	3.	At the tip
D.	Meristemoids	4.	Upper and lower epidermis
E.	Mesophyll	5.	Gives trichomes or stomata

[A] A: 1; B: 2; C: 3; D: 5; E: 4 [B] A: 1; B: 2; C: 3; D: 4; E: 5
[C] A: 4; B: 1; C: 2; D: 3; E: 5 [D] A: 4; B: 2; C: 3; D: 5; E: 1

201. Match the list I with the list II and select the correct answer using the codes given below the lists

	List I [Terms]		List II [Significance]
A.	Lenticels	1.	Porous aggregation
B.	Leucoplasts	2.	Non-pigmented plastid
C.	LFR	3.	Low fluence response
D.	*LFY* gene	4.	For floral meristem identity
E.	*LHCB* gene family	5.	CAB protein

[A] A: 1; B: 3; C: 4; D: 5; E: 2 [B] A: 1; B: 2; C: 3; D: 4; E: 5
[C] A: 4; B: 1; C: 2; D: 3; E: 5 [D] A: 4; B: 2; C: 3; D: 5; E: 1

202. Match the list I with the list II and select the correct answer using the codes given below the lists

	List I [Terms]		List II [Significance]
A.	Light	1.	Radiant energy
B.	Lodging	2.	Bending
C.	Limit dextrinase	3.	Chlorophyll a/b antenna protein
D.	LHC protein	4.	Hydrolysis
E.	Light scattering	5.	Randomization

[A] A: 1; B: 3; C: 4; D: 5; E: 2 [B] A: 1; B: 2; C: 3; D: 4; E: 5
[C] A: 1; B: 2; C: 4; D: 3; E: 5 [D] A: 4; B: 2; C: 3; D: 5; E: 1

203. Match the list I with the list II and select the correct answer using the codes given below the lists

	List I [Terms]		List II [Significance]
A.	Ketose	1.	Terminal ketone
B.	Kinases	2.	Transfer phosphate groups from ATP to other molecules

C.	Kinematics	3.	Motor protein
D.	Kinesins	4.	Applicable to the motion of fluid particle and the shape changes that the fluids undergo
E.	Kinetin	5.	Autoclaved herring sperm DNA

[A] A: 1; B: 3; C: 4; D: 5; E: 2 [B] A: 1; B: 2; C: 3; D: 4; E: 5
[C] A: 4; B: 1; C: 2; D: 3; E: 5 [D] A: 1; B: 2; C: 4; D: 3; E: 5

204. Match the list I with the list II and select the correct answer using the codes given below the lists

	List I [Representative Groups of Hydrophillic Polypeptides Encoded by Cold-Responsive Genes Induced during Cold Acclimation]	List II [Polypeptide(s)]
A.	COR6.6	1. KIN1, KIN2
B.	COR15a	2. BN115, BN26
C.	COR78	3. CAP160
D.	WCS19	4. LTI65
E.	COR47	5. DHN5

[A] A: 1; B: 3; C: 4; D: 5; E: 2 [B] A: 1; B: 2; C: 3; D: 4; E: 5
[C] A: 4; B: 1; C: 2; D: 3; E: 5 [D] A: 1; B: 2; C: 4; D: 3; E: 5

205. Match the Group I with the Group II and select the correct answer using the codes given below the lists

	GROUP I [Hydrophilic polypeptides]		GROUP II [Example]
A.	LEA 1	1.	Em (wheat)
B.	LEAII	2.	WCS120(wheat)
C.	LEAIII	3.	HVA1(barley)
D.	Novel polypeptides	4.	CAS15(alfalfa)

[A] A: 4; B: 3; C: 2; D: 1 [B] A: 3; B: 2; C: 4; D: 1
[C] A: 1; B: 2; C: 3; D: 4 [D] A: 3; B: 2; C: 1; D: 4

206. Match the Group I with the Group II and select the correct answer using the codes given below the lists

	GROUP I [Mutant]		GROUP II [Phenotype]
A.	Sfr	1.	Defective in achieving freezing tolerance
B.	esk1	2.	Constitutively cold acclimated
C.	Sos	3.	Hypersensitive to salt
D.	San	4.	Seed germinate at high NaCl concentration

[A] A: 4; B: 3; C: 2; D: 1 [B] A: 3; B: 2; C: 4; D: 1
[C] A: 1; B: 2; C: 3; D: 4 [D] A: 3; B: 2; C: 1; D: 4

207. Match the Group I with the Group II and select the correct answer using the codes given below the lists

GROUP I [Mutant]		GROUP II [Responses]
A. Sfr	1.	For sensitive to freezing
B. esk1	2.	For eskimol
C. Sos	3.	For salt overly sensitive
D. San	4.	For salty land

[A] A: 4; B: 3; C: 2; D: 1 [B] A: 3; B: 2; C: 4; D: 1
[C] A: 1; B: 2; C: 3; D: 4 [D] A: 3; B: 2; C: 1; D: 4

208. Match the Group I with the Group II and select the correct answer using the codes given below the lists

GROUP I [GA-induced Enzymes in Cereal Aleurone Tissue]		GROUP II [Function]
A. Phosphocholine esterase	1.	Phospholipid metabolism
B. Phosphatase	2.	Phosphorus metabolism
C. Xylanases	3.	Digestion of cell wall
D. Isocitrate lyase	4.	Hydrolysis of lipid

[A] A: 4; B: 3; C: 2; D: 1 [B] A: 3; B: 2; C: 4; D: 1
[C] A: 1; B: 2; C: 3; D: 4 [D] A: 3; B: 2; C: 1; D: 4

209. Match the Group I with the Group II and select the correct answer using the codes given below the lists

GROUP I [GA-induced Enzymes in Cereal Aleurone Tissue]		GROUP II [Function]
A. Cysteine proteinases	1.	Hydrolysis of protein
B. α-Amylases	2.	Hydrolysis of starch
C. RNases	3.	Hydrolysis of nucleic acid
D. Isocitrate lyase	4.	Hydrolysis of lipid

[A] A: 4; B: 3; C: 2; D: 1 [B] A: 3; B: 2; C: 4; D: 1
[C] A: 1; B: 2; C: 3; D: 4 [D] A: 3; B: 2; C: 1; D: 4

210. Match the list I with the list II and select the correct answer using the codes given below the lists

List I [Crop species]		List II [Protein (%)]
A. Barley	1.	12
B. Dent corn (maize)	2.	10

C.	Oats	3.	13
D.	Rye	4.	12
E.	Wheat	5.	12

[A] A: 1; B: 3; C: 4; D: 5; E: 2 [B] A: 1; B: 2; C: 3; D: 4; E: 5
[C] A: 4; B: 1; C: 2; D: 3; E: 5 [D] A: 4; B: 2; C: 3; D: 5; E: 1

211. Match the Group I with the Group II and select the correct answer using the codes given below the lists

	List I [Crop species]		List II [Protein (%)]
A.	Broad bean	1.	23
B.	Garden pea	2.	25
C.	Peanut	3.	31
D.	Soybean	4.	37

[A] A: 4; B: 3; C: 2; D: 1 [B] A: 3; B: 2; C: 4; D: 1
[C] A: 1; B: 2; C: 3; D: 4 [D] A: 3; B: 2; C: 1; D: 4

212. Match the Group I with the Group II and select the correct answer using the codes given below the lists

	List I [Crop species]		List II [Protein (%)]
A.	Castor bean	1.	18
B.	Oil palm	2.	9
C.	Rapeseed	3.	21
D.	Pine	4.	35

[A] A: 4; B: 3; C: 2; D: 1 [B] A: 3; B: 2; C: 4; D: 1
[C] A: 1; B: 2; C: 3; D: 4 [D] A: 3; B: 2; C: 1; D: 4

213. Match the list I with the list II and select the correct answer using the codes given below the lists

	List I [Crop species]		List II [Oil (%)]
A.	Barley	1.	3
B.	Dent corn (maize)	2.	5
C.	Oats	3.	8
D.	Rye	4.	2
E.	Wheat	5.	2

[A] A: 1; B: 3; C: 4; D: 5; E: 2 [B] A: 1; B: 2; C: 3; D: 4; E: 5
[C] A: 4; B: 1; C: 2; D: 3; E: 5 [D] A: 4; B: 2; C: 3; D: 5; E: 1

214. Match the list I with the list II and select the correct answer using the codes given below the lists

List I [Crop species]	List II [Carbohydrate (%)]
A. Barley	1. 76
B. Dent corn (maize)	2. 80
C. Oats	3. 66
D. Rye	4. 76
E. Wheat	5. 75

[A] A: 1; B: 3; C: 4; D: 5; E: 2 [B] A: 1; B: 2; C: 3; D: 4; E: 5
[C] A: 4; B: 1; C: 2; D: 3; E: 5 [D] A: 4; B: 2; C: 3; D: 5; E: 1

215. Match the Group I with the Group II and select the correct answer using the codes given below the lists

List I [Crop species]	List II [Oil (%)]
A. Broad bean	1. 1
B. Garden pea	2. 6
C. Peanut	3. 48
D. Soybean	4. 17

[A] A: 4; B: 3; C: 2; D: 1 [B] A: 3; B: 2; C: 4; D: 1
[C] A: 1; B: 2; C: 3; D: 4 [D] A: 3; B: 2; C: 1; D: 4

216. Match the Group I with the Group II and select the correct answer using the codes given below the lists

List I [Crop species]	List II [Carbohydrate (%)]
A. Broad bean	1. 56
B. Garden pea	2. 52
C. Peanut	3. 12
D. Soybean	4. 26

[A] A: 4; B: 3; C: 2; D: 1 [B] A: 3; B: 2; C: 4; D: 1
[C] A: 1; B: 2; C: 3; D: 4 [D] A: 3; B: 2; C: 1; D: 4

217. Match the Group I with the Group II and select the correct answer using the codes given below the lists

List I [Crop species]	List II [Carbohydrate (%)]
A. Castor bean	1. Negligible
B. Oil palm	2. 28
C. Rapeseed	3. 19
D. Pine	4. 6

[A] A: 4; B: 3; C: 2; D: 1 [B] A: 3; B: 2; C: 4; D: 1
[C] A: 1; B: 2; C: 3; D: 4 [D] A: 3; B: 2; C: 1; D: 4

218. Match the Group I with the Group II and select the correct answer using the codes given below the lists

GROUP I [Classification of plant storage protein type]		GROUP II [Soluble in]	
A.	Albumins	1.	Water or dilute buffer at neutral pH
B.	Globulins	2.	Salt solution
C.	Prolamins	3.	Alcohols (70-90%)
D.	Glutelins	4.	Dilute acids or alkali

[A] A: 4; B: 3; C: 2; D: 1 [B] A: 3; B: 2; C: 4; D: 1
[C] A: 1; B: 2; C: 3; D: 4 [D] A: 3; B: 2; C: 1; D: 4

219. Match the list I with the list II and select the correct answer using the codes given below the lists

List I [Species]		List II [Genes responsive to ABA]	
A.	*Triticum aestivum*	1.	*EM*
B.	*Oryza sativa*	2.	*RAB16*
C.	*Hordenum vulgare*	3.	*HVA22*
D.	*Zea mays*	4.	*C1*
E.	*Craterostigma plantagineum*	5.	*CDeT27-45*

[A] A: 1; B: 3; C: 4; D: 5; E: 2 [B] A: 1; B: 2; C: 3; D: 4; E: 5
[C] A: 4; B: 1; C: 2; D: 3; E: 5 [D] A: 4; B: 2; C: 3; D: 5; E: 1

220. Match the list I with the list II and select the correct answer using the codes given below the lists

List I [Species (Dwarf Phenotype)]		List II [GA Response Mutant]	
A.	*Triticum aestivum*	1.	*Reduced height(Rht)(sd)*
B.	*Zea mays*	2.	*Dwarf(d8)(sd)*
C.	*Arabidopsis thaliana*	3.	*gai*(sd)
D.	*Silene armeria*	4.	*dwarf silence(r)*
E.	*Oryza sativa*	5.	*dwarf1*

[A] A: 1; B: 3; C: 4; D: 5; E: 2 [B] A: 1; B: 2; C: 3; D: 4; E: 5
[C] A: 4; B: 1; C: 2; D: 3; E: 5 [D] A: 4; B: 2; C: 3; D: 5; E: 1

221. Match the Group I with the Group II and select the correct answer using the codes given below the lists

List I [Species (Protein phosphatase]		List II [Cation required]	
A.	PP1	1.	Not known till date
B.	PP2A	2.	Not required
C.	PP2B	3.	Calcium ions
D.	PP2C	4.	Magnesium ions

[A] A: 4; B: 3; C: 2; D: 1 [B] A: 3; B: 2; C: 4; D: 1
[C] A: 1; B: 2; C: 3; D: 4 [D] A: 3; B: 2; C: 1; D: 4

222. Match the Group I with the Group II and select the correct answer using the codes given below the lists

	GROUP I [Genus]		GROUP II [Some of Representative Receptor-like Protein Kinase]
A.	Zea	1.	ZmPK1
B.	Brassica	2.	SRKs
C.	Arabidopsis	3.	ARK1-3
D.	Oryza	4.	OsPK10

[A] A: 4; B: 3; C: 2; D: 1 [B] A: 3; B: 2; C: 4; D: 1
[C] A: 1; B: 2; C: 3; D: 4 [D] A: 3; B: 2; C: 1; D: 4

223. Match the Group I with the Group II and select the correct answer using the codes given below the lists

	GROUP I [Genus]		GROUP II [Some of Representative Receptor-like Protein Kinase]
A.	Phaseolus	1.	PvRK20-1
B.	Catharanthus	2.	CrRLK1
C.	Triticum	3.	WLRK
D.	Peutunia	4.	PRK1

[A] A: 4; B: 3; C: 2; D: 1 [B] A: 3; B: 2; C: 4; D: 1
[C] A: 1; B: 2; C: 3; D: 4 [D] A: 3; B: 2; C: 1; D: 4

224. Match the list I with the list II and select the correct answer using the codes given below the lists

	List I [Different type of light]		List II [Red/FarRed]
A.	Daylight	1.	1.19
B.	Sunset	2.	0.96
C.	Moonlight	3.	0.94
D.	Ivy canopy	4.	0.13
E.	At a depth of 5m in soil	5.	0.88

[A] A: 1; B: 3; C: 4; D: 5; E: 2 [B] A: 1; B: 2; C: 3; D: 4; E: 5
[C] A: 4; B: 1; C: 2; D: 3; E: 5 [D] A: 4; B: 2; C: 3; D: 5; E: 1

225. Match the list I with the list II and select the correct answer using the codes given below the lists

	List I [Different type of light]		List II [Photon flux density ($\mu molm^{-2}s^{-1}$)]
A.	Daylight	1.	1900
B.	Sunset	2.	26.5
C.	Moonlight	3.	0.005
D.	Ivy canopy	4.	17.1
E.	At a depth of 5m in soil	5.	8.6

[A] A: 1; B: 3; C: 4; D: 5; E: 2 [B] A: 1; B: 2; C: 3; D: 4; E: 5
[C] A: 4; B: 1; C: 2; D: 3; E: 5 [D] A: 4; B: 2; C: 3; D: 5; E: 1

226. Match the list I with the list II and select the correct answer using the codes given below the lists

List I [Molecularly Characterized Intermediate in Phytochrome Signaling]			List II [Specific action]
A.	PIF3	1.	A positive regulator of phyB
B.	PKS1	2.	Negatively regulates phyB
C.	NDPK2	3.	Positive regulator of phyA and phyB in cotyledon opening and hook straightening
D.	FAR1	4.	Positively regulates phyA
E.	SPA1	5.	Negative regulator of phyA

[A] A: 1; B: 3; C: 4; D: 5; E: 2 [B] A: 1; B: 2; C: 3; D: 4; E: 5
[C] A: 4; B: 1; C: 2; D: 3; E: 5 [D] A: 4; B: 2; C: 3; D: 5; E: 1

227. Match the list I with the list II and select the correct answer using the codes given below the lists

List I [Molecularly Characterized Intermediate in Phytochrome Signaling]			List II [Characteristic]
A.	PIF3	1.	bHLH transcription factor
B.	PKS1	2.	Substrate for phytochrome activity (cytoplasm location)
C.	NDPK2	3.	Substrate for phytochrome kinase activity (cytochrome and nucleus)
D.	FAR1	4.	Protein with a coiled -coil domain (nucleus)
E.	SPA1	5.	WD repeat protein (nucleus)

[A] A: 1; B: 3; C: 4; D: 5; E: 2 [B] A: 1; B: 2; C: 3; D: 4; E: 5
[C] A: 4; B: 1; C: 2; D: 3; E: 5 [D] A: 4; B: 2; C: 3; D: 5; E: 1

228. Match the list I with the list II and select the correct answer using the codes given below the lists

List I [IAA polar transport mutants in *Arabidopsis*]			List II [Possible site of lesion]
A.	*aux1*	1.	Auxin influx carrier, active in roots
B.	*tir* 3	2.	Possible defect in NPA-binding protein

C. *pin1*	3. Auxin efflux carrier, specific to shoots
D. *Mp*	4. Defective embryo development and defective vasculature in adult plant
E. *lop1*	5. Defective leaf vasculature and morphology

[A] A: 1; B: 3; C: 4; D: 5; E: 2 [B] A: 1; B: 2; C: 3; D: 4; E: 5
[C] A: 4; B: 1; C: 2; D: 3; E: 5 [D] A: 4; B: 2; C: 3; D: 5; E: 1

229. Match the list I with the list II and select the correct answer using the codes given below the lists

List I [Bacteria]	List II [Association type]
A. *Pseudomonas amygdale*	1. Almond canker
B. *P. solanacearum*	2. Bacterial wilt
C. *Erwinia herbicola*	3. Gypsophila gall
D. *Rhodococus fascians*	4. Witches broom
E. *Azotobacrer* Species	5. Rhizosphere

[A] A: 1; B: 3; C: 4; D: 5; E: 2 [B] A: 1; B: 2; C: 3; D: 4; E: 5
[C] A: 4; B: 1; C: 2; D: 3; E: 5 [D] A: 4; B: 2; C: 3; D: 5; E: 1

230. Match the list I with the list II and select the correct answer using the codes given below the lists

List I [Fungi]	List II [Association type]
A. *Rhizopogon roseolus*	1. Ectomyorrhizae
B. *Helminthosporium* sp.	2. Green island on leaves
C. *Dictyostelium*	3. Free living
D. *Sphaceloma maihoticola*	4. Pathogen
E. *Cercospora rosicola*	5. Rose spot

[A] A: 1; B: 3; C: 4; D: 5; E: 2 [B] A: 1; B: 2; C: 3; D: 4; E: 5
[C] A: 4; B: 1; C: 2; D: 3; E: 5 [D] A: 4; B: 2; C: 3; D: 5; E: 1

231. Match the list I with the list II and select the correct answer using the codes given below the lists

List I [Selectable genes]	List II [Enzyme encoded]
A. *hpt* (antibiotic)	1. Hygromycin phosphotransferase
B. *npt II* (antibiotic)	2. Neomycin phosphotransferase
C. *aro A* (herbicide)	3. 5-Enolpyruvylshikimate-3-phosphate synthase
D. *CAT*	4. Chloramphenicol acetyl transferase
E. *GUS*	5. β- Glucuronidase

[A] A: 1; B: 3; C: 4; D: 5; E: 2 [B] A: 1; B: 2; C: 3; D: 4; E: 5
[C] A: 4; B: 1; C: 2; D: 3; E: 5 [D] A: 4; B: 2; C: 3; D: 5; E: 1

232. Match the list I with the list II and select the correct answer using the codes given below the lists

List I [Climacteric fruits]	List II [Production rate of ethylene (nl.g-1.h-1)]
A. Avocado	1. 0.02-130
B. Banana	2. 0.02-20
C. Cerimoya	3. 0.02-80
D. Cantaloupe	4. 0.10-320
E. Tomato	5. 0.02-20

[A] A: 1; B: 3; C: 4; D: 5; E: 2 [B] A: 1; B: 2; C: 3; D: 4; E: 5
[C] A: 4; B: 1; C: 2; D: 3; E: 5 [D] A: 1; B: 2; C: 4; D: 3; E: 5

233. Match the list I with the list II and select the correct answer using the codes given below the lists

List I [Cloned genes that act relatively early in embryogenesis]	List II [Site of Action]
A. *GNOM*	1. Apical-basal polarity
B. *MONOPTEROS*	2. Axialization and vascular development
C. *FACKEL*	3. Proteoderm separation
D. *KNOLLE*	4. Membranes, affected cell division/ growth
E. *SCARECROW*	5. Separation of cortex and endodermis or starch-sheath cells

[A] A: 1; B: 3; C: 4; D: 5; E: 2 [B] A: 1; B: 2; C: 3; D: 4; E: 5
[C] A: 1; B: 2; C: 4; D: 3; E: 5 [D] A: 4; B: 2; C: 3; D: 5; E: 1

234. Match the list I with the list II and select the correct answer using the codes given below the lists

List I [Cloned genes that act relatively early in embryogenesis]	List II [Site of action]
A. *CLAVATA 1*	1. Regulation of shoot meristem size
B. *AINTEGUMENTA*	2. Regulation of size of lateral organs
C. *L1*	3. Specification of tunica 1 layer
D. *GL2*	4. Specification of a trichoblasts in root epidermis
E. *WOL*	5. Specification of vascular tissues

[A] A: 1; B: 3; C: 4; D: 5; E: 2 [B] A: 1; B: 2; C: 3; D: 4; E: 5
[C] A: 4; B: 1; C: 2; D: 3; E: 5 [D] A: 4; B: 2; C: 3; D: 5; E: 1

235. Match the list I with the list II and select the correct answer using the codes given below the lists

List I [Enzymes responsible for modifications in wall architecture]		List II [Action]
A. Cellulase	1.	Break the β-1-4 link in the glucan backbone of xyloglucans(XGs)
B. Xyloglucan endotransglycosylases	2.	Cleave the β-1-4 link in XGs at specific laces, the glycosidic link before the unsubstituted glucose residue
C. Expansins	3.	Thought to brake hydrogen bonds between cellulose fibrils and XGs
D. Pectin methylesterases	4.	Hydrolyze methyl ester of PGA and thus crate free carboxyl groups that can link with Ca^{+2} ions and stabilize two PGA chains
E. Polygalacturonases	5.	Break the α(1-4) bonds between GalA units in PGA and lead to solublization of pectins

[A] A: 1; B: 3; C: 4; D: 5; E: 2 [B] A: 1; B: 2; C: 3; D: 4; E: 5
[C] A: 4; B: 1; C: 2; D: 3; E: 5 [D] A: 4; B: 2; C: 3; D: 5; E: 1

236. Match the list I with the list II and select the correct answer using the codes given below the lists

List I [Common wall protein]		List II [Amount of Protein (%) approximate]
A. Extensins (dicot)	1.	45
B. Extensins (monocot)	2.	70
C. Solanaceous lectins	3.	55
D. Glycine-rich glycoprotein (dicot)	4.	80-100
E. Proline-rich glycoprotein	5.	100

[A] A: 1; B: 2; C: 3; D: 5; E: 4 [B] A: 1; B: 2; C: 3; D: 4; E: 5
[C] A: 4; B: 1; C: 2; D: 3; E: 5 [D] A: 4; B: 2; C: 3; D: 5; E: 1

237. Match the list I with the list II and select the correct answer using the codes given below the lists

	List I [Bacterial Species]		List II [Host plants]
A.	*Sinorhizobium meliloti*	1.	*Trigonella*
B.	*Sinorhizobium fredii*	2.	*Vigna*
C.	*Rhizobium leguminososarum* biovar *trifoli*	3.	*Trifolium*
D.	*Rhizobium tropici*	4.	*Medicago*
E.	*Rhizobium etli*	5.	*Phaseouls*

[A] A: 1; B: 3; C: 4; D: 5; E: 2 [B] A: 1; B: 2; C: 3; D: 4; E: 5
[C] A: 4; B: 1; C: 2; D: 3; E: 5 [D] A: 4; B: 2; C: 3; D: 5; E: 1

238. Match the list I with the list II and select the correct answer using the codes given below the lists

	List I [Source of natural and anthropogenic nitrogen fixation]		List II [Amount of nitrogen fixed]
A.	Lightening	1.	<10 Tg/year
B.	Biological nitrogen fixation in the terrestrial system	2.	90-140 Tg/year
C.	Biological nitrogen fixation in the marine systems	3.	80 Tg/year
D.	N-fertilizers synthesis	4.	30-300 Tg/year
E.	Fossils fuel combustion	5.	>20 Tg/year

[A] A: 1; B: 3; C: 4; D: 5; E: 2 [B] A: 1; B: 2; C: 3; D: 4; E: 5
[C] A: 1; B: 2; C: 4; D: 3; E: 5 [D] A: 4; B: 2; C: 3; D: 5; E: 1

239. Match the list I with the list II and select the correct answer using the codes given below the lists

	List I [Sink]		List II [Osmotic and turgor regulation]
A.	Developmental embryos	1.	Normal development requires constant apoplast osmotic concentration
B.	Legume seed coat	2.	Sucrose release rate is turgor dependent
C.	Corn root tip	3.	Turgor remain constant during sugar accumulation
D.	Sugar beet tap root	4.	Turgor is constant throughout the meristem and is unaffected by changing growth rate
E.	Potato tuber	5.	Starch synthesis and sugar uptake are turgor dependent

[A] A: 1; B: 3; C: 4; D: 5; E: 2 [B] A: 1; B: 2; C: 3; D: 4; E: 5
[C] A: 4; B: 1; C: 2; D: 3; E: 5 [D] A: 1; B: 2; C: 4; D: 3; E: 5

240. Match the list I with the list II and select the correct answer using the codes given below the lists

List I [General characteristic of Phloem exudates]	List II [Phloem exudates]
A. Sugars	1. 100-300 gl^{-1}
B. Amino acids	2. 5-40 $g\,l^{-1}$
C. Inorganic	3. 1-5 gl^{-1}
D. Total solutes	4. 7.3-8.0
E. pH	5. 250-1200 mmol kg^{-1}

[A] A: 1; B: 2; C: 3; D: 5; E: 4 [B] A: 1; B: 2; C: 3; D: 4; E: 5
[C] A: 4; B: 1; C: 2; D: 3; E: 5 [D] A: 4; B: 2; C: 3; D: 5; E: 1

241. Match the list I with the list II and select the correct answer using the codes given below the lists

List I [General characteristic of Xylem exudates]	List II [Xylem exudates]
A. Sugars	1. 0.3 gl^{-1}
B. Amino acids	2. 0.1-2 gl^{-1}
C. Inorganic	3. 0.2-4 $g\,l^{-1}$
D. Total solutes	4. 10-100 mmol kg^{-1}
E. pH	5. 5.0-6.5

[A] A: 1; B: 3; C: 4; D: 5; E: 2 [B] A: 1; B: 2; C: 3; D: 4; E: 5
[C] A: 4; B: 1; C: 2; D: 3; E: 5 [D] A: 4; B: 2; C: 3; D: 5; E: 1

242. Match the list I with the list II and select the correct answer using the codes given below the lists

List I [Stoichiometry of the photosystem in different oxygen-evolving system and in response to change in light quality]		List II [PSII/PSI ratio]
A. Cyanobacteria	1.	0.4
B. Intermittent-light developing plastids	2.	4.1
C. Green algae	3.	1.4
D. Tobacco mutant *Su/Su*	4.	2.7
E. Barley chlorine mutant	5.	3.0

[A] A: 1; B: 3; C: 4; D: 5; E: 2 [B] A: 1; B: 2; C: 3; D: 4; E: 5
[C] A: 4; B: 1; C: 2; D: 3; E: 5 [D] A: 4; B: 2; C: 3; D: 5; E: 1

243. Match the list I with the list II and select the correct answer using the codes given below the lists

List I [Protein subunits of the PSII core complex]	List II [Molecular mass (kDa)]
A. D1	1. 32
B. D2	2. 34
C. CP47	3. 43
D. CP43	4. 51
E. Cytb559 (α-subunit)	5. 9

[A] A: 1; B: 3; C: 4; D: 5; E: 2 [B] A: 1; B: 2; C: 3; D: 4; E: 5
[C] A: 1; B: 2; C: 4; D: 3; E: 5 [D] A: 4; B: 2; C: 3; D: 5; E: 1

244. Match the list I with the list II and select the correct answer using the codes given below the lists

List I [Distribution of the photosynthetic components in chloroplast membrane region]	List II [Appresed]
A. PSII	1. 85
B. PSI	2. 10
C. Cytochrome *b6f* complex	3. 90
D. LHC-II	4. 50
E. ATP synthase	5. 0

[A] A: 1; B: 3; C: 4; D: 5; E: 2 [B] A: 1; B: 2; C: 3; D: 4; E: 5
[C] A: 4; B: 1; C: 2; D: 3; E: 5 [D] A: 1; B: 2; C: 4; D: 3; E: 5

245. Match the list I with the list II and select the correct answer using the codes given below the lists

List I [Distribution of the photosynthetic components in chloroplast membrane region]	List II [Stroma exposed]
A. PSII	1. 15
B. PSI	2. 90
C. Cytochrome *b6f* complex	3. 50
D. LHC-II	4. 100
E. ATP synthase	5. 10

[A] A: 1; B: 2; C: 3; D: 5; E: 4 [B] A: 1; B: 2; C: 3; D: 4; E: 5
[C] A: 4; B: 1; C: 2; D: 3; E: 5 [D] A: 4; B: 2; C: 3; D: 5; E: 1

246. Match the list I with the list II and select the correct answer using the codes given below the lists

List I [Eukaryotic initiation factors (elFs)]		List II [Roles in initiating translation]
A. eIF1	1.	Multiple effects in enhancing initiation of complex formation
B. eIF2	2.	GTP-dependent recognition of Met-tRNA and nucleotide exchange
C. eIF3	3.	Ribosomal subunit dissociation
D. eIF4	4.	Recognizes 5'cap mRNA.
E. eIF5	5.	Promotes elF2 GTPase activity and release of factors from ribosome

[A] A: 1; B: 3; C: 4; D: 5; E: 2 [B] A: 1; B: 2; C: 3; D: 4; E: 5
[C] A: 4; B: 1; C: 2; D: 3; E: 5 [D] A: 4; B: 2; C: 3; D: 5; E: 1

247. Match the list I with the list II and select the correct answer using the codes given below the lists

List I [Lipid composition of plasma membranes from cold-acclimated species]		List II [*Arabidopsis* leaf (% mole)]
A. Phospholipids	1.	47
B. Free sterols	2.	38
C. Steryl glucosides	3.	3
D. Acylated steryl glycosides	4.	5
E. Glucocerebrosides	5.	7

[A] A: 1; B: 3; C: 4; D: 5; E: 2 [B] A: 1; B: 2; C: 3; D: 4; E: 5
[C] A: 1; B: 2; C: 4; D: 3; E: 5 [D] A: 4; B: 2; C: 3; D: 5; E: 1

248. Match the list I with the list II and select the correct answer using the codes given below the lists

List I [Physiologically active alkaloids used in modern medicine]		List II [Plant Source]
A. Ajmaline	1.	*Rauwolfia serpentine*
B. Atropine	2.	*Hyoscyamus niger*
C. Caffeine	3.	*Camptotheca acuminate*
D. Camptothecin	4.	*Coffea Arabica*
E. Cocaine	5.	*Erythroxylon coca*

[A] A: 1; B: 3; C: 4; D: 5; E: 2 [B] A: 1; B: 2; C: 3; D: 4; E: 5
[C] A: 4; B: 1; C: 2; D: 3; E: 5 [D] A: 1; B: 2; C: 4; D: 3; E: 5

249. Match the list I with the list II and select the correct answer using the codes given below the lists

List I [Physiologically active alkaloids used in modern medicine]	List II [Plant Source]
A. Codeine	1. *Papaver somniferum*
B. Coniine	2. *Conium maculatum*
C. Emetine	3. *Uragoga ipecacuanth*
D. Morphine	4. *Nicotiana tabacum*
E. Nicotine	5. *P. somniferum*

[A] A: 1; B: 2; C: 3; D: 5; E: 4 [B] A: 1; B: 2; C: 3; D: 4; E: 5
[C] A: 4; B: 1; C: 2; D: 3; E: 5 [D] A: 4; B: 2; C: 3; D: 5; E: 1

250. Match the list I with the list II and select the correct answer using the codes given below the lists

List I [Physiologically active alkaloids used in modern medicine]	List II [Plant Sources]
A. Pilocarpine	1. *Pilocarpus jaborandi*
B. Quinine	2. *Cinchona officinalis*
C. Sanguinarine	3. *Eschscholzia californica*
D. Scopolamine	4. *H. niger*
E. Strychnine	5. *Strychnos nux-vomica*

[A] A: 1; B: 3; C: 4; D: 5; E: 2 [B] A: 1; B: 2; C: 3; D: 4; E: 5
[C] A: 4; B: 1; C: 2; D: 3; E: 5 [D] A: 4; B: 2; C: 3; D: 5; E: 1

251. Match the list I with the list II and select the correct answer using the codes given below the lists

List I [Physiologically active alkaloids used in modern medicine]	List II [Use]
A. Ajmaline	1. Antiarrythmic
B. Atropine	2. Antidote for nerve gas poisioning
C. Caffeine	3. Potent anticancerous agent
D. Camptothecin	4. Stimulant of central nervous system
E. Cocaine	5. Topical anesthetic

[A] A: 1; B: 3; C: 4; D: 5; E: 2 [B] A: 1; B: 2; C: 3; D: 4; E: 5
[C] A: 1; B: 2; C: 4; D: 3; E: 5 [D] A: 4; B: 2; C: 3; D: 5; E: 1

252. Match the list I with the list II and select the correct answer using the codes given below the lists

List I [Physiologically active alkaloids used in modern medicine]	List II [Use]
A. Codeine	1. Analgesic and antitussive
B. Coniine	2. Used in homeopathy
C. Emetine	3. Adductive drug
D. Morphine	4. Amoebicide
E. Nicotine	5. Highly toxic

[A] A: 1; B: 3; C: 4; D: 5; E: 2 [B] A: 1; B: 2; C: 3; D: 4; E: 5
[C] A: 4; B: 1; C: 2; D: 3; E: 5 [D] A: 1; B: 2; C: 4; D: 3; E: 5

253. Match the list I with the list II and select the correct answer using the codes given below the lists

List I [Physiologically active alkaloids used in modern medicine]	List II [Use]
A. Pilocarpine	1. Used to treat glaucoma
B. Quinine	2. Antimalarial
C. Sanguinarine	3. Antibacterial
D. Scopolamine	4. Powerful narcotic
E. Strychnine	5. Violent titanic poison

[A] A: 1; B: 3; C: 4; D: 5; E: 2 [B] A: 1; B: 2; C: 3; D: 4; E: 5
[C] A: 4; B: 1; C: 2; D: 3; E: 5 [D] A: 4; B: 2; C: 3; D: 5; E: 1

254. Match the list I with the list II and select the correct answer using the codes given below the lists

List I [HSPs Classes]	List II [Size (kDa)]
A. HSP100	1. 100-114
B. HSP90	2. 80-90
C. HSP70	3. 69-71
D. HSP60	4. 15-30
E. smHSP	5. 60

[A] A: 1; B: 2; C: 3; D: 5; E: 4 [B] A: 1; B: 2; C: 3; D: 4; E: 5
[C] A: 4; B: 1; C: 2; D: 3; E: 5 [D] A: 4; B: 2; C: 3; D: 5; E: 1

255. Match the list I with the list II and select the correct answer using the codes given below the lists

	List I [Antioxidative enzymes]		List II [Subcellular locations]
A.	Ascorbate peroxidase	1.	Plastid membrane
B.	Catalase	2.	Perioxisome
C.	Dehydroascorbate reductase	3.	Root nodules
D.	Glutathione reductase	4.	Plastid stroma
E.	Superoxide dismutase (Mn-SOD)	5.	Mitochondria

[A] A: 1; B: 3; C: 4; D: 5; E: 2 [B] A: 1; B: 2; C: 3; D: 4; E: 5
[C] A: 4; B: 1; C: 2; D: 3; E: 5 [D] A: 4; B: 2; C: 3; D: 5; E: 1

256. Match the list I with the list II and select the correct answer using the codes given below the lists

	List I [Sequences and possible functions of some conserved regulatory elements associated with genes that encode seed storage proteins]		List II [Structural gene]
A.	CATGCATG	1.	Seed specific gene
B.	TGTAAAG	2.	Prolamine and legumine gene
C.	CACA	3.	Legume and cereals
D.	GCCAC(C/T)TC	4.	Lectine SP gene
E.	AA/G/CCCA	5.	7S soybean SP gene

[A] A: 1; B: 3; C: 4; D: 5; E: 2 [B] A: 1; B: 2; C: 3; D: 4; E: 5
[C] A: 1; B: 2; C: 4; D: 3; E: 5 [D] A: 4; B: 2; C: 3; D: 5; E: 1

257. Match the list I with the list II and select the correct answer using the codes given below the lists

	List I [Class]		List II [Examples including derived products]
A.	Polyterpens	1.	Rubber
B.	Tetraterpens	2.	Carotenes
C.	Triterpens	3.	Steroids
D.	Diterpens	4.	GA
E.	Sesquiterpens	5.	ABA

[A] A: 1; B: 3; C: 4; D: 5; E: 2 [B] A: 1; B: 2; C: 3; D: 4; E: 5
[C] A: 4; B: 1; C: 2; D: 3; E: 5 [D] A: 1; B: 2; C: 4; D: 3; E: 5

258. Match the list I with the list II and select the correct answer using the codes given below the lists

	List I [Enzyme]		List II [EC]
A.	Ascorbate oxidase	1.	1.10.3.3
B.	Phenol oxidase	2.	1.10.3.1/2
C.	Urate oxidase	3.	1.7.3.3
D.	Glycollate oxidase	4.	1.1.3.1
E.	D-Amino acid oxidase	5.	1.4.3.3

[A] A: 1; B: 3; C: 4; D: 5; E: 2 [B] A: 1; B: 2; C: 3; D: 4; E: 5
[C] A: 1; B: 2; C: 4; D: 3; E: 5 [D] A: 4; B: 2; C: 3; D: 5; E: 1

259. Match the list I with the list II and select the correct answer using the codes given below the lists

	List I [Terms]		List II [Significance]
A.	Prosthetic group	1.	Metal ion or small carbon
B.	Prosystemin	2.	Activation of jasmonic acid
C.	Protease inhibitors	3.	Expression is activated by release of jasmonic acid in target tissues
D.	Proteasome	4.	26S
E.	Protein bodies	5.	Protein storage organelles enclosed by a single unit membrane, found mainly in seed tissues

[A] A: 1; B: 3; C: 4; D: 5; E: 2 [B] A: 1; B: 2; C: 3; D: 4; E: 5
[C] A: 4; B: 1; C: 2; D: 3; E: 5 [D] A: 4; B: 2; C: 3; D: 5; E: 1

260. Match the list I with the list II and select the correct answer using the codes given below the lists

	List I [Terms]		List II [Significance]
A.	Protoderm	1.	Generate epidermis
B.	Protofilaments	2.	Heterodimers
C.	Protomeristem	3.	Become root or shoot meristem upon germination
D.	Proton gradient	4.	Proton motive force
E.	Protonema	5.	Moss spores

[A] A: 1; B: 2; C: 3; D: 5; E: 4 [B] A: 1; B: 2; C: 3; D: 4; E: 5
[C] A: 4; B: 1; C: 2; D: 3; E: 5 [D] A: 4; B: 2; C: 3; D: 5; E: 1

261. Match the list I with the list II and select the correct answer using the codes given below the lists

	List I [Common name]		List II [Trade name]
A.	Gibberellic acid	1.	Gibbrel
B.	Ethephon	2.	Ethrel, Florel
C.	Phenyl Urea	3.	CPPU
D.	Chlormequate	4.	CCC
E.	Dikigulac	5.	Atrinal

[A] A: 1; B: 3; C: 4; D: 5; E: 2 [B] A: 1; B: 2; C: 3; D: 4; E: 5
[C] A: 4; B: 1; C: 2; D: 3; E: 5 [D] A: 4; B: 2; C: 3; D: 5; E: 1

262. Match the list I with the list II and select the correct answer using the codes given below the lists

	List I [Common name]		List II [Important Uses]
A.	Gibberellic acid	1.	Production of parthenocarpic fruits
B.	Ethephon	2.	Sex expression
C.	Phenyl Urea	3.	Senescence control
D.	Chlormequate	4.	Fruit set
E.	Dikigulac	5.	Promotion of lateral branching

[A] A: 1; B: 3; C: 4; D: 5; E: 2 [B] A: 1; B: 2; C: 3; D: 4; E: 5
[C] A: 1; B: 2; C: 4; D: 3; E: 5 [D] A: 4; B: 2; C: 3; D: 5; E: 1

263. Match the list I with the list II and select the correct answer using the codes given below the lists

	List I [Terms]		List II [Significances]
A.	Water uses efficiency	1.	Dry matter production
B.	Waxes	2.	Esters of higher aliphatic alcohols and higher fatty acids of longer chain length than the fatty acids
C.	Weed	3.	Plant out of place
D.	Wilting range	4.	Ultimate wilting point
E.	WSD	5.	For measuring water deficits in plants

[A] A: 1; B: 3; C: 4; D: 5; E: 2 [B] A: 1; B: 2; C: 3; D: 4; E: 5
[C] A: 4; B: 1; C: 2; D: 3; E: 5 [D] A: 4; B: 2; C: 3; D: 5; E: 1

264. Match the list I with the list II and select the correct answer using the codes given below the lists

List I [Terms]	List II [Significances]
A. Tonoplast	1. The cytoplasmic membrane surrounding the vacuole which regulates the passage of materials into and out of the cell sap
B. VP	2. The pressure exerted against the walls of the container and the surface of the water by the water vapour
C. VPD	3. Melchers
D. Vernalin	4. The difference between the actual vapor pressure in the atmosphere and that of the saturated atmosphere at the same temperature
E. Vernalization	5. Early flowering

[A] A: 1; B: 3; C: 4; D: 5; E: 2 [B] A: 1; B: 2; C: 3; D: 4; E: 5
[C] A: 4; B: 1; C: 2; D: 3; E: 5 [D] A: 1; B: 2; C: 4; D: 3; E: 5

265. Match the list I with the list II and select the correct answer using the codes given below the lists

List I [Terms]	List II [Significance]
A. Thermoperiodism	1. Alteration of temperature
B. Thigmotropism	2. Physical contact
C. Thixotropy	3. Reversible sol-gel transformation
D. TLC	4. Stationary phase is thinly coated onto glass, plastic or foil plates
E. Tomato fix	5. Improve fruit set in tomato

[A] A: 1; B: 3; C: 4; D: 5; E: 2 [B] A: 1; B: 2; C: 3; D: 4; E: 5
[C] A: 4; B: 1; C: 2; D: 3; E: 5 [D] A: 4; B: 2; C: 3; D: 5; E: 1

266. Match the list I with the list II and select the correct answer using the codes given below the lists

List I [Terms]	List II [Significance]
A. Symplast	1. The complex of cytoplasm and interconnecting strands
B. Synchronous aging	2. It occurs when the leaves senesce all at once as in deciduous plants in the fall
C. Syndrome	3. A group of factors, sign and symptoms that collectively indicate a disease or disorder

D.	Syneresis	4.	Decreased ability of a gel to hold water with ageing
E.	Synergism	5.	Positive interaction effect of two or more factors which is more than their individuals effects combined

[A] A: 1; B: 3; C: 4; D: 5; E: 2 [B] A: 1; B: 2; C: 3; D: 4; E: 5
[C] A: 4; B: 1; C: 2; D: 3; E: 5 [D] A: 4; B: 2; C: 3; D: 5; E: 1

267. Match the list I with the list II and select the correct answer using the codes given below the lists

	List I [Terms]		List II [Significance]
A.	Single cell protein	1.	Refers to dried cells of micro organisms
B.	Singlet state	2.	Pauli's principle
C.	Sink	3.	Growing towards the darkest sector
D.	Skototropism	4.	Involves in utilization and storage of photosynthesis
E.	Senescence	5.	It refers to aging of plant or its part

[A] A: 1; B: 3; C: 4; D: 5; E: 2 [B] A: 1; B: 2; C: 3; D: 4; E: 5
[C] A: 4; B: 1; C: 2; D: 3; E: 5 [D] A: 1; B: 2; C: 4; D: 3; E: 5

268. Match the list I with the list II and select the correct answer using the codes given below the lists

	List I [Terms]		List II [Significance]
A.	Recalcitrant seeds	1.	Jack fruit, Litchi
B.	Reciprocal crosses	2.	Interchange of male and female parents
C.	Recombinant DNA	3.	Changing the genome of an organism
D.	Red drop	4.	Emerson and Lewis
E.	Redox reaction	5.	One compound is oxidized and the other reduced

[A] A: 1; B: 3; C: 4; D: 5; E: 2 [B] A: 1; B: 2; C: 3; D: 4; E: 5
[C] A: 4; B: 1; C: 2; D: 3; E: 5 [D] A: 4; B: 2; C: 3; D: 5; E: 1

269. Match the list I with the list II and select the correct answer using the codes given below the lists

	List I [Terms]		List II [Significance]
A.	Relaxation time	1.	The time that a plant takes to recover to its original position after the stimulus has been removed
B.	Relative water content	2.	It is the ratio of water content of a tissue to that of the same when the tissue is fully turgid expressed as percentage

C.	Relative retension time	3.	The time of the emergence of the sample relative to that of the standard is determined to give the relative retension time, which is constant for a given column under different conditions
D.	Relative humidity	4.	The actual amount of water vapour present in an atmosphere as percentage of that saturated with water vapour at the same temperature
E.	Repression	5.	The process by which chromosomal DNA is prevented from acting as template

[A] A: 1; B: 3; C: 4; D: 5; E: 2 [B] A: 1; B: 2; C: 3; D: 4; E: 5
[C] A: 4; B: 1; C: 2; D: 3; E: 5 [D] A: 4; B: 2; C: 3; D: 5; E: 1

270. Match the list I with the list II and select the correct answer using the codes given below the lists

	List I [Terms]		List II [Significance]
A.	Proteoglycans	1.	Polysaccharides covalently attached to protein
B.	Protoplasm	2.	It is the basis of life
C.	Protoplast	3.	Nearest to the center
D.	Protoxylem	4.	Protoplasm and the vacuole of the cell are together called protoplast
E.	Proximal	5.	Nearest to the main body of the organism

[A] A: 1; B: 3; C: 4; D: 5; E: 2 [B] A: 1; B: 2; C: 3; D: 4; E: 5
[C] A: 1; B: 2; C: 4; D: 3; E: 5 [D] A: 4; B: 2; C: 3; D: 5; E: 1

271. Match the list I with the list II and select the correct answer using the codes given below the lists

	List I [Terms]		List II [Significance]
A.	Potential energy	1.	Energy of position and bound energy
B.	Potometer	2.	Used for the measuring the transpiration
C.	Precipitation	3.	ACC
D.	Precursor	4.	Colloids of the gelatin and gum arabic
E.	PLC	5.	Objective is the collection of sample components for subsequent identification or use

[A] A: 1; B: 3; C: 4; D: 5; E: 2 [B] A: 1; B: 2; C: 3; D: 4; E: 5
[C] A: 4; B: 1; C: 2; D: 3; E: 5 [D] A: 1; B: 2; C: 4; D: 3; E: 5

272. Match the list I with the list II and select the correct answer using the codes given below the lists

List I [Terms]		List II [Significance]
A. Plane polarized light	1.	Nicol Prism
B. Plant biochemistry	2.	Molecular and chemical basis of plant processes
C. Plant biophysics	3.	Small amount is required
D. PGR	4.	Work and nature of energy
E. Plant physiology	5.	Branch of botany

[A] A: 1; B: 3; C: 4; D: 5; E: 2 [B] A: 1; B: 2; C: 3; D: 4; E: 5
[C] A: 4; B: 1; C: 2; D: 3; E: 5 [D] A: 1; B: 2; C: 4; D: 3; E: 5

273. Match the list I with the list II and select the correct answer using the codes given below the lists

List I [Terms]		List II [Significance]
A. Orthodox seeds	1.	Cereals, legumes
B. Osmolarity	2.	The molarity of an ideal solution such as dextrose having the same osmotic pressure as the unknown solution
C. Osmole	3.	Apparatus
D. Osmometer	4.	Gram molecular weight of osmotically active particles in 1000g of solvent
E. Osmoprotectants	5.	Organic solutes produced as a result of salt stress

[A] A: 1; B: 3; C: 4; D: 5; E: 2 [B] A: 1; B: 2; C: 3; D: 4; E: 5
[C] A: 4; B: 1; C: 2; D: 3; E: 5 [D] A: 1; B: 2; C: 4; D: 3; E: 5

274. Match the list I with the list II and select the correct answer using the codes given below the lists

List I [Terms]		List II [Significance]
A. Nastic movement	1.	Epinasty of leaves in response to ethylene
B. Natrophilic	2.	Sodium loving
C. Necrosis	3.	Local death of cells due to disintegration of the cellular structure with destruction of the nucleus and liquification of the cytoplasm
D. Nephelometry	4.	Measures the intensity of the radiation scattered by a suspension and is commonly used for estimating the concentration of the microorganism
E. NAR	5.	Unit Leaf Rate

[A] A: 1; B: 3; C: 4; D: 5; E: 2 [B] A: 1; B: 2; C: 3; D: 4; E: 5
[C] A: 4; B: 1; C: 2; D: 3; E: 5 [D] A: 4; B: 2; C: 3; D: 5; E: 1

275. Match the list I with the list II and select the correct answer using the codes given below the lists

	List I [Terms]		List II [Significance]
A.	Degree of hydration of protoplasm	1.	Wall pressure
B.	Stretches cell walls and may lead to their enlargement if the walls are sufficiently plastic	2.	Turgor pressure
C.	Increases the free energy of water molecules	3.	Osmotic potential
D.	Indicates the direction of diffusion of water	4.	Water potential
E.	Boron	5.	Terminal leaf necrosis

[A] A: 1; B: 3; C: 4; D: 5; E: 2 [B] A: 1; B: 2; C: 3; D: 4; E: 5
[C] A: 4; B: 1; C: 2; D: 3; E: 5 [D] A: 1; B: 2; C: 4; D: 3; E: 5

276. Match the list I with the list II and select the correct answer using the codes given below the lists

	List I [Essential elements]		List II [Significance]
A.	Boron	1.	Terminal leaf necrosis
B.	Copper	2.	Die-back disease
C.	Iron	3.	Little leaf disease
D.	Zinc	4.	Intervenal chlorosis
E.	Mo	5.	Nitrate reductase

[A] A: 1; B: 3; C: 4; D: 5; E: 2 [B] A: 1; B: 2; C: 3; D: 4; E: 5
[C] A: 4; B: 1; C: 2; D: 3; E: 5 [D] A: 1; B: 2; C: 4; D: 3; E: 5

277. Match the list I with the list II and select the correct answer using the codes given below the lists

	List I [Essential elements]		List II [Significance]
A.	Chlorine	1.	Required for the O_2 evolution in photosynthesis
B.	Copper	2.	Constituent of plastocyanin
C.	Molybdenum	3.	Constituent of nitrate reductase
D.	Zinc	4.	Constituent of alcohol dehydrogenase
E.	Boron	5.	Necessary for the translocation of sugars

[A] A: 1; B: 3; C: 4; D: 5; E: 2 [B] A: 1; B: 2; C: 3; D: 4; E: 5
[C] A: 4; B: 1; C: 2; D: 3; E: 5 [D] A: 4; B: 2; C: 3; D: 5; E: 1

278. Match the list I with the list II and select the correct answer using the codes given below the lists

List I [Terms]		List II [Significance]
A. Field capacity	1.	Water holding capacity of a soil
B. Permanent wilting	2.	Loss of water from plant with little chance of regaining turgidity
C. Water deficit	3.	Forces which hold water to the soil
D. Soil moisture	4.	Combined effect of accelerated water loss and retarded water absorption
E. Boron	5.	Phloem transport

[A] A: 1; B: 3; C: 4; D: 5; E: 2 [B] A: 1; B: 2; C: 3; D: 4; E: 5
[C] A: 4; B: 1; C: 2; D: 3; E: 5 [D] A: 1; B: 2; C: 4; D: 3; E: 5

279. Match the list I with the list II and select the correct answer using the codes given below the lists

List I [Five classes of Heat Shock Proteins found in Plants]		List II [Examples in Prokaryotes]
A. HSP100	1.	ClpB and ClpA/C
B. HSP90	2.	HtpG
C. HSP70	3.	GroES
D. HSP60	4.	Dnak
E. smHSP	5.	AtHSP22

[A] A: 1; B: 3; C: 4; D: 5; E: 2 [B] A: 1; B: 2; C: 3; D: 4; E: 5
[C] A: 4; B: 1; C: 2; D: 3; E: 5 [D] A: 1; B: 2; C: 4; D: 3; E: 5

280. Match the list I with the list II and select the correct answer using the codes given below the lists

List I [Terms]		List II [Significance]
A. Antigibberellin	1.	Cycoel
B. Growth inhibitor	2.	ABA
C. Germination inhibitor	3.	Scopoletin
D. IAA-oxidase inhibitor	4.	Caffeic acid
E. Cell division	5.	Cytokinins

[A] A: 1; B: 3; C: 4; D: 5; E: 2 [B] A: 1; B: 2; C: 3; D: 4; E: 5
[C] A: 4; B: 1; C: 2; D: 3; E: 5 [D] A: 4; B: 2; C: 3; D: 5; E: 1

281. Match the list I with the list II and select the correct answer using the codes given below the lists

List I [Phenomena]		List II [Growth regulators]
A. Root differentiation	1.	Auxin plus gibberellins
B. Vegetative bud differentiation	2.	Auxin followed by kinins
C. Flower bud differentiation	3.	Auxin plus carbohydrate both
D. Fruit development	4.	Kinins followed by auxins
E. Callus development	5.	Both Auxin and Zeatin

[A] A: 1; B: 3; C: 4; D: 5; E: 2 [B] A: 1; B: 2; C: 3; D: 4; E: 5
[C] A: 4; B: 1; C: 2; D: 3; E: 5 [D] A: 1; B: 2; C: 4; D: 3; E: 5

282. Match the list I with the list II and select the correct answer using the codes given below the lists

List I [Compounds]		List II [Mode of Action]
A. Cyanide	1.	Inhibit cytochrome oxidase
B. Antimycin A	2.	Block electron transfer from cytochrome -b to cytochrome -c1
C. Rotenone	3.	Prevent electron transfer from Fe-S center to ubiquione
D. DCMU	4.	Competes with QB for binding size in PSII
E. Oligomycin	5.	Inhibit F0 an CF0

[A] A: 1; B: 3; C: 4; D: 5; E: 2 [B] A: 1; B: 2; C: 3; D: 4; E: 5
[C] A: 4; B: 1; C: 2; D: 3; E: 5 [D] A: 4; B: 2; C: 3; D: 5; E: 1

283. Match the list I with the list II and select the correct answer using the codes given below the lists

List I [Compounds]		List II [Mode of action]
A. DCCD	1.	Blocks proton flow through F0 and CF0
B. FCCP	2.	Hydrophobic proton carrier
C. Valinomycin	3.	In brown fat, forms proton conducting pores in inner mitochondrial membrane
D. Thermogenin	4.	K^+-ionophore
E. Atractyloside	5.	Inhibits Adenine nucleotide translocase

[A] A: 1; B: 3; C: 4; D: 5; E: 2 [B] A: 1; B: 2; C: 3; D: 4; E: 5
[C] A: 4; B: 1; C: 2; D: 3; E: 5 [D] A: 1; B: 2; C: 4; D: 3; E: 5

284. Match the list I with the list II and select the correct answer using the codes given below the lists

	List I [Various enzymes of the Calvin cycle]		List II [Reaction]
A.	Ribulose-5-phosphate kinase	1.	6-Ribulose-5-phosphate + 6ATP to 6 ribulose-1,5-bisphosphate + 6 ADP+17 Pi
B.	Ribose-5-phosphate isomerase	2.	2-Ribose-5-posphate to 2 ribulose-5-phosphate
C.	Ribulose-5-phosphate epimerase	3.	4-Xylulose-5-phosphate to 4 ribulose-5-phosphate
D.	Transketolase	4.	2-Sedoheptulose-7-phosphate + 2 glyceraldehyde-3-phosphate to 2 ribose-5-phosphate + 2 xylulose 5-phosphate
E.	Sedoheptulose-1,7-bisphosphatase	5.	2-Sedoheptulose-7-phosphate + 2 glyceraldehyde-3-phosphate to 2 ribose-5-phospate + 2 xylulose 5 phosphate

[A] A: 1; B: 3; C: 4; D: 5; E: 2 [B] A: 1; B: 2; C: 3; D: 4; E: 5
[C] A: 4; B: 1; C: 2; D: 3; E: 5 [D] A: 4; B: 2; C: 3; D: 5; E: 1

285. Match the list I with the list II and select the correct answer using the codes given below the lists

	List I [Variables]		List II [C3-plants]
A.	Photorespiration	1.	Extensive
B.	Perform Calvin-Benson cycle	2.	Yes
C.	Primary CO_2 acceptor	3.	Rubisco
D.	CO_2-fixing enzymes	4.	RuBP
E.	First product of CO_2 fixation	5.	3PGA

[A] A: 1; B: 3; C: 4; D: 5; E: 2 [B] A: 1; B: 2; C: 3; D: 4; E: 5
[C] A: 4; B: 1; C: 2; D: 3; E: 5 [D] A: 1; B: 2; C: 4; D: 3; E: 5

286. Match the list I with the list II and select the correct answer using the codes given below the lists

	List I [Variables]		List II [C4-plants]
A.	Photorespiration	1.	Minimal
B.	Perform Calvin-Benson cycle	2.	Yes
C.	Primary CO_2 acceptor	3.	PEP carboxylase and Rubisco

D.	CO_2-fixing enzymes	4.	PEP
E.	First product of CO_2 fixation	5.	Oxaloacetate

[A] A: 1; B: 3; C: 4; D: 5; E: 2 [B] A: 1; B: 2; C: 3; D: 4; E: 5
[C] A: 4; B: 1; C: 2; D: 3; E: 5 [D] A: 1; B: 2; C: 4; D: 3; E: 5

287. Match the list I with the list II and select the correct answer using the codes given below the lists

	List I [Inhibitors tools to investigate the electron transport chain of plant mitochondria]		List II [Act on]
A.	Rotenone	1.	Complex I
B.	Antimycin	2.	Complex III
C.	Cyanide	3.	Alternative oxidase
D.	Salicylhydroxamic acid	4.	Complex IV
E.	Atractyloside	5.	Inhibits adenine nucleotide translocase

[A] A: 1; B: 3; C: 4; D: 5; E: 2 [B] A: 1; B: 2; C: 3; D: 4; E: 5
[C] A: 4; B: 1; C: 2; D: 3; E: 5 [D] A: 1; B: 2; C: 4; D: 3; E: 5

288. Match the list I with the list II and select the correct answer using the codes given below the lists

	List I [Free living nitrogen fixing organisms]		List II [Contributions]
A.	Chemotrophic bacteria	1.	*Thiobacillus ferriooxidans*
B.	Photosynthetic bacteria	2.	*Rhodospirillium rubrum*
C.	Cyanobacteria unicellular	3.	*Oscillatoria* sp.
D.	Cyanobacteria filamentous	4.	*Gloeothece* sp.
E.	Cyanobacteria heterocystous	5.	*Anabaena, Nostoc*

[A] A: 1; B: 3; C: 4; D: 5; E: 2 [B] A: 1; B: 2; C: 3; D: 4; E: 5
[C] A: 4; B: 1; C: 2; D: 3; E: 5 [D] A: 1; B: 2; C: 4; D: 3; E: 5

289. Match the list I with the list II and select the correct answer using the codes given below the lists

	List I [Plant Host]		List II [Rhizobial symbiont]
A.	*Parasponia*	1.	*Bradyrhizobium* sp.
B.	Soyabean	2.	*Sinorhizobium fredii*
C.	Alfalfa	3.	*Azorhizobium*
D.	Sesbania	4.	*Sinorhizobium meliloti*
E.	Bean	5.	*Rhizobium leguminosarum*

[A] A: 1; B: 3; C: 4; D: 5; E: 2 [B] A: 1; B: 2; C: 3; D: 4; E: 5
[C] A: 4; B: 1; C: 2; D: 3; E: 5 [D] A: 1; B: 2; C: 4; D: 3; E: 5

290. Match the list I with the list II and select the correct answer using the codes given below the lists

List I [Terms]	List II [Significance]
A. Biotin	1. Carboxylation
B. Coenzyme A	2. Acyl group transfer
C. Thymine pyrophosphate	3. Active aldehyde transfer
D. Uridine diphosphate	4. Sugar transfer
E. Conversion of glucose to pyruvate	5. Glycolysis to yield reduced Co-enzymes

[A] A: 1; B: 3; C: 4; D: 5; E: 2 [B] A: 1; B: 2; C: 3; D: 4; E: 5
[C] A: 4; B: 1; C: 2; D: 3; E: 5 [D] A: 1; B: 2; C: 4; D: 3; E: 5

291. Match the list I with the list II and select the correct answer using the codes given below the lists

List I [Group]	List II [Stage of development and photo- reversible response induced by phytochrome]
A. Angiosperms	1. Seed
B. Gymnosperms	2. Seedlings
C. Pteridophytes	3. Germling
D. Bryophytes	4. Young gametophyte
E. Chlorophyta	5. Mature gametophyte

[A] A: 1; B: 3; C: 4; D: 5; E: 2 [B] A: 1; B: 2; C: 3; D: 4; E: 5
[C] A: 1; B: 2; C: 4; D: 3; E: 5 [D] A: 4; B: 2; C: 3; D: 5; E: 1

292. Match the list I with the list II and select the correct answer using the codes given below the lists

List I [Alkaloid class]	List II [Biosynthetic precursor]
A. Piperridine	1. Lysine
B. Pyrrolizidine	2. Ornithine
C. Quinolizidine	3. Tyrosine
D. Isoquinoline	4. Lysine
E. Indole	5. Tryptophan

[A] A: 1; B: 3; C: 4; D: 5; E: 2 [B] A: 1; B: 2; C: 3; D: 4; E: 5
[C] A: 1; B: 2; C: 4; D: 3; E: 5 [D] A: 4; B: 2; C: 3; D: 5; E: 1

293. Match the list I with the list II and select the correct answer using the codes given below the lists

	List I [Alkaloid class]		List II [Example]
A.	Piperridine	1.	Coline
B.	Pyrrolizidine	2.	Retrorshine
C.	Quinolizidine	3.	Lupinine
D.	Isoquinoline	4.	Codeine Morphine
E.	Indole	5.	Psilocybin

[A] A: 1; B: 3; C: 4; D: 5; E: 2 [B] A: 1; B: 2; C: 3; D: 4; E: 5
[C] A: 4; B: 1; C: 2; D: 3; E: 5 [D] A: 4; B: 2; C: 3; D: 5; E: 1

294. Match the list I with the list II and select the correct answer using the codes given below the lists

	List I [Alkaloid class]		List II [Human Use]
A.	Piperridine	1.	Paralyzes motor neurons
B.	Pyrrolizidine	2.	None
C.	Quinolizidine	3.	Restoration of heart rhythm
D.	Isoquinoline	4.	Analgesic
E.	Indole	5.	Hallucinogen

[A] A: 1; B: 3; C: 4; D: 5; E: 2 [B] A: 1; B: 2; C: 3; D: 4; E: 5
[C] A: 4; B: 1; C: 2; D: 3; E: 5 [D] A: 4; B: 2; C: 3; D: 5; E: 1

295. Match the list I with the list II and select the correct answer using the codes given below the lists

	List I [Process]		List II [Protein(s)/Reaction]
A.	Wounding	1.	Proteinase inhibitors
B.	Pathogen attack	2.	Thionins
C.	Elicitor application	3.	Phytoalexin-producing enzymes
D.	Contact	4.	Tendril movement
E.	Drought stress	5.	Dehydrins

[A] A: 1; B: 3; C: 4; D: 5; E: 2 [B] A: 1; B: 2; C: 3; D: 4; E: 5
[C] A: 4; B: 1; C: 2; D: 3; E: 5 [D] A: 4; B: 2; C: 3; D: 5; E: 1

296. Match the list I with the list II and select the correct answer using the codes given below the lists

	List I [Stress]		List II [Enzyme]
A.	Osmotic	1.	Delta-pyrrolling -5-carboxylate synthetase

B.	Drought and salinity	2.	Mannitol-1-phosphate dehydrogenase
C.	Cold and salt	3.	Choline dehydrogenase (bet A)
D.	Salt	4.	Choline oxidase (cod A)
E.	Cold	5.	Omega-3-fatty acid desaturase

[A] A: 1; B: 3; C: 4; D: 5; E: 2 [B] A: 1; B: 2; C: 3; D: 4; E: 5
[C] A: 1; B: 2; C: 4; D: 3; E: 5 [D] A: 4; B: 2; C: 3; D: 5; E: 1

297. Match the list I with the list II and select the correct answer using the codes given below the lists

	List I [Terms]		List II [Significance]
A.	Ecology	1.	The study of living organisms in relation to environment and their effects on each other
B.	Economic yield	2.	Economically useful part of total dry matter, producing by a plant or crop in unit land area
C.	Ecosystem	3.	It includes not only the biotic communities but also its physical and chemical environments such as lithosphere, hydrosphere and atmosphere
D.	Ecotype	4.	A variety or strain within a given species adapted to a specific environment
E.	Elasticity	5.	Reversible extension

[A] A: 1; B: 3; C: 4; D: 5; E: 2 [B] A: 1; B: 2; C: 3; D: 4; E: 5
[C] A: 4; B: 1; C: 2; D: 3; E: 5 [D] A: 4; B: 2; C: 3; D: 5; E: 1

298. Match the list I with the list II and select the correct answer using the codes given below the lists

	List I [Terms]		List II [Significance]
A.	Endosperm	1.	The reserve food in seeds, which develops from the fertilization of the polar nuclei of the ovule by the second male nucleus and nourishes the embryo
B.	Endothermic reaction	2.	Those reactions that proceed at the expense of energy supplied from outside are endothermic reactions
C.	Energy	3.	Heat content of a system
D.	Enthalpy	4.	Capacity to do work
E.	Entropy	5.	The randomness of the energy of a system that is less capable of doing work then the initial non random energy

[A] A: 1; B: 3; C: 4; D: 5; E: 2 [B] A: 1; B: 2; C: 3; D: 4; E: 5
[C] A: 1; B: 2; C: 4; D: 3; E: 5 [D] A: 4; B: 2; C: 3; D: 5; E: 1

299. Match the list I with the list II and select the correct answer using the codes given below the lists

List I [Terms]		List II [Significance]
A. Epistasis	1.	One gene influencing the phenotypic expression of another gene
B. Epithem	2.	Mass of thin walled, loosely arranged cells bordering the chamber of hydathodes
C. Epitope	3.	Is the ratio of antibody and antigen
D. Equivalency	4.	Structure is predictable by variable region antibody binding sites
E. Escape	5.	Because of timing sequence in development and growth, some plants are not subjected to stress

[A] A: 1; B: 3; C: 4; D: 5; E: 2 [B] A: 1; B: 2; C: 3; D: 4; E: 5
[C] A: 1; B: 2; C: 4; D: 3; E: 5 [D] A: 4; B: 2; C: 3; D: 5; E: 1

300. Match the list I with the list II and select the correct answer using the codes given below the lists

List I [Terms]		List II [Significance]
A. Fluorescence	1.	Excited singlet state to ground state
B. Foliar nutrition	2.	Nutrition take up by leaves
C. Foot candle	3.	Required for initiation of flowering
D. Fractional photoperiodic inductioh	4.	An unit of illumination
E. Free energy	5.	Fraction of the total energy of a system that can do work at constant temperature and pressure

[A] A: 1; B: 3; C: 4; D: 5; E: 2 [B] A: 1; B: 2; C: 3; D: 4; E: 5
[C] A: 1; B: 2; C: 4; D: 3; E: 5 [D] A: 4; B: 2; C: 3; D: 5; E: 1

301. Match the list I with the list II and select the correct answer using the codes given below the lists

List I [Terms]		List II [Significance]
A. Girdling or ringing	1.	Removal of a ring of bark
B. Glass electrode	2.	Solution of standard pH
C. Global warming	3.	Formation of glucose from non-carbohydrate precursors, mainly through reversal of citric acid cycle and glycolysis

D.	Gluconeogenesis	4.	Industrialization
E.	Glycolipids	5.	Substituted lipids containing carbohydrates and nitrogen

[A] A: 1; B: 3; C: 4; D: 5; E: 2 [B] A: 1; B: 2; C: 3; D: 4; E: 5
[C] A: 1; B: 2; C: 4; D: 3; E: 5 [D] A: 4; B: 2; C: 3; D: 5; E: 1

302. Match the list I with the list II and select the correct answer using the codes given below the lists

	List I [Terms]		List II [Significance]
A.	Derived lipid	1.	Compounds obtained from lipids, generally their break down products
B.	Desiccation tolerance	2.	Protoplasmic resistance to injury under extreme water stress conditions
C.	Detoxification	3.	Series of successive and orderly transformation
D.	Development	4.	Living cells can counter the harmful effects of toxic substance
E.	Devernalization	5.	High temperature

[A] A: 1; B: 3; C: 4; D: 5; E: 2 [B] A: 1; B: 2; C: 3; D: 4; E: 5
[C] A: 1; B: 2; C: 4; D: 3; E: 5 [D] A: 4; B: 2; C: 3; D: 5; E: 1

303. Match the list I with the list II and select the correct answer using the codes given below the lists

	List I [Terms]		List II [Significance]
A.	Lectins	1.	A group of plant proteins which bind to specific carbohydrate
B.	Ligands	2.	A substance that interact specifically and reversibly with a biological substance
C.	Ligate	3.	Opening in the bark through which gas exchange can take place in the stems of woody plants
D.	Lenticels	4.	The biological substances that is specifically bound by an immobilized to a solid support
E.	LAI	5.	Watson

[A] A: 1; B: 3; C: 4; D: 5; E: 2 [B] A: 1; B: 2; C: 3; D: 4; E: 5
[C] A: 1; B: 2; C: 4; D: 3; E: 5 [D] A: 4; B: 2; C: 3; D: 5; E: 1

304. Match the list I with the list II and select the correct answer using the codes given below the lists

	List I [Terms]		List II [Significance]
A.	Light quality	1.	Composition of light in terms of wave length of monochromatic radiation
B.	Light reaction	2.	Photolysis of water
C.	Light saturation intensity	3.	The point in light intensity at which an increase in it fails to speed up photosynthesis
D.	Lignin	4.	Constituent of woody tissue
E.	Longevity	5.	Life span of a living plant or its part

[A] A: 1; B: 3; C: 4; D: 5; E: 2 [B] A: 1; B: 2; C: 3; D: 4; E: 5
[C] A: 4; B: 1; C: 2; D: 3; E: 5 [D] A: 4; B: 2; C: 3; D: 5; E: 1

305. Match the list I with the list II and select the correct answer using the codes given below the lists

	List I [Terms]		List II [Significance]
A.	Immunogen	1.	Used to induce an immune response
B.	Immunoglobulin	2.	A class of globular protein
C.	Immunology	3.	Loss of turgidity by the leaf cells is only partial
D.	Incipient wilting	4.	Study of immune system
E.	Incomplete dominance	5.	F1 phenotype

[A] A: 1; B: 3; C: 4; D: 5; E: 2 [B] A: 1; B: 2; C: 3; D: 4; E: 5
[C] A: 1; B: 2; C: 4; D: 3; E: 5 [D] A: 4; B: 2; C: 3; D: 5; E: 1

306. Match the list I with the list II and select the correct answer using the codes given below the lists

	List I [Terms]		List II [Significance]
A.	Primary growth	1.	Growth, which is initiated by apical stem and root meristem
B.	Primary root system	2.	The primary root and its branches are collectively called the primary root system
C.	Primary wall	3.	It is the first production in cell wall synthesis
D.	Primer	4.	It is an organ in the early stages of development
E.	Primordium	5.	A short DNA fragment used for the initiation of DNA synthesis with DNA polymerase

[A] A: 1; B: 2; C: 3; D: 5; E: 4 [B] A: 1; B: 2; C: 3; D: 4; E: 5
[C] A: 4; B: 1; C: 2; D: 3; E: 5 [D] A: 4; B: 2; C: 3; D: 5; E: 1

307. Match the list I with the list II and select the correct answer using the codes given below the lists

List I [Terms]		List II [Significance]
A. Protoplast fusion	1.	Fusion of two protoplasts beginning with firm adhesion between the bounding membranes of the two protoplasts
B. Protoplast	2.	Cell without cell membrane
C. Protoplasmic particulates	3.	The bodies in protoplasm are sometimes referred to as protoplasmic particulates
D. Protoplasm	4.	Higher molecular weight
E. Protopectins	5.	Basis of life

[A] A: 1; B: 2; C: 3; D: 5; E: 4 [B] A: 1; B: 2; C: 3; D: 4; E: 5
[C] A: 4; B: 1; C: 2; D: 3; E: 5 [D] A: 4; B: 2; C: 3; D: 5; E: 1

308. Match the list I with the list II and select the correct answer using the codes given below the lists

List I [Terms]		List II [Significance]
A. Spherosomes	1.	Equivalent to the lysosomes
B. Spontaneous reactions	2.	These can take place without a net supply of energy
C. Staling products	3.	Products of the metabolism of an organ or an organism which accumulate in the medium in which growth is taking place and which depress the activity of the growing cells
D. Starch	4.	Coloum packing material which is coated to a solid support in GLC
E. Stationary phase	5.	Composed of two polysaccharids

[A] A: 1; B: 2; C: 3; D: 5; E: 4 [B] A: 1; B: 2; C: 3; D: 4; E: 5
[C] A: 4; B: 1; C: 2; D: 3; E: 5 [D] A: 4; B: 2; C: 3; D: 5; E: 1

309. Match the list I with the list II and select the correct answer using the codes given below the lists

List I [Terms]		List II [Significance]
A. Haulm	1.	The above ground portion of the a tuber crop like potato
B. Heat rigor	2.	A plant is said to have passes into a state of heat rigor in the range of temperature in which it neither grows nor exhibits growth movements

C.	Heredity	3.	Transmission of the characteristics from one generation to successive generation
D.	Heteroblastic development of leaves	4.	Plants having different forms of leaves
E.	Heterophyllus plants	5.	Production of leaves with varying length and breadth at different nodes reflecting the various stages of development

[A] A: 1; B: 2; C: 3; D: 5; E: 4 [B] A: 1; B: 2; C: 3; D: 4; E: 5

[C] A: 4; B: 1; C: 2; D: 3; E: 5 [D] A: 4; B: 2; C: 3; D: 5; E: 1

310. Match the list I with the list II and select the correct answer using the codes given below the lists

	List I [Terms]		List II [Significance]
A.	Dormin	1.	A bud dormancy inducing substance isolated from the leaves of Acer
B.	Drip culture	2.	Nutrient solution are added to the solid culture by dripping on the surface
C.	Drought avoidance	3.	Prevention of loss of water by closing stomata resulting in the maintenance of turgor during periods conducive for high transpiration rates
D.	Drought escape	4.	The capacity of surviving conditions of drought with little or no injury
E.	Drought tolerance	5.	Complete their life cycle before the onset of drought and thus escape drought

[A] A: 1; B: 2; C: 3; D: 5; E: 4 [B] A: 1; B: 2; C: 3; D: 4; E: 5

[C] A: 4; B: 1; C: 2; D: 3; E: 5 [D] A: 4; B: 2; C: 3; D: 5; E: 1

311. Match the list I with the list II and select the correct answer using the codes given below the lists

	List I [Terms]		List II [Significance]
A.	Growth analysis	1.	Method of analyzing growth
B.	Growth correlation	2.	The controlling effect exerted by one part over the other
C.	Growth curve	3.	Some characteristic is plotted against time or age

D. Growth movement	4. The ratio of the increase in size to the time
E. Growth rate	5. Tropisms

[A] A: 1; B: 2; C: 3; D: 5; E: 4 [B] A: 1; B: 2; C: 3; D: 4; E: 5
[C] A: 4; B: 1; C: 2; D: 3; E: 5 [D] A: 4; B: 2; C: 3; D: 5; E: 1

312. Match the list I with the list II and select the correct answer using the codes given below the lists

List I [Terms]	List II [Significance]
A. Hybridization of DNA	1. DNA-DNA association
B. Hybrid vigour	2. Heterosis
C. Hydration	3. Association of water
D. Hydathodes	4. Breakdown of a compound with the addition of water
E. Hydrolysis	5. Pore-like structures in the leaves

[A] A: 1; B: 2; C: 3; D: 5; E: 4 [B] A: 1; B: 2; C: 3; D: 4; E: 5
[C] A: 4; B: 1; C: 2; D: 3; E: 5 [D] A: 4; B: 2; C: 3; D: 5; E: 1

313. Match the list I with the list II and select the correct answer using the codes given below the lists

List I [Terms]		List II [R/FR ratio]
A. Wheat canopy filtered light	1.	0.5
B. Maize canopy filtered light	2.	0.20
C. Oak woodland canopy filtered light	3.	0.12
D. Maple wood canopy filtered light	4.	0.15-0.33
E. Spruce forest woodland canopy	5.	0.14-0.28

[A] A: 1; B: 2; C: 3; D: 5; E: 4 [B] A: 1; B: 2; C: 3; D: 4; E: 5
[C] A: 4; B: 1; C: 2; D: 3; E: 5 [D] A: 4; B: 2; C: 3; D: 5; E: 1

314. Match the list I with the list II and select the correct answer using the codes given below the lists

List I [Endangered Medicinal Plants]		List II [Part Used]
A. *Aconitum balfourli*	1.	Roots
B. *Acorus calamus*	2.	Rhizomes
C. *Aquilaria malaccensis*	3.	Wood
D. *Amebia benethamii*	4.	Danti
E. *Baliospermum*	5.	Leaves, flower

[A] A: 1; B: 2; C: 3; D: 5; E: 4 [B] A: 1; B: 2; C: 3; D: 4; E: 5
[C] A: 4; B: 1; C: 2; D: 3; E: 5 [D] A: 4; B: 2; C: 3; D: 5; E: 1

315. Match the list I with the list II and select the correct answer using the codes given below the lists

List I [Endangered Medicinal Plants]		List II [Part used]
A. *Berberia petiolaris*	1.	Bark of root
B. *Bergenia ciliate*	2.	Rhizome
C. *Bunium persicum*	3.	Fruits
D. *Butea monosperma*	4.	Bulb
E. *Fritillaria royei*	5.	Leaves, Root and Bark

[A] A: 1; B: 2; C: 3; D: 5; E: 4 [B] A: 1; B: 2; C: 3; D: 4; E: 5
[C] A: 4; B: 1; C: 2; D: 3; E: 5 [D] A: 4; B: 2; C: 3; D: 5; E: 1

316. Match the list I with the list II and select the correct answer using the codes given below the lists

List I [Medical terms related to Human]		List II [Significance]
A. Abortifacient	1.	Drug or agent inducing expulsions of a non-viable fetus
B. Abscess	2.	Localized collection of pus formed as reaction to pyegenic organisms
C. Acne	3.	A skin condition common in adolescence
D. Acute	4.	Any disease or pathological condition
E. Affection	5.	Short and severe, not long drawn out or chronic

[A] A: 1; B: 2; C: 3; D: 5; E: 4 [B] A: 1; B: 2; C: 3; D: 4; E: 5
[C] A: 4; B: 1; C: 2; D: 3; E: 5 [D] A: 4; B: 2; C: 3; D: 5; E: 1

317. Match the list I with the list II and select the correct answer using the codes given below the lists

List I [Medical term related to Human]		List II [Significance]
A. Ague	1.	Malarial fever with successive stages of fever and chills
B. Alopecia	2.	Baldness which can be congenital, premature or senile
C. Alterative	3.	Able to restore health
D. Amenorrhoea	4.	Infection of large intestine by the protozoa *Entamoeba histolytica*
E. Amoebiasis	5.	Absence of the menses

[A] A: 1; B: 2; C: 3; D: 5; E: 4 [B] A: 1; B: 2; C: 3; D: 4; E: 5
[C] A: 4; B: 1; C: 2; D: 3; E: 5 [D] A: 4; B: 2; C: 3; D: 5; E: 1

318. Match the list I with the list II and select the correct answer using the codes given below the lists

List I [Scientific name]		List II [Chromosome number (2n)]
A. *Haplopapuus gracilis*	1.	4
B. *Crepis capillaries*	2.	6
C. *Trillium grandiflorum*	3.	10
D. *Spinacea oleracea*	4.	14
E. *Hordenum vulgare*	5.	12

[A] A: 1; B: 2; C: 3; D: 5; E: 4 [B] A: 1; B: 2; C: 3; D: 4; E: 5
[C] A: 4; B: 1; C: 2; D: 3; E: 5 [D] A: 4; B: 2; C: 3; D: 5; E: 1

319. Match the list I with the list II and select the correct answer using the codes given below the lists

List I [Terms]	List II [Significance]
A. Salicylic acid	1. A benzoic acid derivative believed to be an endogenous signal for *SAR*
B. Huxley	2. Collected Essays
C. Weismann	3. Studies in the Theory of Descent
D. Haeckel	4. Materials for the study of variations
E. Bateson	5. General Morphology of Organism

[A] A: 1; B: 2; C: 3; D: 5; E: 4 [B] A: 1; B: 2; C: 3; D: 4; E: 5
[C] A: 4; B: 1; C: 2; D: 3; E: 5 [D] A: 4; B: 2; C: 3; D: 5; E: 1

320. Match the list I with the list II and select the correct answer using the codes given below the lists

List I [Terms]	List II [Examples]
A. Periderm	1. Cork cells
B. Collenchyma	2. Collenchyma cells
C. Sclerenchyma	3. Fibers
D. Xylem	4. Trichomes
E. Epidermis	5. Vessels membrane

[A] A: 1; B: 2; C: 3; D: 5; E: 4 [B] A: 1; B: 2; C: 3; D: 4; E: 5
[C] A: 4; B: 1; C: 2; D: 3; E: 5 [D] A: 4; B: 2; C: 3; D: 5; E: 1

321. Match the list I with the list II and select the correct answer using the codes given below the lists

List I [Reductant/Oxidant]		List II [Em(mV)]
A. Ferredoxin reduction/oxidation	1.	-430
B. $H_2/2H^+$	2.	-420
C. NADH + H^+/NAD	3.	-320
D. Succinate/fumarate	4.	+40
E. Ubiquione red/ox	5.	+30

[A] A: 1; B: 2; C: 3; D: 5; E: 4 [B] A: 1; B: 2; C: 3; D: 4; E: 5
[C] A: 4; B: 1; C: 2; D: 3; E: 5 [D] A: 4; B: 2; C: 3; D: 5; E: 1

322. Match the list I with the list II and select the correct answer using the codes given below the lists

List I [Radiation of Principle Interest to Biologists]		List II [Wavelength Range(nm)]
A. UV-C	1.	100-280
B. UV-B	2.	280-320
C. UV-A	3.	320-400
D. Ultraviolet	4.	700-740
E. Far-red	5.	100-400

[A] A: 1; B: 2; C: 3; D: 5; E: 4 [B] A: 1; B: 2; C: 3; D: 4; E: 5
[C] A: 4; B: 1; C: 2; D: 3; E: 5 [D] A: 4; B: 2; C: 3; D: 5; E: 1

323. Match the list I with the list II and select the correct answer using the codes given below the lists

List I [Significant features]		List II [C3-plant]
A. Photorespiration	1.	Yes
B. CO_2 compensation	2.	20-100
C. Photosynthesis optimum temp	3.	20-25
D. Quantum yield as a function of temperature	4.	500-1000
E. Transpiration ratio	5.	Declining

[A] A: 1; B: 2; C: 3; D: 5; E: 4 [B] A: 1; B: 2; C: 3; D: 4; E: 5
[C] A: 4; B: 1; C: 2; D: 3; E: 5 [D] A: 4; B: 2; C: 3; D: 5; E: 1

324. Match the list I with the list II and select the correct answer using the codes given below the lists

List I [Plant Species]		List II [(Approximate lipid content (% dry weight)]
A. Macadamia nut	1.	75
B. Hazal nut	2.	65
C. Safflower	3.	50
D. Sun flower	4.	45
E. Canola	5.	40

[A] A: 1; B: 2; C: 3; D: 5; E: 4 [B] A: 1; B: 2; C: 3; D: 4; E: 5
[C] A: 4; B: 1; C: 2; D: 3; E: 5 [D] A: 4; B: 2; C: 3; D: 5; E: 1

325. Match the list I with the list II and select the correct answer using the codes given below the lists

List I [Molecules]		List II [Molecular mass (Da)]
A. Water	1.	18
B. Hydrogen sulphide	2.	34
C. Ammonia	3.	17
D. Carbon dioxide	4.	44
E. Methane	5.	16

[A] A: 1; B: 3; C: 4; D: 5; E: 2 [B] A: 1; B: 2; C: 3; D: 4; E: 5
[C] A: 4; B: 1; C: 2; D: 3; E: 5 [D] A: 4; B: 2; C: 3; D: 5; E: 1

326. Match the list I with the list II and select the correct answer using the codes given below the lists

List I [Molecules]		List II [Dielectric constants]
A. Water	1.	78.4
B. Methanol	2.	33.6
C. Ethanol	3.	24.3
D. Benzene	4.	2.3
E. Hexane	5.	1.9

[A] A: 1; B: 3; C: 4; D: 5; E: 2 [B] A: 1; B: 2; C: 3; D: 4; E: 5
[C] A: 4; B: 1; C: 2; D: 3; E: 5 [D] A: 4; B: 2; C: 3; D: 5; E: 1

327. Match the list I with the list II and select the correct answer using the codes given below the lists

List I [Plants]		List II [Duration of the juvenile phase (Years)]
A. Apple	1.	6-8
B. Scotch pine	2.	5-10

C.	Pear	3.	8-12
D.	Perilla	4.	<0.25
E.	Bryophyllum	5.	1-2

[A] A: 1; B: 3; C: 4; D: 5; E: 2 [B] A: 1; B: 2; C: 3; D: 4; E: 5
[C] A: 4; B: 1; C: 2; D: 3; E: 5 [D] A: 4; B: 2; C: 3; D: 5; E: 1

328. Match the list I with the list II and select the correct answer using the codes given below the lists

	List I [Rhythm]		List II [Organism]
A.	Sleep movement	1.	Many species
B.	Stomatal opening	2.	Banana, Tobacco, Viccia
C.	Stem growth	3.	Tomato
D.	CO_2 production	4.	Dry onion seed
E.	Gas uptake	5.	Orchid flower

[A] A: 1; B: 2; C: 3; D: 5; E: 4 [B] A: 1; B: 2; C: 3; D: 4; E: 5
[C] A: 4; B: 1; C: 2; D: 3; E: 5 [D] A: 4; B: 2; C: 3; D: 5; E: 1

329. Match the list I with list II and select the correct answer using the codes given below the lists

	List I [Chilling resistant plant]		List II [Ratio of unsaturated/ saturated fatty acids]
1.	*Brassica oleracea* (bud)	P.	3.2
2.	*Brassica campestris* (turnip)	Q.	80
3.	*Pisum sativum* (pea)	R.	98

[A] 1: Q; 2: R; 3: P [B] 1: S; 2: P; 3: R
[C] 1: P; 2: Q; 3: R [D] None of the above

330. Match the Group I with Group II and select the correct answer using the codes given below

	Group I (Chilling-sensitive tissues)		Group II (Ratio of unsaturated and saturated fatty acid)
A.	*Phaseolus vulgaris*	1.	2.8
B.	*Ipomoea batatas*	2.	1.7
C.	*Zea mays*	3.	2.1
D.	*Lycopersicon esculentum*	4.	2.8

[A] A: 3; B: 1; C: 2; D:4 [B] A: 2; B: 1; C: 3; D: 4
[C] A: 1; B: 2; C: 3; D: 4 [D] A: 4; B: 3; C: 2; D: 1
[E] None of the above

331. Match the Group I with Group II and select the correct answer using the codes given below

Group I (Plant Source)	Group II (Gene)
A. *Arabidopsis*	1. *cor15a*
B. *Alfalfa*	2. *cas15*
C. *Wheat*	3. *wcs120*
D. *Barley*	4. *bva1*

[A] A: 3; B: 1; C: 2; D:4
[B] A: 2; B: 1; C: 3; D: 4
[C] A: 1; B: 2; C: 3; D: 4
[D] A: 4; B: 3; C: 2; D: 1
[E] None of the above

332. Match the Group I with Group II and select the correct answer using the codes given below

Group I (Plant Source)	Group II (Polypeptides)
A. *Arabidopsis*	1. cor15a
B. *Alfalfa*	2. cas15
C. *Wheat*	3. wcs120
D. *Barley*	4. bva1

[A] A: 3; B: 1; C: 2; D:4
[B] A: 2; B: 1; C: 3; D: 4
[C] A: 1; B: 2; C: 3; D: 4
[D] A: 4; B: 3; C: 2; D: 1
[E] None of the above

333. Match the Group I with Group II and select the correct answer using the codes given below

Group I (Plant Source)	Group II [Molecular mass of polypeptides (kD)]
A. *Arabidopsis*	1. 15
B. *Alfalfa*	2. 15
C. *Wheat*	3. 39
D. *Barley*	4. 22

[A] A: 3; B: 1; C: 2; D:4
[B] A: 2; B: 4; C: 3; D: 1
[C] A: 1; B: 2; C: 3; D: 4
[D] A: 4; B: 3; C: 2; D: 1
[E] None of the above

334. Match the Group I with Group II and select the correct answer using the codes given below

Group I (HSP family)	Group II (Probable function)
A. HSP110	1. Unknown
B. HSP90	2. Protecting receptor proteins
C. HSP 70	3. ATP-dependent protein assembly
D. HSP 60	4. Molecular chaperons

[A] A: 3; B: 1; C: 2; D:4 [B] A: 2; B: 1; C: 3; D: 4
[C] A: 1; B: 2; C: 3; D: 4 [D] A: 4; B: 3; C: 2; D: 1
[E] None of the above

335. Match the Group I with Group II and select the correct answer using the codes given below

Group I (Plant Source)	Group II (Gene)
A. *Arabidopsis*	1. *cor15a*
B. *Alfalfa*	2. *cas15*
C. *Wheat*	3. *wcs120*
D. *Barley*	4. *bva1*

[A] A: 3; B: 1; C: 2; D:4 [B] A: 2; B: 1; C: 3; D: 4
[C] A: 1; B: 2; C: 3; D: 4 [D] A: 4; B: 3; C: 2; D: 1
[E] None of the above

336. Match the list I with the list II and select the correct answer using the codes given below the lists

List I [Class of herbicides]	List II [Year First Detected]
A. Bypyridilium	1. 1976
B. Substituted ureas	2. 1983
C. Triazines	3. 1968
D. Uracils	4. 1986
E. ALS inhibitors	5. 1988

[A] A: 1; B: 2; C: 3; D: 5; E: 4 [B] A: 1; B: 2; C: 3; D: 4; E: 5
[C] A: 4; B: 1; C: 2; D: 3; E: 5 [D] A: 4; B: 2; C: 3; D: 5; E: 1

337. Match the list I with the list II and select the correct answer using the codes given below the lists

List I [Class of Herbicides]	List II [Number of species with Resistance]
A. Bypyridilium	1. 18
B. Substituted ureas	2. 7
C. Triazines	3. 57
D. Uracils	4. 8
E. ALS inhibitors	5. 2

[A] A: 1; B: 2; C: 3; D: 5; E: 4 [B] A: 1; B: 2; C: 3; D: 4; E: 5
[C] A: 4; B: 1; C: 2; D: 3; E: 5 [D] A: 4; B: 2; C: 3; D: 5; E: 1

338. Match the list I with the list II and select the correct answer using the codes given below the lists

List I [Common name of short day plants]	List II [Scientific name]
A. Red goosefoot	1. *Chenopodium rubrum*
B. Cocklebur	2. *Xanthium strumarium*
C. Yellow osmosis	3. *Cosmos sulphureus*
D. Poinsettia	4. *Glycine max*
E. Soybean	5. *Euphorbia pulcherrima*

[A] A: 1; B: 2; C: 3; D: 5; E: 4
[B] A: 1; B: 2; C: 3; D: 4; E: 5
[C] A: 4; B: 1; C: 2; D: 3; E: 5
[D] A: 4; B: 2; C: 3; D: 5; E: 1

339. Match the list I with the list II and select the correct answer using the codes given below the lists

List I [Common name of the long day plants]	List II [Scientific name]
A. Dill	1. *Anethum graveolens*
B. Rye grass	2. *Lolium* sp.
C. Radish	3. *Raphanus sativus*
D. White mustard	4. *Spinacea oleracea*
E. Spinach	5. *Sinapis alba*

[A] A: 1; B: 2; C: 3; D: 5; E: 4
[B] A: 1; B: 2; C: 3; D: 4; E: 5
[C] A: 4; B: 1; C: 2; D: 3; E: 5
[D] A: 4; B: 2; C: 3; D: 5; E: 1

340. Match the list I with the list II and select the correct answer using the codes given below the lists

List I [Common name of the day neutral plants]	List II [Scientific name]
A. Cucumber	1. *Cucumis sativus*
B. Sunflower	2. *Helianthus annus*
C. Common bean	3. *Phaseolus vulgarus*
D. Garden pea	4. *Zea mays*
E. Corn	5. *Pisum sativum*

[A] A: 1; B: 2; C: 3; D: 5; E: 4
[B] A: 1; B: 2; C: 3; D: 4; E: 5
[C] A: 4; B: 1; C: 2; D: 3; E: 5
[D] A: 4; B: 2; C: 3; D: 5; E: 1

341. Match the list I with the list II and select the correct answer using the codes given below the lists

List I [Some values for water potential (ψ) as a function of relative humidity (RH)]	List II [ψ(MPa)=1.06 T log(RH/100)]
A. 100	1. 0.0
B. 95	2. -6.9
C. 90	3. -14.2
D. 50	4. -217.1
E. 20	5. -93.5

[A] A: 1; B: 2; C: 3; D: 5; E: 4 [B] A: 1; B: 2; C: 3; D: 4; E: 5
[C] A: 4; B: 1; C: 2; D: 3; E: 5 [D] A: 4; B: 2; C: 3; D: 5; E: 1

342. Match the list I with list II and select the correct answer using the codes given below the lists

List I [Molecules]	List II [Molecular mass (Da)]
1. Ethane	P. 30
2. Methanol	Q. 32
3. Ethanol	R. 46

[A] 1: Q; 2: R; 3: P [B] 1: S; 2: P; 3: R
[C] 1: P; 2: Q; 3: R [D] None of the above

343. Match the list I with list II and select the correct answer using the codes given below the lists

List I [Bacterium]	List II [Host]
1. *Azorhizobium*	P. Sesbania
2. *Bradyrhizobium*	Q. Glycine
3. *Rhizobium meliloti*	R. Medicago

[A] 1: Q; 2: R; 3: P [B] 1: S; 2: P; 3: R
[C] 1: P; 2: Q; 3: R [D] None of the above

344. Match the list I with list II and select the correct answer using the codes given below the lists

List I [Bacterium]	List II [Host Plants]
1. *Rhizobium loti*	P. Lotus
2. *Rhizobium leguminosarum*	Q. Lens
3. *Rhizobium mililoti*	R. Medicago

[A] 1: Q; 2: R; 3: P [B] 1: S; 2: P; 3: R
[C] 1: P; 2: Q; 3: R [D] None of the above

345. Match the list I with list II and select the correct answer using the codes given below the lists

List I [Plants]		List II [Approximate specific dark respiration rates (μmol CO_2 evolved G^{-1} dry mass h^{-1} at 20°C]
1. Crops	P.	70-180
2. Sun leaves	Q.	70-90
3. Shade leaves	R.	20-45

[A] 1: Q; 2: R; 3: P
[B] 1: S; 2: P; 3: R
[C] 1: P; 2: Q; 3: R
[D] None of the above

346. Match the list I with list II and select the correct answer using the codes given below the lists

List I [Chemical composition of phloem exudates (organic) from stems of actively growing castor bean]		List II [mg^{-1}]
1. Sucrose	P.	80-106
2. Protein	Q.	1.45-2.20
3. Malic acid	R.	2.0-3.2

[A] 1: Q; 2: R; 3: P
[B] 1: S; 2: P; 3: R
[C] 1: P; 2: Q; 3: R
[D] None of the above

347. Match the list I with list II and select the correct answer using the codes given below the lists

List I [Chemical composition of phloem exudates (Inorganic) from stems of actively growing castor bean]		List II [mg^{-1}]
1. Anions	P.	20-30
2. Cations	Q.	74-138
3. Total dry matter	R.	100-125

[A] 1: Q; 2: R; 3: P
[B] 1: S; 2: P; 3: R
[C] 1: P; 2: Q; 3: R
[D] None of the above

348. Match the list I with list II and select the correct answer using the codes given below the lists

List I [Genus]		List II [Number of stomata (mm^{-2}) at upper surface]
1. *Allium*	P.	175
2. *Hordenum*	Q.	70
3. *Triticum*	R.	50

[A] 1: Q; 2: R; 3: P
[B] 1: S; 2: P; 3: R
[C] 1: P; 2: Q; 3: R
[D] None of the above

349. Match the list I with list II and select the correct answer using the codes given below the lists

	List I [Genus]		List II [Number of stomata (mm^{-2}) at upper surface]
1.	*Helianthus*	P.	120
2.	*Medicago*	Q.	169
3.	*Pelargonium*	R.	29

[A] 1: Q; 2: R; 3: P [B] 1: S; 2: P; 3: R
[C] 1: P; 2: Q; 3: R [D] None of the above

350. Match the list I with list II and select the correct answer using the codes given below the lists

	List I [Genus]		List II [Number of stomata (mm^{-2}) at lower surface]
1.	*Allium*	P.	175
2.	*Hordenum*	Q.	85
3.	*Triticum*	R.	40

[A] 1: Q; 2: R; 3: P [B] 1: S; 2: P; 3: R
[C] 1: P; 2: Q; 3: R [D] None of the above

351. Match the list I with list II and select the correct answer using the codes given below the lists

	List I [Genus]		List II [Number of stomata (mm^{-2}) at lower surface]
1.	*Helianthus*	P.	175
2.	*Medicago*	Q.	188
3.	*Pelargonium*	R.	179

[A] 1: Q; 2: R; 3: P [B] 1: S; 2: P; 3: R
[C] 1: P; 2: Q; 3: R [D] None of the above

352. Match the list I with list II and select the correct answer using the codes given below the lists

	List I [Genus]		List II [Number of stomata (mm^{-2}) at lower surface]
1.	*Aesculus*	P.	210
2.	*Quercus*	Q.	340
3.	*Tilia*	R.	370

[A] 1: Q; 2: R; 3: P [B] 1: S; 2: P; 3: R
[C] 1: P; 2: Q; 3: R [D] None of the above

353. Match the list I with list II and select the correct answer using the codes given below the lists

	List I [Anthocyanidin]		List II [Source]
1.	Pelargonidin	P.	Geranium petals
2.	Cyanidin	Q.	Red rose petals
3.	Malvidin	R.	Blue grapes

[A] 1: Q; 2: R; 3: P [B] 1: S; 2: P; 3: R
[C] 1: P; 2: Q; 3: R [D] None of the above

354. Match the list I with list II and select the correct answer using the codes given below the lists

	List I [Common name of saturated fatty acid]		List II [Melting point (°C)]
1.	Lauric acid	P.	44
2.	Myristic acid	Q.	52
3.	Palmitic acid	R.	63

[A] 1: Q; 2: R; 3: P [B] 1: S; 2: P; 3: R
[C] 1: P; 2: Q; 3: R [D] None of the above

355. Match the list I with the list II and select the correct answer using the codes given below the lists

	List I [Terms]		List II [Significance]
A.	Aberrant	1.	A growth that deviated from the normal growth
B.	A23187	2.	An Ionophore which transport divalent cations
C.	Acarpous	3.	Devoid of fruit
D.	Accession	4.	Plant or seed sample, strain or population held in a genebank or breeding programme for conservation and use
E.	Acclimate	5.	Adjust physiologically to a new environment

[A] A: 1; B: 3; C: 4; D: 5; E: 2 [B] A: 1; B: 2; C: 3; D: 4; E: 5
[C] A: 4; B: 1; C: 2; D: 3; E: 5 [D] A: 4; B: 2; C: 3; D: 5; E: 1

356. Match the list I with the list II and select the correct answer using the codes given below the lists

	List I [Terms]		List II [Significance]
A.	Aceticlastic	1.	Able to catabolize acetate
B.	Acetogenesis	2.	Acetate formation
C.	Acetosyringone	3.	One of the phenolic compounds released by wounded plant cells

D. Achlamydeous	4. Growth-promoting effect of auxins in higher plants
E. Acid growth theory	5. Possessing neither calyx nor corolla

[A] A: 1; B: 2; C: 3; D: 5; E: 4 [B] A: 1; B: 2; C: 3; D: 4; E: 5
[C] A: 4; B: 1; C: 2; D: 3; E: 5 [D] A: 4; B: 2; C: 3; D: 5; E: 1

357. Match the list I with the list II and select the correct answer using the codes given below the lists

List I [Terms]	List II [Significance]
A. Acidulent	1. A substance used to increase acidity
B. Aconta	2. Algae that do not have a flagellate stage in their life cycle
C. Acridine dye	3. Organic molecules
D. Acropetal development	4. Graph of biological activity versus wavelength
E. Action spectra	5. Development from the base or point of attachment

[A] A: 1; B: 2; C: 3; D: 5; E: 4 [B] A: 1; B: 2; C: 3; D: 4; E: 5
[C] A: 4; B: 1; C: 2; D: 3; E: 5 [D] A: 4; B: 2; C: 3; D: 5; E: 1

358. Match the list I with the list II and select the correct answer using the codes given below the lists

List I [Terms]	List II [Significance]
A. Active transport	1. Joint assay with Darwin
B. Activity-range	2. The area within which individuals of a single generation may move
C. Adaptive enzyme	3. Formed in response to an outside stimulus during adaptation
D. Addendum	4. Supplement
E. Additive genes	5. Interact but that show no dominance

[A] A: 1; B: 3; C: 4; D: 5; E: 2 [B] A: 1; B: 2; C: 3; D: 4; E: 5
[C] A: 4; B: 1; C: 2; D: 3; E: 5 [D] A: 4; B: 2; C: 3; D: 5; E: 1

359. Match the list I with the list II and select the correct answer using the codes given below the lists

List I [Terms]	List II [Significance]
A. Adenylate cyclase	1. cAMP
B. Adenylate kinase	2. Trimeric in nature
C. Adermin	3. Vitamin

D. Adhesin	4. Polymers
E. ADRY reagents	5. Stimulate the photooxidation

[A] A: 1; B: 3; C: 4; D: 5; E: 2 [B] A: 1; B: 2; C: 3; D: 4; E: 5
[C] A: 4; B: 1; C: 2; D: 3; E: 5 [D] A: 4; B: 2; C: 3; D: 5; E: 1

360. Match the list I with the list II and select the correct answer using the codes given below the lists

List I [Terms]	List II [Significance]
A. Adsorption	1. Binding of a gas to the surface of a solid
B. Aerate	2. Supply with or mix with air of gas
C. Aerobic	3. Molecular oxygen
D. Aerosols	4. Causation study
E. Aetiology	5. Suspended in air

[A] A: 1; B: 2; C: 3; D: 5; E: 4 [B] A: 1; B: 2; C: 3; D: 4; E: 5
[C] A: 4; B: 1; C: 2; D: 3; E: 5 [D] A: 4; B: 2; C: 3; D: 5; E: 1

361. Match the list I with the list II and select the correct answer using the codes given below the lists

List I [Terms]	List II [Significance]
A. Agamic complex	1. A group of closely resembling and asexually reproducing plants which cannot be
B. Agamic	2. Having no pistils
C. Agamogenesis	3. Asexual reproduction by buds
D. Agar	4. A gelatinous polysaccharide
E. Ageing of lakes	5. Eutrophication

[A] A: 1; B: 3; C: 4; D: 5; E: 2 [B] A: 1; B: 2; C: 3; D: 4; E: 5
[C] A: 4; B: 1; C: 2; D: 3; E: 5 [D] A: 4; B: 2; C: 3; D: 5; E: 1

362. Match the list I with the list II and select the correct answer using the codes given below the lists

List I [Terms]	List II [Significance]
A. Aging	1. Biochemical and genetic changes that leads to gradual senescence and death of a cell or organism
B. Agonisis	2. Competition as between pollen grains of different genotypes, in the rapidity with which they can grow down the style
C. Agrenet	3. Communication network
D. Albescent	4. A white seedling due to a recessive gene
E. Albino seedling	5. Seedling emerge green and later become whitish, with variable expression

[A] A: 1; B: 2; C: 3; D: 5; E: 4 [B] A: 1; B: 2; C: 3; D: 4; E: 5
[C] A: 4; B: 1; C: 2; D: 3; E: 5 [D] A: 4; B: 2; C: 3; D: 5; E: 1

363. Match the list I with the list II and select the correct answer using the codes given below the lists

List I [Terms]		List II [Significance]
A. Albinotic	1.	Affected with albinism
B. Albomycin	2.	Antibiotic
C. Albugo	3.	Obligatory plant-parasitic fungi
D. ADH	4.	Catalysing the conversion of alcohols into aldehydes and ketons
E. Alcian blue	5.	Water soluble stain

[A] A: 1; B: 3; C: 4; D: 5; E: 2 [B] A: 1; B: 2; C: 3; D: 4; E: 5
[C] A: 4; B: 1; C: 2; D: 3; E: 5 [D] A: 4; B: 2; C: 3; D: 5; E: 1

364. Match the list I with the list II and select the correct answer using the codes given below the lists

List I [Terms]		List II [Significance]
A. Aldonic acids	1.	Monocarboxylic acid
B. Aldosterone	2.	A steroid hormone
C. Aleurone layer	3.	Outermost layer of the endosperm in some seeds
D. AMV	4.	Phaeophyta
E. Alga	5.	ssRNA-containing virus

[A] A: 1; B: 2; C: 3; D: 5; E: 4 [B] A: 1; B: 2; C: 3; D: 4; E: 5
[C] A: 4; B: 1; C: 2; D: 3; E: 5 [D] A: 4; B: 2; C: 3; D: 5; E: 1

365. Match the list I with the list II and select the correct answer using the codes given below the lists

List I [Terms]		List II [Significance]
A. Algicide	1.	A specific chemical highly toxic to algae
B. Alginate	2.	A salt of alginic acid
C. Allantoin	3.	A substance which stimulate cell growth
D. Allautogamia	4.	A little-used terms referring to substance effecting allelopathic reactions
E. Allelochemical	5.	The state of having a facultative method of pollination in addition to a normal method

[A] A: 1; B: 2; C: 3; D: 5; E: 4 [B] A: 1; B: 2; C: 3; D: 4; E: 5
[C] A: 4; B: 1; C: 2; D: 3; E: 5 [D] A: 4; B: 2; C: 3; D: 5; E: 1

366. Match the list I with the list II and select the correct answer using the codes given below the lists

	List I [Terms]		List II [Significance]
A.	Allocarpy	1.	The production of fruit following cross-fertilization
B.	Allogamy	2.	Spontaneous cross-pollination
C.	Allogenous flora	3.	Relic plants of an earlier prevalling flora and environment
D.	Allograft	4.	Homograft
E.	Allobiogenesis	5.	An alteration of a sexual with an asexual form

[A] A: 1; B: 3; C: 4; D: 5; E: 2 [B] A: 1; B: 2; C: 3; D: 4; E: 5
[C] A: 4; B: 1; C: 2; D: 3; E: 5 [D] A: 4; B: 2; C: 3; D: 5; E: 1

367. Match the list I with the list II and select the correct answer using the codes given below the lists

	List I [Terms]		List II [Significance]
A.	Allopatric	1.	Existing in different environments or geographical regions
B.	Alloplasma	2.	Highly differentiated protoplasm
C.	Allosteric effect	3.	A compound that induce a conformational changes in a protein
D.	Allosteric effect	4.	A flower so shaped that its nectar is easy of access to insect
E.	Allotropous flower	5.	An enzyme that shows unusual kinetic properties in response to changes in substrate concentration/or in response to the action of the allosteric effectors that are not substrates

[A] A: 1; B: 2; C: 3; D: 5; E: 4 [B] A: 1; B: 2; C: 3; D: 4; E: 5
[C] A: 4; B: 1; C: 2; D: 3; E: 5 [D] A: 4; B: 2; C: 3; D: 5; E: 1

368. Match the list I with the list II and select the correct answer using the codes given below the lists

	List I [Terms]		List II [Significance]
A.	Alpha amylase	1.	Wheat seed at germination
B.	Alsike clover	2.	A species of clover
C.	Alternate complement pathway	3.	Pathway in which the complement cascade is activated at the C3 levels
D.	Amaranthus	4.	Inhibitor of acetycholinesterase
E.	Amaryllidaceae alkaloids	5.	A red dye belonging to the betacyanin group

[A] A: 1; B: 2; C: 3; D: 5; E: 4 [B] A: 1; B: 2; C: 3; D: 4; E: 5
[C] A: 4; B: 1; C: 2; D: 3; E: 5 [D] A: 4; B: 2; C: 3; D: 5; E: 1

369. Match the list I with the list II and select the correct answer using the codes given below the lists

List I [Terms]	List II [Significance]
A. Amatoxins	1. Toxic cyclic peptides
B. Amber	2. UAG
C. Ambient	3. The environment at a particular time
D. Ambiguity	4. Hermaphrodite
E. Ambisporangiate	5. The coding for more than one amino acid

[A] A: 1; B: 2; C: 3; D: 5; E: 4 [B] A: 1; B: 2; C: 3; D: 4; E: 5
[C] A: 4; B: 1; C: 2; D: 3; E: 5 [D] A: 4; B: 2; C: 3; D: 5; E: 1

370. Match the list I with the list II and select the correct answer using the codes given below the lists

List I [Terms]	List II [Significance]
A. Ament	1. A catkin
B. AAA pathway	2. Lysine biosynthesis
C. Aminopeptidases	3. Containing metal ions
D. Amixia	4. Cross sterility
E. Amphibolic pathway	5. TCA cycle

[A] A: 1; B: 3; C: 4; D: 5; E: 2 [B] A: 1; B: 2; C: 3; D: 4; E: 5
[C] A: 4; B: 1; C: 2; D: 3; E: 5 [D] A: 4; B: 2; C: 3; D: 5; E: 1

371. Match the list I with the list II and select the correct answer using the codes given below the lists

List I [Terms]	List II [Significance]
A. Amphigony	1. Reproduction in which two individuals participate
B. Amphimict	2. An individual which reproduces by amphimixis
C. Amphiplasty	3. The loss of a satellite from a chromosome
D. Amphitoky	4. Parthenogenetic
E. Anaboly	5. An evolutionary changes

[A] A: 1; B: 3; C: 4; D: 5; E: 2 [B] A: 1; B: 2; C: 3; D: 4; E: 5
[C] A: 4; B: 1; C: 2; D: 3; E: 5 [D] A: 4; B: 2; C: 3; D: 5; E: 1

372. Match the list I with the list II and select the correct answer using the codes given below the lists

	List I [Terms]		List II [Significance]
A.	Anachromasis	1.	The changes which take place in a nucleus during prophase
B.	Anaerobe	2.	Absence of air or oxygen
C.	Anaerobic fixed film digester	3.	An alternative term for an anaerobic filter
D.	Anagenesis	4.	Having no stamens
E.	Anandrous	5.	The evolutionary changes of a single lineage in the course

[A] A: 1; B: 2; C: 3; D: 5; E: 4 [B] A: 1; B: 2; C: 3; D: 4; E: 5
[C] A: 4; B: 1; C: 2; D: 3; E: 5 [D] A: 4; B: 2; C: 3; D: 5; E: 1

373. Match the list I with the list II and select the correct answer using the codes given below the lists

	List I [Terms]		List II [Significance]
A.	Anchondroplasia	1.	One type of inherited dwarfism
B.	Balance of nature	2.	The natural tendency of plants and animals population resulting from natural regulative processes in an undistributed environment to neither decline in number to extinction nor increases to infinite density
C.	Basement membrane	3.	A thin layer of jelly or cement that is found at the base of an epithelium
D.	Batonette	4.	Part of Golgi apparatus
E.	Biennial	5.	Complete its life cycle within two years

[A] A: 1; B: 3; C: 4; D: 5; E: 2 [B] A: 1; B: 2; C: 3; D: 4; E: 5
[C] A: 4; B: 1; C: 2; D: 3; E: 5 [D] A: 4; B: 2; C: 3; D: 5; E: 1

374. Match the list I with the list II and select the correct answer using the codes given below the lists

	List I [Terms]		List II [Significance]
A.	Bioassay	1.	Assessment procedure
B.	Biocatalyst	2.	Derived from a living organism
C.	Biocatalyzer	3.	Catalyst derived from living organism
D.	BOD	4.	Development of chromatography
E.	Biochemical taxonomy	5.	Measures for determining the level of organic pollution in a sample of water

[A] A: 1; B: 2; C: 3; D: 5; E: 4 [B] A: 1; B: 2; C: 3; D: 4; E: 5
[C] A: 4; B: 1; C: 2; D: 3; E: 5 [D] A: 4; B: 2; C: 3; D: 5; E: 1

375. Match the list I with the list II and select the correct answer using the codes given below the lists

	List I [Terms]		List II [Significance]
A.	Bioprobe	1.	A type of biosensor in which the catalyst is immobilized with a detection device
B.	Biorythum	2.	A physiological or behavioral event displaying a periodic pattern of occurrence in a living organism or population
C.	Biosensor	3.	A device that uses an agent of biological origin
D.	Biota	4.	All of the organism, including animals, plants, fungi, and microorganisms, found in a given area
E.	Biotin	5.	A water soluble vitamin

[A] A: 1; B: 3; C: 4; D: 5; E: 2 [B] A: 1; B: 2; C: 3; D: 4; E: 5
[C] A: 4; B: 1; C: 2; D: 3; E: 5 [D] A: 4; B: 2; C: 3; D: 5; E: 1

376. Match the list I with the list II and select the correct answer using the codes given below the lists

	List I [Terms]		List II [Significance]
A.	Biuret test	1.	Protein
B.	Blade	2.	Expanded part
C.	Blastema	3.	A group of cells in organism that will develop into an organized structure during regeneration
D.	Bract	4.	A leaf like structure subtending flower or inflorescence
E.	Bradyauxesis	5.	A growth relationship in which the growth of a part of an individual's is slower than that of the body as a whole

[A] A: 1; B: 3; C: 4; D: 5; E: 2 [B] A: 1; B: 2; C: 3; D: 4; E: 5
[C] A: 4; B: 1; C: 2; D: 3; E: 5 [D] A: 4; B: 2; C: 3; D: 5; E: 1

377. Match the list I with the list II and select the correct answer using the codes given below the lists

	List I [Terms]		List II [Significance]
A.	Bubble column	1.	A type of fermentor or bioreactor used for the growth of plant cell culture in liquid suspension

B.	Bud primordium	2.	Meristematic tissue
C.	Bud sport	3.	Mutation
D.	Burgeon	4.	Sprout or grow
E.	Bulbil	5.	A modified bud capable of developing into a new plant when separated from the parent plant

[A] A: 1; B: 3; C: 4; D: 5; E: 2 [B] A: 1; B: 2; C: 3; D: 4; E: 5
[C] A: 4; B: 1; C: 2; D: 3; E: 5 [D] A: 4; B: 2; C: 3; D: 5; E: 1

378. Match the list I with the list II and select the correct answer using the codes given below the lists

	List I [Terms]		List II [Significance]
A.	C/N ratio	1.	Culture medium
B.	Cabernet sauvignon	2.	A variety of purple grape used in the production of high- quality wines
C.	Cabinet array	3.	Biological safety
D.	Pantnothenic acid	4.	Calcium salt of vitamin B5
E.	Calyptra	5.	A hood or cap found partially covering the apex of the sporophyte capsule in many moss species, formed from the expanded wall and neck of the archegonium

[A] A: 1; B: 3; C: 4; D: 5; E: 2 [B] A: 1; B: 2; C: 3; D: 4; E: 5
[C] A: 4; B: 1; C: 2; D: 3; E: 5 [D] A: 4; B: 2; C: 3; D: 5; E: 1

379. Match the list I with the list II and select the correct answer using the codes given below the lists

	List I [Terms]		List II [Significance]
A.	Canker	1.	Disease
B.	Calyx	2.	All of the sepals of the flower, collectively
C.	Cannula	3.	A small tube for insertion into tissue
D.	Capping	4.	A process by which cell surface molecules are caused to aggregate on the cell membrane
E.	Capsomers	5.	Protein unit

[A] A: 1; B: 3; C: 4; D: 5; E: 2 [B] A: 1; B: 2; C: 3; D: 4; E: 5
[C] A: 4; B: 1; C: 2; D: 3; E: 5 [D] A: 4; B: 2; C: 3; D: 5; E: 1

380. Match the list I with the list II and select the correct answer using the codes given below the lists

List I [Terms]	List II [Significance]
A. Carbon budget	1. The amount of atmospheric carbon incorporated into organic molecules by a plant
B. Carbon source	2. A source of the non-metallic elements
C. Cassava	3. Any of several plants of the spurge family grown in the tropics for their fleshy edible rootstocks, which yield a nutritional starch
D. Catabolic repression	4. The decrease synthesis of many enzymes that tend to provide glucose for a cell; caused by the presence of excellent carbon sources, particularly glucose
E. Caudal	5. Tail

[A] A: 1; B: 3; C: 4; D: 5; E: 2 [B] A: 1; B: 2; C: 3; D: 4; E: 5
[C] A: 4; B: 1; C: 2; D: 3; E: 5 [D] A: 4; B: 2; C: 3; D: 5; E: 1

381. Match the list I with the list II and select the correct answer using the codes given below the lists

List I [Terms]	List II [Significance]
A. Caulogenesis	1. Shoot formation; as de novo shoot development from callus
B. Cell count	2. The number of cells per unit suspension volume or callus weight
C. Cell line	3. Developmental history or descent, through cell division from a single original cell
D. Cell number	4. The absolute number or approximation of the number of cells per unit area of a culture or medium volume
E. Cell selection	5. Selection within a group of genetically different cells; involves competition between cells often under some stress

[A] A: 1; B: 3; C: 4; D: 5; E: 2 [B] A: 1; B: 2; C: 3; D: 4; E: 5
[C] A: 4; B: 1; C: 2; D: 3; E: 5 [D] A: 4; B: 2; C: 3; D: 5; E: 1

382. Match the list I with the list II and select the correct answer using the codes given below the lists

	List I [Terms]		List II [Significance]
A.	Cell specific	1.	A condition in which the effect produced is directed to the cells of a particular species
B.	Cell strain	2.	Having specific properties
C.	Cell suspension	3.	Liquid medium
D.	Cell theory	4.	Well established
E.	Cell transformation	5.	Permanent alteration of the cell phenotype

[A] A: 1; B: 3; C: 4; D: 5; E: 2 [B] A: 1; B: 2; C: 3; D: 4; E: 5
[C] A: 4; B: 1; C: 2; D: 3; E: 5 [D] A: 4; B: 2; C: 3; D: 5; E: 1

383. Match the list I with the list II and select the correct answer using the codes given below the lists

	List I [Terms]		List II [Significance]
A.	Cellulose	1.	A straight chain polymer of glucose molecule, used by plants as a structure supporting material
B.	Centriole	2.	A cylindrical organelle in animal and some protest cells, containing microtubules in a characteristic arrangement
C.	Chelating agent	3.	Usually organic
D.	Chardonnay	4.	A variety of green grape used in the production of white wine
E.	Chemotaxis	5.	Increased directional migration of cells in response to concentration gradient of certain chemotactic factors

[A] A: 1; B: 3; C: 4; D: 5; E: 2 [B] A: 1; B: 2; C: 3; D: 4; E: 5
[C] A: 4; B: 1; C: 2; D: 3; E: 5 [D] A: 4; B: 2; C: 3; D: 5; E: 1

384. Match the list I with the list II and select the correct answer using the codes given below the lists

	List I [Years]		List II [Significance]
A.	1878	1.	Vochting studied plant polarity and tried to grow small lumps of tissue *in vitro*
B.	1893	2.	Rechniger placed segments of polar stem and pieces of beet and candeloin roots on a wet filter paper

C. 1902	3.	Haberlandt tried to obtain plant tissue from the isolated mesophyll cells of Lamium on artificial medium
D. 1925	4.	The utility of embryo culture technology became apparent
E. 1934	5.	Successful culture of tomato roots was demonstrated by white

[A] A: 1; B: 3; C: 4; D: 5; E: 2 [B] A: 1; B: 2; C: 3; D: 4; E: 5
[C] A: 4; B: 1; C: 2; D: 3; E: 5 [D] A: 4; B: 2; C: 3; D: 5; E: 1

385. Match the list I with the list II and select the correct answer using the codes given below the lists

List I [Years]		List II [Significance]
A. 1937	1.	Using a synthetic medium containing inorganic salts
B. 1941	2.	Use of coconut milk to meet nutritional requirements of the younger embryo
C. 1944	3.	First in vitro cultures of tobacco were used to study adventitious root formation by Skoog
D. 1952	4.	Virus-free dahlias were obtained by Morel & martin using meristematic culture
E. 1953	5.	Isolation and culture of single cell was impossible until Muir devised the technology of agitating callus tissue in liquid medium through paper-raft nurse technique

[A] A: 1; B: 3; C: 4; D: 5; E: 2 [B] A: 1; B: 2; C: 3; D: 4; E: 5
[C] A: 4; B: 1; C: 2; D: 3; E: 5 [D] A: 4; B: 2; C: 3; D: 5; E: 1

386. Match the list I with the list II and select the correct answer using the codes given below the lists

List I [Years]		List II [Significance]
A. 1955	1.	Discovery of kinetin, a cell division hormone, by Miller
B. 1957	2.	Skoog and Miller obtained roots and stems from callus treated with auxin and kinetin
C. 1958	3.	Regeneration of somatic embryo in vitro from the nucellus of citrus ovules
D. 1959	4.	Publication of first extensive handbook on plant tissue culture by Gautheret

E.	1960	5.	Cocking showed enzymatic degradation of the cell walls to get large number of protoplast

[A] A: 1; B: 3; C: 4; D: 5; E: 2 [B] A: 1; B: 2; C: 3; D: 4; E: 5
[C] A: 4; B: 1; C: 2; D: 3; E: 5 [D] A: 4; B: 2; C: 3; D: 5; E: 1

387. Match the list I with the list II and select the correct answer using the codes given below the lists

	List I [Years]		List II [Significance]
A.	1962	1.	Murashige & Skoog medium developed
B.	1964	2.	First haploid Datura plants obtained from pollen grains by Sipra Guha & Maheshwari
C.	1967	3.	Haploids plant obtained from tobacco pollen grains by Bourgin and Nitsch
D.	1969	4.	First successful isolation of protoplasts from a suspension culture of *Hapopappus gracilis*
E.	1970	5.	Selection of biochemical mutants *in vitro* by Carlson

[A] A: 1; B: 3; C: 4; D: 5; E: 2 [B] A: 1; B: 2; C: 3; D: 4; E: 5
[C] A: 4; B: 1; C: 2; D: 3; E: 5 [D] A: 4; B: 2; C: 3; D: 5; E: 1

388. Match the list I with the list II and select the correct answer using the codes given below the lists

	List I [Years]		List II [Significance]
A.	1971	1.	First plant regenerated from protoplasts
B.	1972	2.	Interspecific hybridization
C.	1973	3.	Cytokinin found potential of breaking dormancy in excised capitulum explants of the Gerbera by Pierik
D.	1974	4.	Discovery of Ti-plasmid by Larebeke
E.	1976	5.	Shoot initiation from cryopreserved shoot apices of carnation demonstrated by Siebert

[A] A: 1; B: 3; C: 4; D: 5; E: 2 [B] A: 1; B: 2; C: 3; D: 4; E: 5
[C] A: 4; B: 1; C: 2; D: 3; E: 5 [D] A: 4; B: 2; C: 3; D: 5; E: 1

389. Match the list I with the list II and select the correct answer using the codes given below the lists

	List I [Years]		List II [Significance]
A.	1978	1.	Somatic hybridization of tomato and potato by Melchers
B.	1981	2.	Introduction of the term somaclonal Variation
C.	1982	3.	Protoplasts able to incorporate naked DNA
D.	1983	4.	Production of the transformed tobacco plants following single cell transformation or gene insertion, shown by Chilton
E.	1984	5.	Vasil emphasized for the development of cell culture and somatic genetics

[A] A: 1; B: 3; C: 4; D: 5; E: 2 [B] A: 1; B: 2; C: 3; D: 4; E: 5
[C] A: 4; B: 1; C: 2; D: 3; E: 5 [D] A: 4; B: 2; C: 3; D: 5; E: 1

390. Match the list I with list II and select the correct answer using the codes given below the lists

	List I [Examples of root culture plants]		List II [Response observed]
1.	*Albizia lebbeck*	P.	Plantlets
2.	*Dalbergia sissoo*	Q.	Shoot buds
3.	*Vicia faba*	R.	Rooted shoots

[A] 1: Q; 2: R; 3: P [B] 1: S; 2: P; 3: R
[C] 1: P; 2: Q; 3: R [D] None of the above

391. Match the list I with list II and select the correct answer using the codes given below the lists

	List I [Examples of stem culture plants]		List II [Response observed]
1.	*Antirhinum majus*	P.	Callus
2.	*Cajanus cajan*	Q.	Multiple shoots
3.	*Citrus grandis*	R.	Plantlets

[A] 1: Q; 2: R; 3: P
[B] 1: S; 2: P; 3: R
[C] 1: P; 2: Q; 3: R
[D] None of the above

392. Match the list I with list II and select the correct answer using the codes given below the lists

List I [Examples of stem culture plants]	List II [Explant used]
1. *Antirhinum majus*	P. Shoots
2. *Cajanus cajan*	Q. Epicotyl segment
3. *Psidium guajava*	R. Nodal segment

[A] 1: Q; 2: R; 3: P
[B] 1: S; 2: P; 3: R
[C] 1: P; 2: Q; 3: R
[D] None of the above

393. Match the list I with list II and select the correct answer using the codes given below the lists

List I [Examples of stem tip/meristem culture plants]	List II [Response observed]
1. *Allium sativum*	P. Plantlets
2. *Artocarpus heterophyllus*	Q. Multiple shooting
3. *Morus indica*	R. Embryoids

[A] 1: Q; 2: R; 3: P
[B] 1: S; 2: P; 3: R
[C] 1: P; 2: Q; 3: R
[D] None of the above

394. Match the list I with list II and select the correct answer using the codes given below the lists

List I [Examples of leaf culture plants]	List II [Response observed]
1. *Azadirachta indica*	P. Shoot differentiation
2. *Citrus sp*	Q. Plantlets
3. *Zea mays*	R. Somatic embryoids

[A] 1: Q; 2: R; 3: P
[B] 1: S; 2: P; 3: R
[C] 1: P; 2: Q; 3: R
[D] None of the above

395. Match the list I with list II and select the correct answer using the codes given below the lists

List I [Examples of micropro-pagation as employed in medicinal plants]	List II [Explant used]
1. *Artimisia annua*	P. Nodes
2. *Datura innoxia*	Q. Anther
3. *Catharanthus roseus*	R. Stem

[A] 1: Q; 2: R; 3: P
[B] 1: S; 2: P; 3: R
[C] 1: P; 2: Q; 3: R
[D] None of the above

396. Match the list I with list II and select the correct answer using the codes given below the lists

	List I [Examples of micropro-pagation as employed in medicinal plants]		List II [Explant used]
1.	*Moringa oleifera*	P.	Hypocotyl
2.	*Delonix regia*	Q.	Leaves
3.	*Panax ginseng*	R.	Anther

[A] 1: Q; 2: R; 3: P
[B] 1: S; 2: P; 3: R
[C] 1: P; 2: Q; 3: R
[D] None of the above

397. Match the list I with list II and select the correct answer using the codes given below the lists

	List I [Examples of micropro-pagation as employed in medicinal plants]		List II [Explant used]
1.	*Taraxacum officinale*	P.	Root
2.	*Rauwolfia serpentine*	Q.	Plant parts
3.	*Hyoscyamus niger*	R.	Seedlings

[A] 1: Q; 2: R; 3: P
[B] 1: S; 2: P; 3: R
[C] 1: P; 2: Q; 3: R
[D] None of the above

398. Match the list I with list II and select the correct answer using the codes given below the lists

	List I [Sources of the plant tissue from which protoplast can be isolated]		List II [Part used]
1.	Plant tissue	P.	Young leaves, roots, stem
2.	Flower parts	Q.	Petal, reproductive organs
3.	Callus/cells	R.	Friable callus tissues/cells

[A] 1: Q; 2: R; 3: P
[B] 1: S; 2: P; 3: R
[C] 1: P; 2: Q; 3: R
[D] None of the above

399. Match the list I with the list II and select the correct answer using the codes given below the lists

	List I [Constituents of the Modified B5 medium]		List II [Concentration]
A.	$CaCl_2.2H_2O$	1.	0.0875%
B.	Mannitol	2.	4.5% (w/v)
C.	d-glucose	3.	0.25% (w/v)
D.	D-ribose	4.	0.125% (w/v)
E.	NAA	5.	10^{-6} M

[A] A: 1; B: 3; C: 4; D: 5; E: 2 [B] A: 1; B: 2; C: 3; D: 4; E: 5
[C] A: 4; B: 1; C: 2; D: 3; E: 5 [D] A: 4; B: 2; C: 3; D: 5; E: 1

400. Match the list I with the list II and select the correct answer using the codes given below the lists

	List I [Examples of the plants regenerated from protoplasts]		List II [Plant source]
A.	*Helianthus annuus*	1.	Hypocotyl
B.	*Populus sp.*	2.	Leaf
C.	*H. vulgare*	3.	ECS
D.	*Ipomea batata*	4.	Petiole and stem
E.	*Prunus avium*	5.	Cell suspension (root)

[A] A: 1; B: 3; C: 4; D: 5; E: 2 [B] A: 1; B: 2; C: 3; D: 4; E: 5
[C] A: 4; B: 1; C: 2; D: 3; E: 5 [D] A: 4; B: 2; C: 3; D: 5; E: 1

401. Match the list I with list II and select the correct answer using the codes given below the lists

	List I [Some examples of plant regeneration through direct DNA and macromolecule uptake in the protoplast]		List II [Promoter/gene]
1.	*O. sativa*	P.	Promoter and the first intron of the alcohol dehydrogenase gene of the maize/p-glucuronidase gene (GUS)
2.	*S. tuberosum*	Q.	Nopaline synthase gene of *A. fumefaciens*/neomycin
3.	*A. thaliana*	R.	35s/hph

[A] 1: Q; 2: R; 3: P [B] 1: S; 2: P; 3: R
[C] 1: R; 2: P; 3: Q [D] None of the above

402. Match the list I with list II and select the correct answer using the codes given below the lists

	List I [Examples of selected hybrids showing potential characteristic useful in crop cultivation]		List II [Trait transferred]
1.	*Solanum pinnatisectum + S. tuberosum*	P.	Late blight resistance
2.	*Brassica napus + B. nigra*	Q.	Black leg resistance
3.	*Cucumis melo + C. sativum*	R.	Disease resistance

[A] 1: Q; 2: R; 3: P [B] 1: S; 2: P; 3: R
[C] 1: R; 2: P; 3: Q [D] None of the above

403. Match the list I with list II and select the correct answer using the codes given below the lists

	List I [Examples of selected hybrids showing potential characteristic useful in crop cultivation]		List II [Trait transferred]
1.	*Brassica napus* + *Raphanus sativus*	P.	Beet cyst nematode resistance
2.	*Brassica napus* + *Moricandia arvensis*	Q.	Resistance to fungal disease; krantz leaf anatomy
3.	*Citrus reticulata* + *Citropsis gilletiana*	R.	Overcomes barrier to sexual hybridization

[A] 1: Q; 2: R; 3: P
[B] 1: S; 2: P; 3: R
[C] 1: R; 2: P; 3: Q
[D] None of the above

404. Match the list I with list II and select the correct answer using the codes given below the lists

	List I [Different gels for encapsulation of somatic embryos]		List II [Complexing agent]
1.	Sodium algenate	P.	Calcium salts
2.	Sodium algenate with gelatine	Q.	Calcium chloride
3.	Carrageenan with Locust Bean Gum	R.	Potassium or Ammonium chloride

[A] 1: Q; 2: R; 3: P
[B] 1: S; 2: P; 3: R
[C] 1: R; 2: P; 3: Q
[D] None of the above

405. Match the list I with list II and select the correct answer using the codes given below the lists

	List I [Some examples of haploid plants obtained by excised anther/ pollen]		List II [Response observed]
1.	*Asparagus sp.*	P.	Plantlet
2.	*Brassica napus*	Q.	Embryos
3.	*Nicotiana tabacum*	R.	Plants

[A] 1: Q; 2: R; 3: P
[B] 1: S; 2: P; 3: R
[C] 1: R; 2: P; 3: Q
[D] None of the above

406. Match the list I with the list II and select the correct answer using the codes given below the lists

List I [Examples of meristem cultures to eliminate virus]	List II [Virus eliminated]
A. *Allium sativum*	1. Mosaic virus
B. *Brassica oleracea*	2. Turnip mosaic virus, CaMV
C. *Chrysanthemum* sp.	3. Virus B
D. *Dianthus barbatus*	4. Motle virus
E. *Ipomoea batata*	5. Rugos/mosaic virus

[A] A: 1; B: 3; C: 4; D: 5; E: 2
[B] A: 1; B: 2; C: 3; D: 4; E: 5
[C] A: 4; B: 1; C: 2; D: 3; E: 5
[D] A: 4; B: 2; C: 3; D: 5; E: 1

407. Match the list I with list II and select the correct answer using the codes given below the lists

List I [Examples of meristem culture to eliminate virus]	List II [Virus elimination]
1. *Musa* sp.	P. Cucumber mosaic virus
2. *Solanum tuberosum*	Q. Potato Virus-X
3. *Petunia sp.*	R. TMV

[A] 1: Q; 2: R; 3: P
[B] 1: S; 2: P; 3: R
[C] 1: R; 2: P; 3: Q
[D] None of the above

408. Match the list I with list II and select the correct answer using the codes given below the lists

List I [Examples of somaclonal variation in some crops]	List II [Traits]
1. Sugarcane	P. Resistance to *Helminthosporium sacchari*
2. Potato	Q. Resistant to *Phytophthora infestans*
3. Tobacco	R. Resistance to *Pseudomonas* sp.

[A] 1: Q; 2: R; 3: P
[B] 1: S; 2: P; 3: R
[C] 1: R; 2: P; 3: Q
[D] None of the above

409. Match the list I with list II and select the correct answer using the codes given below the lists

List I [Examples of somaclonal variation in some crops]		List II [Traits]	
1.	Tomato	P.	Resistance to *Fusarium oxysporum*
2.	Wheat	Q.	Adh isozyme
3.	Rice	R.	Resistance to *Xanthomonas oryzae*

[A] 1: Q; 2: R; 3: P
[B] 1: S; 2: P; 3: R
[C] 1: R; 2: P; 3: Q
[D] None of the above

410. Match the list I with list II and select the correct answer using the codes given below the lists

List I [Somaclonal variation induced in morphological traits in some crops]		List II [Traits]	
1.	Tobacco	P.	Yield, plant height, leaf number, inflorescence type
2.	Brassica	Q.	Abnormal leaves, slow growth, large pollens, multiple stem branching
3.	Rice	R.	Number of tillers per plant, frequency of fertile seeds

[A] 1: Q; 2: R; 3: P
[B] 1: S; 2: P; 3: R
[C] 1: R; 2: P; 3: Q
[D] None of the above

411. Match the list I with list II and select the correct answer using the codes given below the lists

List I [Somaclonal variants having improved traits in some crop plants]		List II [Improved traits]	
1.	Wheat	P.	Tolerance to drought stress
2.	Rice	Q.	Lysine content
3.	Potato	R.	Resistance to *Fusarium* sp.

[A] 1: Q; 2: R; 3: P
[B] 1: S; 2: P; 3: R
[C] 1: R; 2: P; 3: Q
[D] None of the above

412. Match the list I with list II and select the correct answer using the codes given below the lists

	List I [Somaclonal variants having improved traits in some crop plants]		List II [Improved traits]
1.	Brassica	P.	Salt tolerance
2.	Sugarcane	Q.	Resistant to Eyespot
3.	Tobacco	R.	Herbicide resistant

[A] 1: Q; 2: R; 3: P
[B] 1: S; 2: P; 3: R
[C] 1: R; 2: P; 3: Q
[D] None of the above

413. Match the list I with list II and select the correct answer using the codes given below the lists

	List I [Examples of *in vitro* raised stress tolerant plants]		List II [Improved traits]
1.	*Medicago sativa*	P.	Salt tolerance cells require salt for plant growth
2.	*Brassica juncea*	Q.	Salt tolerant plant
3.	*Cicer areitinum*	R.	Stress tolerant callus selected

[A] 1: Q; 2: R; 3: P
[B] 1: S; 2: P; 3: R
[C] 1: R; 2: P; 3: Q
[D] None of the above

414. Match the list I with list II and select the correct answer using the codes given below the lists

	List I [Examples of *in vitro* raised stress tolerant plants]		List II [Improved traits]
1.	*Cymboogon meatinii*	P.	NaCl tolerant plants
2.	*Saccharum officinarum*	Q.	Fe-efficient plants
3.	*Datura innoxia*	R.	Zn-tolerant plants

[A] 1: Q; 2: R; 3: P
[B] 1: S; 2: P; 3: R
[C] 1: R; 2: P; 3: Q
[D] None of the above

415. Match the list I with the list II and select the correct answer using the codes given below the lists

	List I [Terms]		List II [Significance]
A.	Saturation	1.	A maximum state
B.	*SAUR* genes	2.	In soybean
C.	Sclereid	3.	A type of non elongated sclerenchyma cell commonly found in hard structures such as seed coats

D.	Sclerenchyma	4.	Plant tissue composed of cells, often dead at maturity with thick, lignified secondary cell walls
E.	SCR gene	5.	*Arabidopsis SCARECROW* gene that controls tissue organization and cell differentiation in the embryo, hypocotyls, primary roots, and secondary roots

[A] A: 1; B: 3; C: 4; D: 5; E: 2
[B] A: 1; B: 2; C: 3; D: 4; E: 5
[C] A: 4; B: 1; C: 2; D: 3; E: 5
[D] A: 4; B: 2; C: 3; D: 5; E: 1

416. Match the list I with the list II and select the correct answer using the codes given below the lists

	List I [Terms]		List II [Significance]
A.	S lignin	1.	Sinapyl alcohol
B.	S phase	2.	DNA replication
C.	S-type channels	3.	A type of gated channels
D.	SAG genes	4.	Senescence associated gene
E.	Salicylic acid	5.	Endogenous signal for SAR

[A] A: 1; B: 3; C: 4; D: 5; E: 2
[B] A: 1; B: 2; C: 3; D: 4; E: 5
[C] A: 4; B: 1; C: 2; D: 3; E: 5
[D] A: 4; B: 2; C: 3; D: 5; E: 1

417. Match the list I with the list II and select the correct answer using the codes given below the lists

	List I [Terms]		List II [Significance]
A.	Salicylhydroxamic acid	1.	Specific inhibitor of the alternative oxidase
B.	Salinity	2.	Contrast with sodicity
C.	Salinization	3.	The accumulation of mineral ions, particularly sodium chloride and sodium sulfate, in soil often due to irrigation
D.	Salt stress	4.	The adverse effects of excess minerals on plants
E.	Salt-tolerant plants	5.	Plants that can survive or even thrive in high salt soils

[A] A: 1; B: 3; C: 4; D: 5; E: 2
[B] A: 1; B: 2; C: 3; D: 4; E: 5
[C] A: 4; B: 1; C: 2; D: 3; E: 5
[D] A: 4; B: 2; C: 3; D: 5; E: 1

418. Match the list I with the list II and select the correct answer using the codes given below the lists

List I [Terms]	List II [Significance]
A. *scr* mutant	1. Arabidopsis mutant in which hypocotyls and inflorescence are agravitropic and lack both endodermis and starch sheath
B. Scutellum	2. The single cotyledon of the grass
C. Second messenger	3. Intracellular molecules
D. Secondary active transport	4. Proton motive forces
E. Secondary cross-regulation	5. Involves the output of one signal pathways regulating the abundance or perception of a second signal

[A] A: 1; B: 3; C: 4; D: 5; E: 2
[B] A: 1; B: 2; C: 3; D: 4; E: 5
[C] A: 4; B: 1; C: 2; D: 3; E: 5
[D] A: 4; B: 2; C: 3; D: 5; E: 1

419. Match the list I with the list II and select the correct answer using the codes given below the lists

List I [Scientist]	List II [Contributions]
A. Aristole	1. Father of Biology
B. Theophrastus	2. Historia plantarum
C. Jenseen and Jenssen	3. Compound microscope
D. Robert Hooke	4. Bottle cork
E. Leeuwenhoek	5. Observed bacteria

[A] A: 1; B: 3; C: 4; D: 5; E: 2
[B] A: 1; B: 2; C: 3; D: 4; E: 5
[C] A: 4; B: 1; C: 2; D: 3; E: 5
[D] A: 4; B: 2; C: 3; D: 5; E: 1

420. Match the list I with the list II and select the correct answer using the codes given below the lists

List I [Scientist]	List II [Contributions]
A. Malpighi	1. Studied plant anatomy
B. N. Grew	2. Described reproductive parts in flowers for the first time
C. R. Camerarius	3. Gave the impotence of pollination
D. Stephan Hales	4. Father of Plant Physiology
E. Carolus Linnaeus	5. Father of Plant Taxonomy

[A] A: 1; B: 3; C: 4; D: 5; E: 2
[B] A: 1; B: 2; C: 3; D: 4; E: 5
[C] A: 4; B: 1; C: 2; D: 3; E: 5
[D] A: 4; B: 2; C: 3; D: 5; E: 1

421. Match the list I with the list II and select the correct answer using the codes given below the lists

	List I [Scientist]		List II [Contributions]
A.	Fontana	1.	Nucleoli
B.	Goethe	2.	Proposed flower is the condensed metamorphic shoot
C.	Lamarck	3.	Inheritance of acquired character
D.	AP Condolle	4.	Introduce the term Taxonomy
E.	Amici	5.	Observed pollen tube

[A] A: 1; B: 3; C: 4; D: 5; E: 2 [B] A: 1; B: 2; C: 3; D: 4; E: 5
[C] A: 4; B: 1; C: 2; D: 3; E: 5 [D] A: 4; B: 2; C: 3; D: 5; E: 1

422. Match the list I with the list II and select the correct answer using the codes given below the lists

	List I [Scientist]		List II [Contributions]
A.	Robert Brown	1.	Brownian movement
B.	Ehrenberg	2.	The term bacterium used for the first time
C.	Felix Dujardin	3.	Named the living substance of cell-Sarcode
D.	Hugo von Mohl	4.	Emphasized importance of protoplasm and described cell division
E.	J.E. Purkinije	5.	Introduced the term protoplasm

[A] A: 1; B: 3; C: 4; D: 5; E: 2 [B] A: 1; B: 2; C: 3; D: 4; E: 5
[C] A: 4; B: 1; C: 2; D: 3; E: 5 [D] A: 4; B: 2; C: 3; D: 5; E: 1

423. Match the list I with the list II and select the correct answer using the codes given below the lists

	List I [Scientist]		List II [Contributions]
A.	Mayer	1.	During photosynthesis light energy is converted into chemical energy
B.	De Bary	2.	Protoplsmic theory
C.	Kollicker	3.	Mitochondria
D.	W. Hofmeister	4.	Nuclear division
E.	R. Virchow	5.	Omnis cellula e cellula

[A] A: 1; B: 3; C: 4; D: 5; E: 2 [B] A: 1; B: 2; C: 3; D: 4; E: 5
[C] A: 4; B: 1; C: 2; D: 3; E: 5 [D] A: 4; B: 2; C: 3; D: 5; E: 1

424. Match the list I with the list II and select the correct answer using the codes given below the lists

	List I [Scientist]		List II [Contributions]
A.	George Bentham	1.	Genera Plantarum
B.	A Kollicker	2.	Cytoplasm
C.	Louis Pasteur	3.	Germ theory of disease
D.	E. Haeckel	4.	Coined the term Protista
E.	J. Huxley	5.	Protoplasm in the physical basis of life

[A] A: 1; B: 3; C: 4; D: 5; E: 2 [B] A: 1; B: 2; C: 3; D: 4; E: 5
[C] A: 4; B: 1; C: 2; D: 3; E: 5 [D] A: 4; B: 2; C: 3; D: 5; E: 1

425. Match the list I with the list II and select the correct answer using the codes given below the lists

	List I [Scientist]		List II [Contributions]
A.	W. Kuhne	1.	Joint assay with Darwin
B.	Joseph Lister	2.	Collected Essays
C.	W. Flemming	3.	Studies in the Theory of Descent
D.	E. Strasburger	4.	General Morphology of Organism
E.	E.G. Balbiani	5.	Discovered larval salivary gland

[A] A: 1; B: 3; C: 4; D: 5; E: 2 [B] A: 1; B: 2; C: 3; D: 4; E: 5
[C] A: 4; B: 1; C: 2; D: 3; E: 5 [D] A: 4; B: 2; C: 3; D: 5; E: 1

426. Match the list I with the list II and select the correct answer using the codes given below the lists

	List I [Authors]		List II [Contributions]
A.	Wallace	1.	Joint assay with Darwin
B.	Huxley	2.	Collected Essays
C.	Weismann	3.	Studies in the Theory of Descent
D.	Haeckel	4.	General Morphology of Organism
E.	Bateson	5.	Materials for the study of variations

[A] A: 1; B: 3; C: 4; D: 5; E: 2 [B] A: 1; B: 2; C: 3; D: 4; E: 5
[C] A: 4; B: 1; C: 2; D: 3; E: 5 [D] A: 4; B: 2; C: 3; D: 5; E: 1

427. Match the list I with the list II and select the correct answer using the codes given below the lists

	List I [Scientist]		List II [Contributions]
A.	C. Benda	1.	Mitochondria
B.	E. Buchner	2.	Zymase
C.	S.G. Nawaschin	3.	Double fertilization

D.	Dixon and Jolly	4.	Ascent of sap
E.	Ivanowaski	5.	Provide proof for the existence of the virus

[A] A: 1; B: 3; C: 4; D: 5; E: 2 [B] A: 1; B: 2; C: 3; D: 4; E: 5
[C] A: 4; B: 1; C: 2; D: 3; E: 5 [D] A: 4; B: 2; C: 3; D: 5; E: 1

428. Match the list I with the list II and select the correct answer using the codes given below the lists

	List I [Scientist]		List II [Contributions]
A.	R. Altman	1.	Nucleic acid
B.	E. Von Tschemak	2.	Rediscover Mandel work
C.	E. Strasburger	3.	Plasmodesmata
D.	C.E. McLung	4.	Showed the importance of reduction division
E.	W.S. Sutton	5.	Identified sex chromosome in grasshopper

[A] A: 1; B: 2; C: 3; D: 5; E: 4 [B] A: 1; B: 2; C: 3; D: 4; E: 5
[C] A: 4; B: 1; C: 2; D: 3; E: 5 [D] A: 4; B: 2; C: 3; D: 5; E: 1

429. Match the list I with the list II and select the correct answer using the codes given below the lists

	List I [Scientist]		List II [Contributions]
A.	C.de Duve	1.	Joint assay with Darwin
B.	C.O. Miller	2.	Discovery of Kinetin
C.	D.I. Arnon	3.	Photophosphorylation
D.	Robinson and Brown	4.	First observed ribosome in Bean roots
E.	F.A. Lipman	5.	Discovered coenzyme A and its role in metabolism

[A] A: 1; B: 3; C: 4; D: 5; E: 2 [B] A: 1; B: 2; C: 3; D: 4; E: 5
[C] A: 4; B: 1; C: 2; D: 3; E: 5 [D] A: 4; B: 2; C: 3; D: 5; E: 1

430. Match the list I with the list II and select the correct answer using the codes given below the lists

	List I [Scientist]		List II [Contributions]
A.	T. H. Morgan	1.	Joint assay with Darwin
B.	F. Kogl, Haagen-Smit and Exleben	2.	First to isolate IAA from human urine but called it hetroauxin
C.	Otto H. Warburg	3.	Gave action of enzymes
D.	M. Knoll and E. Ruska	4.	Mass flow hypothesis
E.	Munch	5.	Invented electron microscope

[A] A: 1; B: 2; C: 3; D: 5; E: 4 [B] A: 1; B: 2; C: 3; D: 4; E: 5
[C] A: 4; B: 1; C: 2; D: 3; E: 5 [D] A: 4; B: 2; C: 3; D: 5; E: 1

431. Match the list I with the list II and select the correct answer using the codes given below the lists

List I [Scientist]		List II [Contributions]
A. Rosenberg	1.	Joint assay with Darwin
B. Winkler	2.	Collected Essays
C. A. Kossel	3.	Studies in the Theory of Descent
D. W. L. Johannsen	4.	General Morphology of Organism
E. P. Rous	5.	Rous sarcoma virus

[A] A: 1; B: 3; C: 4; D: 5; E: 2 [B] A: 1; B: 2; C: 3; D: 4; E: 5
[C] A: 4; B: 1; C: 2; D: 3; E: 5 [D] A: 4; B: 2; C: 3; D: 5; E: 1

432. Match the list I with the list II and select the correct answer using the codes given below the lists

List I [Scientist]		List II [Contributions]
A. A. Stoll	1.	Gave chemistry and structure of chlorophyll
B. W.W. Garner and Allard	2.	Photoperiodism
C. R. Feulgen	3.	Feulgen reaction
D. J.C. Bose	4.	Phylogenic system of classification of angiosperm
E. J. Hutchinson	5.	Ascent of sap

[A] A: 1; B: 2; C: 3; D: 5; E: 4 [B] A: 1; B: 2; C: 3; D: 4; E: 5
[C] A: 4; B: 1; C: 2; D: 3; E: 5 [D] A: 4; B: 2; C: 3; D: 5; E: 1

433. Match the list I with the list II and select the correct answer using the codes given below the lists

List I [Scientist]		List II [Contributions]
A. T. Svedberg	1.	Developed Ultracentrifuge
B. F.W. Went	2.	Auxin
C. H. J. Muller	3.	Produced mutations by X-rays in animals
D. F. Griffiths	4.	Discovered Penicillin
E. Alexander Fleming	5.	Transformation experiment

[A] A: 1; B: 2; C: 3; D: 5; E: 4 [B] A: 1; B: 2; C: 3; D: 4; E: 5
[C] A: 4; B: 1; C: 2; D: 3; E: 5 [D] A: 4; B: 2; C: 3; D: 5; E: 1

434. Match the list I with the list II and select the correct answer using the codes given below the lists

	List I [Scientist]		List II [Contributions]
A.	Palade	1.	Ribosomes
B.	S. Ochoa	2.	In vitro synthesis of polyribonucleotides
C.	Korenberg	3.	In vitro synthesis of Deoxyribonucleotides
D.	H. Fraenkel Conart	4.	RNA in genetic material of TMV
E.	J. Lederberg	5.	Genetic recombination in bacteria

[A] A: 1; B: 3; C: 4; D: 5; E: 2 [B] A: 1; B: 2; C: 3; D: 4; E: 5
[C] A: 4; B: 1; C: 2; D: 3; E: 5 [D] A: 4; B: 2; C: 3; D: 5; E: 1

435. Match the list I with the list II and select the correct answer using the codes given below the lists

	List I [Scientist]		List II [Contributions]
A.	F. Sanger	1.	Amino acid sequence of Insulin
B.	Robertson	2.	Give the concept of unit membrane
C.	A. Korenberg	3.	Synthesis of DNA and RNA
D.	Park and Pon	4.	Glyoxysomes
E.	Beevers	5.	Quantasome

[A] A: 1; B: 2; C: 3; D: 5; E: 4 [B] A: 1; B: 2; C: 3; D: 4; E: 5
[C] A: 4; B: 1; C: 2; D: 3; E: 5 [D] A: 4; B: 2; C: 3; D: 5; E: 1

436. Match the list I with the list II and select the correct answer using the codes given below the lists

	List I [Scientist]		List II [Contributions]
A.	Nass and Nass	1.	DNA in mitochondria
B.	Chance and Parsons	2.	Discovered elementary particle in mitochondria
C.	Menke	3.	Reported thyalokoids in chloroplast
D.	Malvin Calvin	4.	Path of carbon in photosynthesis
E.	Nirenberg and Mathii	5.	Broken genetic code

[A] A: 1; B: 3; C: 4; D: 5; E: 2 [B] A: 1; B: 2; C: 3; D: 4; E: 5
[C] A: 4; B: 1; C: 2; D: 3; E: 5 [D] A: 4; B: 2; C: 3; D: 5; E: 1

437. Match the list I with the list II and select the correct answer using the codes given below the lists

	List I [Scientist]		List II [Contributions]
A.	Roger Beachy	1.	Demonstrate that plant can acquire resistant to viral pathogens
B.	E. Fischer, E. Krebs	2.	Alteration of enzyme activity
C.	R. Roberts, P Sharp	3.	Structure and function of GTP binding proteins
D.	A Gilman	4.	Split genes and RNA processing
E.	EB Lewis	5.	Genetic control of embryonic development

[A] A: 1; B: 3; C: 4; D: 5; E: 2 [B] A: 1; B: 2; C: 3; D: 4; E: 5
[C] A: 4; B: 1; C: 2; D: 3; E: 5 [D] A: 1; B: 2; C: 4; D: 3; E: 5

438. Match the list I with the list II and select the correct answer using the codes given below the lists

	List I [Scientists]		List II [Contributions]
A.	W. Waldeyer	1.	Chromosome
B.	T. Boveri	2.	Centriole
C.	R. Altman	3.	Bioblast
D.	Ivanowaski	4.	Virus
E.	W. Roux	5.	Chromosome contains the unit of heredity

[A] A: 1; B: 3; C: 4; D: 5; E: 2 [B] A: 1; B: 2; C: 3; D: 4; E: 5
[C] A: 4; B: 1; C: 2; D: 3; E: 5 [D] A: 4; B: 2; C: 3; D: 5; E: 1

439. Match the list I with the list II and select the correct answer using the codes given below the lists

	List I [Scientist]		List II [Contributions]
A.	Melchers	1.	Vernalin
B.	F. Gustafsan	2.	Produce parthenocarpy fruit
C.	Hans Krebs	3.	GA
D.	Yabuta, Hayashi, Sumiki	4.	Citric acid Cycle
E.	R. Hill	5.	In light reaction of the photosynthesis Oxygen is released and a reducing substance is produced

[A] A: 1; B: 3; C: 4; D: 5; E: 2 [B] A: 1; B: 2; C: 3; D: 4; E: 5
[C] A: 4; B: 1; C: 2; D: 3; E: 5 [D] A: 1; B: 2; C: 4; D: 3; E: 5

440. Match the list I with the list II and select the correct answer using the codes given below the lists

List I [Scientist]		List II [Contributions]
A. S.A. Waksman	1.	Streptomycin
B. K.R. Porter	2.	ER
C. J.B. Summner	3.	Crystalized first enzyme
D. H.J. Muller	4.	Production of mutation by X-rays
E. N.D. Jinder and Lederberg	5.	Transduction

[A] A: 1; B: 3; C: 4; D: 5; E: 2 [B] A: 1; B: 2; C: 3; D: 4; E: 5
[C] A: 4; B: 1; C: 2; D: 3; E: 5 [D] A: 4; B: 2; C: 3; D: 5; E: 1

441. Match the list I with the list II and select the correct answer using the codes given below the lists

List I [Terms]		List II [Significance]
A. Phytochromobilin	1.	The linear tetrapyrrole chromophore of phytochrome
B. Phytoecdysones	2.	Group of plant steroids that are toxic to insects because of chemical similarity to the insect molting hormone
C. Phytoferritin	3.	A minor constituents of starch having a branched structure
D. Phytoglycogen	4.	An iron-protein complex which stores surplus iron in the plant cells
E. Phytol	5.	The long hydrocarbon chain found on the chlorophyll

[A] A: 1; B: 3; C: 4; D: 5; E: 2 [B] A: 1; B: 2; C: 3; D: 4; E: 5
[C] A: 4; B: 1; C: 2; D: 3; E: 5 [D] A: 1; B: 2; C: 4; D: 3; E: 5

442. Match the list I with the list II and select the correct answer using the codes given below the lists

List I [Terms]		List II [Significance]
A. Phytomere	1.	A developmental unit consisting of one or more leaves
B. Phytosiderophores	2.	A special class of iron chelators produced by the grasses and made of amino acids that are not found in proteins
C. PIF3	3.	Transcriptional factor

D. PILs	4. DNA binding proteins that selectively interact with phytochromes in their active Pfr conformations
E. PIN1	5. Membrane protein that transport auxin out of the basal ends

[A] A: 1; B: 3; C: 4; D: 5; E: 2 [B] A: 1; B: 2; C: 3; D: 4; E: 5
[C] A: 4; B: 1; C: 2; D: 3; E: 5 [D] A: 4; B: 2; C: 3; D: 5; E: 1

443. Match the list I with the list II and select the correct answer using the codes given below the lists

List I [Terms]	List II [Significance]
A. Pit	1. Microscopic regions where the secondary wall of treachery elements is absent
B. Pit fields	2. Depression in the primary cell wall
C. Pit membrane	3. The adjacent pits of adjoined tracheid cells
D. Pit pairs	4. The porous layer between pit pairs, consisting of two thinned primary walls and a middle lamella
E. Plant growth substance	5. For Basic metabolism

[A] A: 1; B: 3; C: 4; D: 5; E: 2 [B] A: 1; B: 2; C: 3; D: 4; E: 5
[C] A: 4; B: 1; C: 2; D: 3; E: 5 [D] A: 1; B: 2; C: 4; D: 3; E: 5

444. Match the list I with the list II and select the correct answer using the codes given below the lists

List I [Terms]	List II [Significance]
A. Axis	1. The hypothetical central line
B. *axr*1 mutant	2. *Arabidopsis* auxin resistant mutant
C. Bacillus thuringiensis	3. A soil bacterium
D. Bacteriochlorophyll	4. Light-absorbing pigments
E. Bacteroids	5. Nitrogen-fixing organelles that develop from endosymbiotic bacteria upon a signal from the host plant

[A] A: 1; B: 3; C: 4; D: 5; E: 2 [B] A: 1; B: 2; C: 3; D: 4; E: 5
[C] A: 4; B: 1; C: 2; D: 3; E: 5 [D] A: 4; B: 2; C: 3; D: 5; E: 1

445. Match the list I with the list II and select the correct answer using the codes given below the lists

	List I [Terms]		List II [Significance]
A.	BAK1	1.	A second LRR-receptor kinase
B.	Bark	2.	Tissues outside the cambium
C.	basal	3.	Contrast with apical
D.	Basal cell	4.	Embryogenesis
E.	Basipetal	5.	Growing tip

[A] A: 1; B: 3; C: 4; D: 5; E: 2
[B] A: 1; B: 2; C: 3; D: 4; E: 5
[C] A: 4; B: 1; C: 2; D: 3; E: 5
[D] A: 4; B: 2; C: 3; D: 5; E: 1

446. Match the list I with the list II and select the correct answer using the codes given below the lists

	List I [Terms]		List II [Significance]
A.	Biennial	1.	Two growing season
B.	Bidirectional transport	2.	Simultaneous transport in two direction
C.	BIN2	3.	Measurement of a suspected biologically active substance
D.	Bioassy	4.	A repressor of brassinosteroid-induced gene expression
E.	Biolistics	5.	Gen gun

[A] A: 1; B: 3; C: 4; D: 5; E: 2
[B] A: 1; B: 2; C: 3; D: 4; E: 5
[C] A: 4; B: 1; C: 2; D: 3; E: 5
[D] A: 1; B: 2; C: 4; D: 3; E: 5

447. Match the list I with the list II and select the correct answer using the codes given below the lists

	List I [Terms]		List II [Significance]
A.	Bleached	1.	The loss of chlorophyll properties
B.	Blue-light response	2.	400 to 500 nm
C.	Bolting	3.	A thin file of still air at the leaf surface
D.	Boundary layer	4.	GA
E.	Bowen ratio	5.	The ratio of sensible heat loss to evaporational heat loss

[A] A: 1; B: 3; C: 4; D: 5; E: 2
[B] A: 1; B: 2; C: 3; D: 4; E: 5
[C] A: 4; B: 1; C: 2; D: 3; E: 5
[D] A: 1; B: 2; C: 4; D: 3; E: 5

448. Match the list I with the list II and select the correct answer using the codes given below the lists

List I [Terms]	List II [Significance]
A. Hybrid vigour	1. Heterosis
B. Hydathodes	2. Also a site of auxin synthesis in immature leaves of *Arabidopsis*
C. Hydrulic conductivity	3. Closing of stomata in response to a closing signal
D. Hydroactive stomatal closer	4. Describes how readily water can move across a membrane
E. Hydrogen bond	5. A weak chemical bond formed between a hydrogen atom and an oxygen or nitrogen atom

[A] A: 1; B: 3; C: 4; D: 5; E: 2 [B] A: 1; B: 2; C: 3; D: 4; E: 5
[C] A: 4; B: 1; C: 2; D: 3; E: 5 [D] A: 1; B: 2; C: 4; D: 3; E: 5

449. Match the list I with the list II and select the correct answer using the codes given below the lists

List I [Terms]	List II [Significance]
A. Hypoxia	1. Refers to oxygen concentrations lower than normal
B. Indole	2. Precursor
C. Induced defenses	3. Flower that lack either the male or female
D. Imperfect flower	4. Plant defenses that are initiated only after actual damage occurs
E. Secondary plasmodesmata	5. Plasmodesmata that forms and permit symplastic transport between non clonally related cells

[A] A: 1; B: 3; C: 4; D: 5; E: 2 [B] A: 1; B: 2; C: 3; D: 4; E: 5
[C] A: 4; B: 1; C: 2; D: 3; E: 5 [D] A: 1; B: 2; C: 4; D: 3; E: 5

30

SEED PHYSIOLOGY OF CROPS

1. **Botanically seed is**
   [A] A ripened ovule [B] Containing embryo
   [C] Developed after fertilization [D] All of the above
2. **In agriculture, seed is**
   [A] Any part of organ of a plant [B] Maintaining the intrinsic quality
   [C] Both [D] None of the above
3. **Dormancy in a general sense is the inactive period of seed embryo which has attended maturity, up to when it begins to germinate. Among the following, find out the which one shows the dormancy?**
   [A] Vegetative buds and Tubers [B] Bulbs and Rhizomes
   [C] Bulbils [D] All of the above
4. **Dormancy is**
   [A] An endogenous process [B] Environmentally imposed
   [C] Temporary suspension [D] All of the above
5. **Dormancy that results from some characteristic of seed is called**
   [A] Organic dormancy [B] Inorganic dormancy
   [C] Both [D] None of the above
6. **Exogenous dormancy is related to**
   [A] Endosperm [B] Seed coat
   [C] Fruit walls [D] All of the above
7. **Endogenous dormancy is related to**
   [A] Endosperm [B] Seed coat
   [C] Fruit walls [D] Embryo
8. **Non-deep, intermediate and deep dormancy are the examples of**
   [A] Physiological dormancy [B] Morphological dormancy
   [C] Morpho-physiological dormancy
   [D] None of the above
9. **Non-Deep physiological dormancy is common in seeds of**
   [A] Most of weeds [B] Vegetables
   [C] Garden flowers [D] All of the above

10. Seeds of Acer, Corylus, Fagus, Fraxinus, Melampyrum, Polygonum have

[A] Intermediate physiological dormancy
[B] Physiological dormancy
[C] Morphological dormancy
[D] None of the above

11. Species such as *Impatiens parviflora, Malus domestica, Sorbus* spp. *Rhodotypos kerrioides, Prunus persica, Crataegus* have

[A] Intermediate physiological dormancy
[B] Deep physiological dormancy
[C] Morphological dormancy
[D] None of the above

12. Morpho-physiological dormancy (MPD) occurs in seeds with

[A] Rudimentary embryo [B] Linear embryo
[C] Both [D] None of the above

13. Hard Seed coat dormancy occurs in the seeds of

[A] Anacardiaceae [B] Cannanaceae
[C] Curcurbitaceae [D] All of the above

14. The impermeability of the seed coat is due to

[A] Suberin [B] Cutin
[C] Lignin [D] All of the above

15. In physical dormancy, germination is prevented by lack of

[A] Imbibition of water [B] pH of water
[C] EC of water [D] All of the above

16. Anatomical studies of several spp. have shown that a specific region of the seed coat, rather than the whole seed coat becomes permeable. In Malvaceae, water enters through the opening in the

[A] Palisade layer of the chalazal region
[B] The dense paranchymatous plug moves and water enters after the palisade cells surrounding the plug down
[C] When palisade cells pull away from subpalisade cells in the chalazal region
[D] None of the above

17. In cotton, water enters through the opening in the

[A] When palisade cells pull away from subpalisade cells in the chalazal region
[B] Palisade layer of the chalazal region
[C] The dense paranchymatous plug moves and water enters after the palisade cells surrounding the plug down
[D] None of the above

18. In *Sida spinosa*, the seed coat becomes permeable when

[A] Palisade cells pull away from subpalisade cells in the chalazal region

[B] Palisade layer of the chalazal region

[C] The dense paranchymatous plug moves and water enters after the palisade cells surrounding the plug down

[D] None of the above

19. The strophiole is the site of water entry into seeds of

[A] Leguminosae [B] Pappilionoidae

[C] Malvaceae [D] None of the above

20. In the subfamiliy, ..., there is a zone of weakness in the seed coat near hilum where macrosclerids dull apart.

[A] Leguminosae [B] Pappilionoidae

[C] Malvaceae [D] None of the above

21. In ...., water enters the seed when a small plug like structure near the micropyle is dislodged or removed.

[A] Convolvulaceae [B] Pappilionoidae

[C] Malvaceae [D] None of the above

22. As a consequence of seed coat limiting the amount of oxygen available to the embryo, the respiration needed for germination would also be affected, enforcing the embryo to remain dormant. In some seeds, phenolic compounds in embryo covers fix oxygen by oxidation, thus making it unavailable for embryo. Among the following which one is the correct statement ?

[A] If we increase the oxygen concentration, it will increase the germination of seeds

[B] If we increase the oxygen concentration, it will decrease the germination of seeds

[C] If we increase the oxygen concentration, it will not affect the germination of seeds

[D] None of the above

23. Alternate freezing and thawing had little or no effect on softening the seeds of

[A] *Medicago* sativa [B] *Cuscuta* sp.

[C] *Sesbania* sp. [D] All of the above

24. High temperature causes hard seed coats to become permeable. Drastic treatments such as immersion into boiling water for short periods or exposure to dry heat at 60°C to 100°C results in 100% permeability. In nature fire is the source of high temperature, resulting in germination of hard seeded spp. The above said statement is

[A] True [B] False

[C] Not known [D] None of the above

25. Some seeds have a combination of physiological and physical dormancy. They have dormant embryo and impermeable seed coats. In these seeds, high fluctuating temperature is required to render the seed coat permeable and either low winter temperature or high summer temperature is required to overcome dormancy of the embryo. For seeds of *Tilia ameriana, Cercis Canadensis, Parkia* sp. low temperature is needed to overcome the embryo dormancy. This statement is

[A] False [B] True
[C] Not known [D] None of the above

26. The presence of physical and physiological dormancy in a single seed, sometimes is referred to as

[A] Combined dormancy [B] Single dormancy
[C] Alone dormancy [D] None of the above

27. In ...., physical dormancy is broken before physiological dormancy

[A] Tilia [B] Stylosanthus
[C] Both [D] None of the above

28. In ...., physiological dormancy is broken before physical dormancy.

[A] Tilia [B] Stylosanthus
[C] Both [D] None of the above

29. Match the species in Group I with their location of inhibitors listed in Group II

	Group I		Group II
P.	*Gossypium* sp.	I.	Pericarp, testa
Q.	*Coriandrum sativum*	II.	Pericarp
R.	*Helianthus* sp.,	III.	Pericarp, testa
S.	*Oryza sativa*	IV.	Hull

[A] P-I, Q-II, R-III, S-IV [B] P-IV, Q-II, R-III, S-I
[C] P-III, Q-IV, R-III, S-I [D] P-II, Q-IV, R-III, S-I

30. Match the species in group I with their name of inhibitors listed in group II

	Group I		Group II
P.	*Gossypium* sp.	I.	ABA
Q.	*Coriandrum sativum*	II.	Coumarin
R.	*Helianthus* sp.,	III.	Hydrocyanic acid
S.	*Oryza sativa*	IV.	Probably ABA

[A] P-I, Q-II, R-III, S-IV [B] P-IV, Q-II, R-III, S-I
[C] P-III, Q-IV, R-III, S-I [D] P-II, Q-IV, R-III, S-I

31. Match the species in group I with their name of inhibitors listed in group II

	Group I		Group II
P.	*Triticum* spp.	I.	Catechin, Tannins
Q.	*Hordenum vulgare*	II.	Phenolic acid
R.	*Elaegnus angustifolia*	III.	Coumarin
S.	*Beta vulgare*	IV.	High concentration of inorganic ions

[A] P-I, Q-II, R-III, S-IV [B] P-IV, Q-II, R-III, S-I
[C] P-III, Q-IV, R-III, S-I [D] P-II, Q-IV, R-III, S-I

32. Match the family in group I with their seed type listed in group II

	Group I		Group II
P.	Anacardiaceae	I.	Bent
Q.	Bombacaceae	II.	Spatulate
R.	Cannaceae	III.	Linear
S.	Convolvulaceae	IV.	Folded

[A] P-I, Q-II, R-III, S-IV [B] P-IV, Q-II, R-III, S-I
[C] P-III, Q-IV, R-III, S-I [D] P-II, Q-IV, R-III, S-I

33. Match the family in group I with their specialized seed structure listed in group II

	Group I		Group II
P.	Anacardiaceae	I.	Endocarp slit
Q.	Bombacaceae	II.	Micropyle
R.	Cannaceae	III.	Lid on raphe
S.	Convolvulaceae	IV.	Plug near micropyle

[A] P-I, Q-II, R-III, S-IV [B] P-IV, Q-II, R-III, S-I
[C] P-III, Q-IV, R-III, S-I [D] P-II, Q-IV, R-III, S-I

34. Match the family in group I with their specialized seed structure listed in group II

	Group I		Group II
P.	Geraniaceae	I.	Hilum
Q.	Malvaceae	II.	Chalaza plug
R.	Musaceae	III.	Micropylar lid-like structure
S.	Nelumbonaceae	IV.	Protuberance

[A] P-I, Q-II, R-III, S-IV [B] P-IV, Q-II, R-III, S-I
[C] P-III, Q-IV, R-III, S-I [D] P-II, Q-IV, R-III, S-I

35. Stony endocarps are found in

[A] Anacardiaceae [B] Apocynaceae
[C] Arecaceae [D] All of the above

36. Among the following, which one has stony endocarps?

[A] Burseraceae [B] Cornaceae
[C] Elaeagnaceae [D] All of the above

37. Stony endocarp is found in

[A] Juglandaceae [B] Meliaceae
[C] Tiliaceae [D] All of the above

38. Dormant seeds exhibit dormancy at one particular temperature, but not at another. e.g. wheat grains are usually dormant for a short while after harvest and will not germinate at 20°C temperature. If the temperature is lowered to 15°C nearly all the grains germinate. After the loss of dormancy, germination occurs at both temperatures. This is called

[A] Relative dormancy [B] After ripening
[C] Primary dormancy [D] None of the above

39. A condition of the seed when it fails to germinate because of the suitable environmental conditions normally required for the growth are not present is known as

[A] Relative dormancy [B] Quiescence
[C] Primary dormancy [D] None of the above

40. Physiological process that occurs in the seed after harvest that enables germination to take place is known as

[A] Relative dormancy [B] After ripening
[C] Primary dormancy [D] None of the above

41. Seeds whose germination has been inhibited fail to recover even when the inhibitory factor is removed. This dormancy is known as

[A] Induced dormancy [B] Secondary dormancy
[C] Both [D] None of the above

42. Among the following find out the methods of breaking of seed dormancy

[A] Scarification [B] Acid Scarification
[C] Mechanical Scarification [D] All of the above

43. Among the following find out the methods of breaking of seed dormancy

[A] Stream scarification [B] Stratification
[C] Cold stratification [D] All of the above

44. Among the following find out the methods of breaking of seed dormancy

[A] Water stratification [B] Hot water treatment
[C] Removal of structure around the seed
[D] All of the above

45. Among the following find out the methods of breaking of seed dormancy

[A] Leaching of metabolites

[B] Low temperature treatments

[C] High temperature treatments

[D] All of the above

46. Among the following find out the methods of breaking of seed dormancy

[A] Dry storage

[B] Alternate temperature treatments

[C] Fire treatment

[D] All of the above

47. Among the following find out the methods of breaking of seed dormancy

[A] Light and Phytochrome

[B] Light intensity

[C] Promoter *vs* Inhibitor concept

[D] All of the above

48. Among the following find out the methods of breaking of seed dormancy

[A] Seed treatments with growth regulators/chemicals

[B] Infra Red Treatment

[C] Pressure Treatment

[D] Magnetic speed treatments

[e] All of the above

49. Any treatment, may be physical or chemical, that weakens or softens the seed coat for early germination is known as

[A] Scarification [B] Stratification

[C] Both [D] None of the above

50. Scarification is more applicable to

[A] Malvaceae [B] Leguminosae

[C] Both [D] None of the above

51. By using concentrated H_2SO_4 @ 100ml/kg of seed for 2-3 minutes treatment can overcome dormancy in seeds. The duration of treatment will vary and it depends on the type and nature of seed coat. This treatment is known as

[A] Mechanical scarification

[B] Acid scarification

[C] Stream scarification

[D] None of the above

52. Seeds are rubbed on a sand paper or with the help of mechanical scarifier or by puncturing on seed coat with the help of needle to enhance/increase the moisture absorption by seeds. This treatment is known as

[A] Mechanical scarification [B] Acid scarifications

[C] Stream scarification [D] None of the above

53. The treatment can be given for about 15-20 minutes. Seed can be kept in the way of boiled steam and dormancy can be broken. This treatment is known as

[A] Mechanical scarification [B] Acid scarifications

[C] Stream scarification [D] None of the above

54. The seed is incubated at low temperature of 0-5°C over a moist substratum for 2-3 days to several months. It depends on the nature of seed and kind of dormancy. This treatment is known as

[A] Cold stratification [B] Water stratification

[C] Hot water treatment [D] None of the above

55. Some seed requires a temperature of 40-50°C water for few days for early germination. This treatment is known as

[A] Cold stratification [B] Water stratification

[C] Hot water treatment [D] None of the above

56. The seeds should be soaked in boiling water for 1-5 minutes to 60-80 minutes. It is effective in case of leguminous tree crop seeds. This treatment is known as

[A] Cold stratification [B] Water stratification

[C] Hot water treatment [D] None of the above

57. For water stratification in lentil, onion, cauliflower, cabbage, radish, opium, turnip and cole crops, the required temperature is 5-10°C for 7 days. This statement is

[A] True [B] False

[C] Not known [D] None of the above

58. Plants which grow in temperate and cooler climates require a period of chilling for breakage of dormancy. For example, apple seed dormancy can be released by low temperature treatment by storing the seeds at 5°C. This statement is

[A] True [B] False

[C] Not known [D] None of the above

59. Most of the plant species which grow in .... and .... regions require alternate temperature for breakage of dormancy.

[A] Temperate, cool temperate [B] Tropical, hot climate

[C] Temperate, hot temperate [D] None of the above

60. **Seeds can be kept in the magnetic field for about 1 to 10 days for breaking dormancy is known as**
[A] Magnetic seed treatment [B] Pressure treatment
[C] Infra Red treatment [D] None of the above

61. **Dormant seeds can be kept in an autoclave and required pressure can be employed for breaking dormancy is known as**
[A] Magnetic seed treatment [B] Pressure treatment
[C] Infra Red treatment [D] None of the above

62. **In Jussiae, Pavlonia and Kalanchoe blossfeldiana, light in the 750-1250 lux intensity from cool white lamps is illuminated for 8 hours in every 24 hour during high temperature period to break the dormancy of seeds with alternate temperature. This statement is**
[A] True [B] False
[C] Not Known [D] None of the above

63. **Infra Red rays can be used for removal of dormancy in the seeds. This statement is**
[A] True [B] False
[C] Both [D] None of the above

64. **If the endogenous dormancy is due to the presence of inhibitors, we can apply growth regulators at the low level to overcome the dormancy. GA & cytokinin and kinetin can be used at concentration of 100-1000 ppm to break the dormancy. This statement is**
[A] False [B] Ture
[C] Both [D] None of the above

65. **........................ can be used for breaking dormancy of both light and chilling treatment requiring seeds e.g. Lettuce.**
[A] Thiourea [B] Ethrel
[C] Hydroxylamine [D] None of the above

66. **................................ can be used for breaking the dormancy of groundnut and cotton seed. The dormancy in cotton seed is due to the presence of ABA in pericarp of seed.**
[A] Thiourea [B] Ethrel
[C] Hydroxylamine [D] None of the above

67. **Among the following, find out the nitrogenous compounds that can be used for breaking the dormancy.**
[A] Nitrate [B] Nitric acid
[C] Hydroxylamine [D] All of the above

68. **Among the following find out the biological significance of dormancy**
[A] Dormancy is important for preventing vivipary and precocious germination
[B] Survival
[C] Synchronise germination
[D] All of the above

69. Bud dormancy of many woody perennials can be overcomed by providing chilling treatment before growth can be resumed. This statement is

[A] True [B] False

[C] Not known [D] None of the above

70. ...., and .... can be used to overcome the bud dormancy.

[A] Ethylene [B] Thiourea

[C] Dinitrophenol [D] All of the above

71. .... and .... are two important environmental factors regulating bud dormancy.

[A] Temperature [B] Light

[C] Both [D] None of the above

72. The seed quality factors refers to

[A] Genetic purity [B] Physical quality

[C] Physiological quality [D] All of the above

73. Seed is the basic input and plays a vital role in agriculture. Quality seed alone can increase the yield up to

[A] 1 to 5 % [B] 5 to 10%

[C] 15 to 20 % [D] None of the above

74. Seed quality can be considered as the summation of all factors that contribute to seed performance. This statement is

[A] True [B] False

[C] Not known [D] None of the above

75. Quality seed means

[A] The most economic planting rate

[B] A higher percentage of seed germination

[C] Emergence in the field

[D] Minimum replanting

[E] All of the above

76. Quality seeds include

[A] Vigorous seedling establishment

[B] Fast growth rate

[C] Greater resistance to pest, disease and stress

[D] Uniformity in ripening

[E] All of the above

77. Seed quality can be maintained which plays a major role in increasing the seed yield. It covers

[A] Pre sowing seed treatment

[B] Storage seed treatment

[C] Foliar treatment

[D] All of the above

78. Colouring seed is an important treatment for increasing market value. More of discolouration of seed with good germination is of less market value. Discoloured seeds which may otherwise be viable with recommended germination as per seed certification standards, many a time may not be acceptable as seed for sowing because of poor physical appearance and high expected incidence of seed borne fungi.

Match the crop in group I with their specialized seed colour listed in group II

	Group I		Group II
P.	Paddy	I.	Yellow
Q.	Maize	II.	Pink
R.	Soybean	III.	Yellow
S.	Sunflower	IV.	Black

[A] P-I, Q-II, R-III, S-IV
[B] P-IV, Q-II, R-III, S-I
[C] P-III, Q-IV, R-III, S-I
[D] P-II, Q-IV, R-III, S-I

79. Match the crop in group I with their specialized seed colour listed in group II

	Group I		Group II
P.	Tomato	I.	Red
Q.	Bhendi	II.	Green
R.	Sorghum	III.	Pink
S.	Paddy	IV.	Yellow

[A] P-I, Q-II, R-III, S-IV
[B] P-IV, Q-II, R-III, S-I
[C] P-III, Q-IV, R-III, S-I
[D] P-II, Q-IV, R-III, S-I

80. The purpose of acid delinting of cotton seeds is to remove fuzz and improve seed quality characteristic . Addition of commercial sulphuric acid (100 ml/kg) to fuzzy cotton seeds and stirring for 1-2 minutes, removes the fuzz, facilitates separation of filled seeds. Among the following find out what are the benefits of the same.

[A] High germination percentage

[B] Quick and uniform seedling emergence

[C] Vigourous seedlings

[D] High tillering rate

[E] Enhanced seed yield

[F] All of the above

81. The term "vigour" was used for the first time by a German Plant Pathologist, ..., who observed the cereal seeds infected by Fusarium disease were capable of germination but the resulting seedlings were incapable of penetrating through a thick layer of brick grit under a specified condition of temperature and moisture.

[A] Hiltner [B] McDaniel

[C] Whalley [D] None of the above

82. In rice, seedling vigour is strongly influenced by the

[A] Method of crop establishment

[B] Temperature at sowing

[C] Both

[D] None of the above

83. Among the following, find out the factors affecting seed and seedling vigour

[A] Seed aging during storage

[B] Cultivar response

[C] Environment during maturation

[D] All of the above

84. Find out the factors affecting seedling vigour

[A] Stage of harvesting

[B] Seed size and weight

[C] Mechanical injury

[D] All of the above

85. Among the following find out the factors affecting seedling vigour

[A] Pest management

[B] Seed protein contents

[C] Post harvest storage

[D] Rate of respiration and oxygen uptake

[E] All of the above

86. What is the full name of ISTA

[A] International Seed Testing Association

[B] Internal Seed Testing Association

[C] International Soil Testing Association

[D] None of the above

87. Among the following find out the enzymes related with seed and seedlings vigour

[A] Glutamic acid decarboxylase (GADA)

[B] Dehydrogenase

[C] Peroxidase

[D] α-amylase

[E] All of the above

88. Ageing in all organisms is the sum total of the deterioative process that eventually lead to death. This statement is

[A] True [B] False

[C] Not known [D] None of the above

89. Castor seeds are highly toxic to humans and many animals, ingestion of two beans can be lethal to humans. This occurs due to the presence of

[A] Ricin [B] Aicin

[C] Picin [D] None of the above

90. A section of cell from endoperm of castor seed is depicted here. Protein in solid form occurs frequently as aleurons grains. The aleurone of cereal seeds form small granules but in castor they are up to 60 micron size and show a complete structure. These granules consist of an amorphous protein mixture are embedded a protein crystalloid and globoid consisting of a double phosphate of calcium and magnesium.

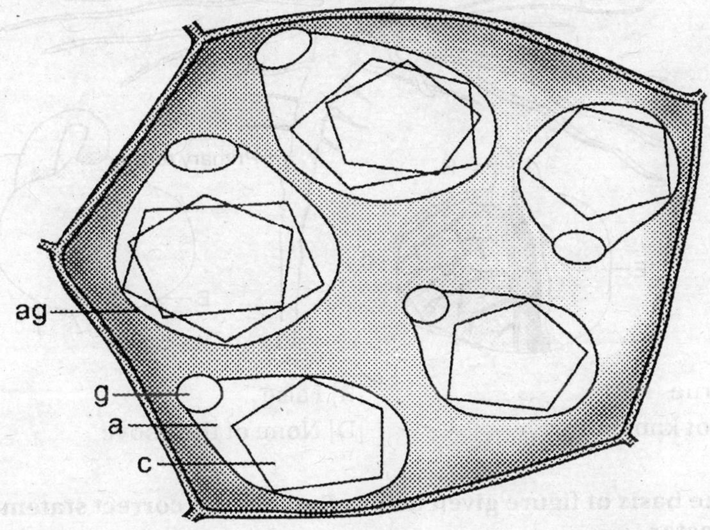

In the given figure, identify the ag, g, a and c. Match the symbols in group I with their specialized seed structure in the above figure listed in group II

	Group I		Group II
P.	ag	I.	aleuron grain
Q.	g	II.	globoid
R.	a	III.	amorphousprotein
S.	c	IV.	crystal protein

[A] P-I, Q-II, R-III, S-IV [B] P-IV, Q-II, R-III, S-I

[C] P-III, Q-IV, R-III, S-I [D] P-II, Q-IV, R-III, S-I

91. In the figure given below, it is clear that castor seeds are endospermic, the embryo is spat late axile and the seed has physiological dormancy. The cotyledons are thin and broad and the endosperm is the major storage tissue. Castor seed is a classical system for studying endosperm reserve breakdown especially for lipid and protein mobilization. Indicate if this statement is true or false in relation to castor seed.

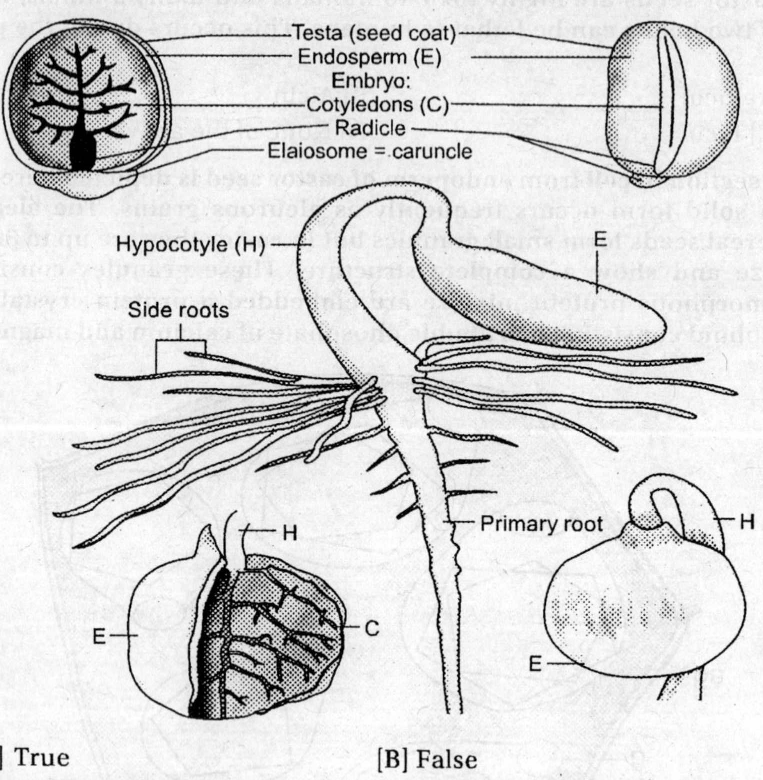

[A] True
[B] False
[C] Not known
[D] None of the above

92. On the basis of figure given below, find out the correct statement about the castor

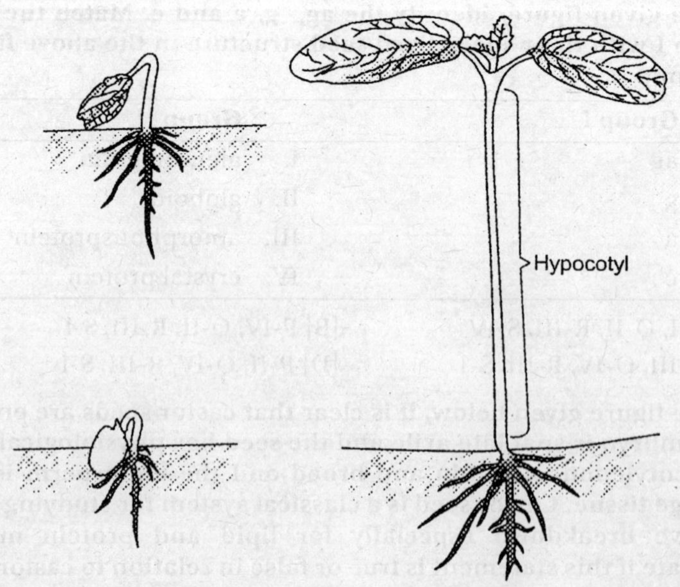

[A] Epigeal germination of castor with cotyledons carried above the soil and long brittle hypocotyls

[B] Hypogeal germination of castor with cotyledons carried below the soil and long brittle hypocotyls

[C] Both

[D] None of the above

93. Match the vegetable crops in group I with their common visual symptoms of maturation listed in group II

	Group I		Group II
P.	Brinjal	I.	Yellow colouration of fruits
Q.	Tomato	II.	Reddish colouration of fruits
R.	Bhendi	III.	Hairline crack formation in dried fruits
S.	Lab lab	IV.	Pods turn light straw colour and seeds to brown colour

[A] P-I, Q-II, R-III, S-IV [B] P-IV, Q-II, R-III, S-I

[C] P-III, Q-IV, R-III, S-I [D] P-II, Q-IV, R-III, S-I

94. Match the vegetable crops in group I with their common visual symptoms of maturation listed in group II

	Group I		Group II
P.	Cluster bean	I.	Pod colour changed to dark yello-wish brown and seed colour to dark brown
Q.	Pea	II.	Change of pod colour from greenish yellow to yellow
R.	Cowpea	III.	Pods turn light straw colour and seeds to brown colour
S.	Ashgourd	IV.	Ashy coating over the fruit

[A] P-I, Q-II, R-III, S-IV [B] P-IV, Q-II, R-III, S-I

[C] P-III, Q-IV, R-III, S-I [D] P-II, Q-IV, R-III, S-I

95. Match the vegetable crops in group I with their common visual symptoms of maturation listed in group II

	Group I		Group II
P.	Pumpkin	I.	Yellowish browning of fruit rind
Q.	Ribbed gourd	II.	Rattling sound of seed in fruit. Browning of dried fruits
R.	Snakergourd	III.	Change of fruit colour yellow
S.	Bittergourd	IV.	Changes of fruit colour yellow/ red at distal end or any part of fruit

[A] P-I, Q-II, R-III, S-IV [B] P-IV, Q-II, R-III, S-I

[C] P-III, Q-IV, R-III, S-I [D] P-II, Q-IV, R-III, S-I

96. Match the vegetable crops in group I with their common visual symptoms of maturation listed in group II

	Group I		Group II
P.	Bottle gourd	I.	Browning of dried fruits. Ratting sound of seed in fruit
Q.	Onion	II.	Umbel colour changes into yellow and turning of seed colour from white to deep black
R.	Carrot	III.	The colour of umbel turn brown and secondary umbel starts to brown
S.	Amaranthus	IV.	Black colour seed

[A] P-I, Q-II, R-III, S-IV [B] P-IV, Q-II, R-III, S-I
[C] P-III, Q-IV, R-III, S-I [D] P-II, Q-IV, R-III, S-I

97. The common name of *Manihot esculenta* Crantz is

[A] Tapioca [B] Yuca
[C] Manioc [D] All of the above

98. Cassava can be propagated by

[A] Asexual method [B] Sexual method
[C] Both [D] None of the above

99. The Central Tube Crops Research Institute (CTCRI), is located at

[A] Thiruvananthapuram [B] West Bengal
[C] Punjab [D] None of the above

100. A mature seed has four components which are significant both physiologically and ecologically for survival

[A] The seed coat, a protective envelope
[B] The embryo, an embryonic plant or sporophyte
[C] Stored food and mineral reserves that nourish the new sporophyte
[D] Enzymes and hormones required for digestion of food reserves
[E] All of the above

101. .... is the most common carbohydrate or polysaccharide, stored in seeds.

[A] Starch [B] Protein
[C] Lipid [D] None of the above

102. The two glucosan starches, ... and ...., are common in seed.

[A] Amylose [B] Amylopectine
[C] Both [D]None of the above

103. Amylose is a straight chain of .... to .... glucose molecules.

[A] 300 to 400 [B] 900 to 1000
[C] 1200 to 1500 [D] None of the above

104. Amylopectine may contain more than a .... glucose molecule units, consequently it has a higher molecular weight and different chemical and physical properties than amylose.

[A] >100 [B] >500

[C] >700 [D] >1000

105. In the iodine test for starch, amylopectin stains .... colour.

[A] Red [B] Blue

[C] Black [D] None of the above

106. In the iodine test of for starch, amylose stains .... colour

[A] Red [B] Blue

[C] Black [D] None of the above

107. .... is more viscous upon wetting.

[A] Amylopectin [B] Amylose

[C] Both [D] None of the above

108. Amylose is ....% digestible by α-amylase.

[A] 10% [B] 50%

[C] 75% [D] 100%

109. Amylopectin is approximately ....% digestible by α-amylase.

[A] 10% [B] 50%

[C] 75% [D] 100%

110. In the given figure, straight chain of glucose is depicted. Find the correct answer.

$(--O-O-O-O-O-O--)_n$

[A] It is amylose [B] It is amylopectine

[C] Both [D] None of the above

111. On the basis of figure, find the correct answer. The branch chain of glucose is depicted as under.

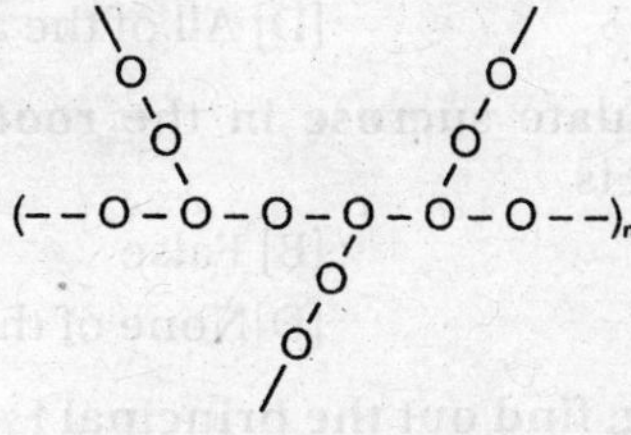

[A] It is amylose [B] It is amylopectin

[C] Both [D] None of the above

112. Corn starch is about .... % amylose and .... % of amylopectin.

[A] 25%, 75% [B] 75%, 25%

[C] 50%, 50% [D] 90%, 10%

113. Hydrolysis of gluosan starches yields

[A] Monosaccharide [B]Disaccharide

[C] Both [D]None of the above

114. Hydrolysis of starches yields

[A] Glucose [B] Maltose

[C] Both [D] None of the above

115. ...., a relatively small starch molecule composed of fructose sugar molecules, is the principle food reserve in barley and certain other temperate grasses.

[A] Inulin [B] Mannans

[C] Xylose [D] None of the above

116. Seeds of some legumes are rich in mannans and this is

[A] Long-chain polymers of mannose sugar

[B] Short-chain polymers of mannose sugar

[C] Both

[D] None of the above

117. The mucilages are a complex family of carbohydrates composed principally of

[A] Polyuromides [B] Galactomides

[C] Both [D] None of the above

118. Mucilage may serve as food reserves, but they also function as a seed coat covering, becoming sticky when wet. This statement is

[A] True [B] False

[C] Not known [D] None of the above

119. Pectins, carbohydrate that are long-chain polymers of galacturonic acid, are bonding between cell walls (middle lamella) of seeds. Pectins are composed principally of

[A] Pectic acid [B] Propectin

[C] Ca and Mg salt [D] All of the above

120. Sugar crops accumulate sucrose in the root or stem, but not in the seed. This statement is

[A] True [B] False

[C] Not known [D]None of the above

121. Among the following, find out the principal fatty acids in oil seeds.

[A] Oleic acid [B] Linoleic acid

[C] Linolenic acid [D] All of the above

122. The dominant fatty acid depends on plant species, ... being the primary in soybean.

[A] Linoleic acid [B] Lauric acid

[C] Linolenic acid [D] None of the above

123. In castor bean .... is the dominant fatty acid.

[A] Lesquerolic acid [B] Lauric acid
[C] Oleic acid [D] None of the above

124. Oils are converted to fats commercially by saturating the double bonds of the fatty acids with hydrogen, a process called

[A] Saponification [B] Hydrogenation
[C] Both [D] None of the above

125. Strong alkalis are used commercially to hydrolyze fats, a process called

[A] Saponification [B] Hydrogenation
[C] Both [D] None of the above

126. Legume seed are generally rich in

[A] Albumins [B] Globulins
[C] Glutelins [D] None of the above

127. Glutelins are water ....

[A] Soluble [B] Insoluble
[C] Not known [D] None of the above

128. The bread wheats are high in ...., a glutelin protein. It imparts the capacity of wheat dough to stretch or rise.

[A] Glutenin [B] Prolamins
[C] Albumins [D] None of the above

129. Zein protein in maize is the example of

[A] Albumins [B] Globulins
[C] Glutelins [D] Prolamins

130. In general, cereals are high in ....

[A] Prolamin [B] Glutelin
[C] Both [D] None of the above

131. Germination includes the following physiological and morphological events. These are

[A] Imbibition and absorption of water
[B] Hydration of tissues
[C] Absorption of oxygen
[D] Activation of enzymes and digestion
[E] Transport of the hydrolyzed molecules to the embryo axis
[F] Increases in respiration and assimilation
[G] Initiation of cell division and enlargement
[G] Embryo emergence
[I] All of the above

132. Seeds of cotton are very sensitive to .... during germination, especially during imbibitions.

[A] Chilling [B] Heat
[C] Rain [D] None of the above

133. Potassium nitrate, which is used routinely in germination tests of many grass seeds and generally are photoblastic seeds. This statement is

[A] True [B] False
[C] Not known [D] None of the above

134. Thiourea, which is not used extensively but does stimulate germination in certain species. It cannot substitute for light or temperature requirements. This statement is

[A] True [B] False
[C] Both [D] None of the above

135. Hydrogen peroxide is effective in the seeds of certain legumes, tomato and barley. This statement is

[A] True [B] False
[C] Both [D] None of the above

136. Prolamine is deficient in .... and ....

[A] Lysine and Tryptophan [B] Methionine and Tryptophan
[C] Lysine and Trypotophan [D] None of the above
[D] None of the above

31

CELL BIOLOGY

1. **Which part of the compound microscope is responsible for gathering and focusing light rays in order to view the specimen under the microscope?**

[A] Condenser lens [B] Eyepiece lens

[C] Objective lens [D] Magnifying lens

2. **How far does the eye have to be from any object before it can focus on it?**

[A] 9 cm [B] 18 cm

[C] 25 cm [D] 37 cm

3. **A microscope's resolving power is determined by**

[A] Wavelength of light

[B] Numerical aperture of the lens system

[C] Refractive Index

[D] Both [A] and [B]

4. **It is possible to achieve the highest resolution in light microscopy by using**

[A] An objective with a minimum numerical aperture

[B] Longest wavelength of visible light used

[C] Shortest wavelength of visible light used

[D] Both [A] and [C]

5. **Oil immersion objective lens has an NA [NA= refractive index X sine (half-aperture angle) value of**

[A] 1.33 [B] 1.01

[C] 2.09 [D] 3.8

6. **According to fluorescence microscopy, which of the following objects is responsible for removing all light except for the blue light in the experiment**

[A] Barrier filter [B] Mercury arc lamp

[C] Exciter filter [D] Dichroic mirror

7. **A total magnification of the image can be obtained by**

   [A] Magnifying power of the objective lens

   [B] Magnifying power of the eyepiece

   [C] Magnifying power of condenser lens

   [D] Both [A] and [B]

8. **What is the first fixative used in light microscopy before staining technique is used in order to prepare the sample for examination?**

   [A] Glutaraldehyde [B] Heat

   [C] Salinity [D] Water

9. **In phase contrast microscopy, the rate at which light enters through an object is measured as the ratio of the**

   [A] Directly Proportional to their refractive indices

   [B] Exponentially related to their refractive indices

   [C] Inversely proportional to their refractive indices

   [D] None of these

10. **.................... of the light microscope controls the intensity of light entering the viewing area?**

    [A] Diaphragm [B] Fine Adjustment

    [C] Coarse adjustment [D] Lens

11. **The negative staining method is used as a method for examining**

    [A] Virus particles [B] Protein particles

    [C] Bacterial flagella [D] All of the above

12. **In order to get a three-dimensional representation of the specimen, which of the following techniques is most useful?**

    [A] SEM [B] TEM

    [C] Compound microscope [D] All of the above

13. **In SEM, the secondary electrons that are radiated back to the detector are collected by**

    [A] Cathode [B] Anode

    [C] Chamber [D] All of the above

14. **How does TEM work to examine the cellular structure of cells by using one of the following techniques?**

    [A] Negative staining [B] Shadow casting

    [C] Ultra-thin sectioning [D] All of the above

15. **Cellular activities are mainly concentrated in the following areas of the cell:**

    [A] Mitochondria [B] Nucleus

    [C] Cytoplasm [D] Plasma membrane

16. Some animals have been observed to have cholesterol present in their plasma membranes which

[A] Enables the membrane to stay fluid more easily when the temperature of the cell drops

[B] Makes the membrane less flexible permitting it to sustain greater pressure from within the cell

[C] Enables animals to remove hydrogen atoms from saturated phospholipids

[D] Enables animals to add hydrogen atoms tounsaturated phospholipids

17. Viruses are cellular organisms. This statement is

[A] True [B] False

[C] Not Known [D] None of the above

18. What is the basic unit of functional and structural organization of organisms?

[A] DNA [B] Nucleus

[C] Gene [D] None of the above

19. What is the name of the hollow sphere formed by bilayers of lipids that resembles a disc

[A] Lipid raft [B] Liposome

[C] Micelle [D] Cholesterol

20. Identify the component that does not form a part of the bilayer of lipids.

[A] Phosphate [B] Fatty acids

[C] Methionine [D] Glycerol

21. In the ABC transport system, which of the following proteins transports lipids in the opposite direction from what transports protein?

[A] Scramblase [B] Flippase

[C] Floppases [D] ATPase

22. What is the one lipid transporter in the body that is not an ATP-dependent transporter?

[A] V-type ATPase [B] Scramblase

[C] Flippase [D] Floppases

23. Which of the following techniques is used to visualize the lateral movement of lipids?

[A] FRAP [B] Microscopy

[C] Spectroscopy [D] None of the above

24. The gram-positive bacteria and the gram-negative bacteria can be classified according to their cell walls, which makes them into two main classes. This statement is:

[A] True [B] False

[C] Not Known [D] None of the above

25. **Is there any organism without a cell wall?**
[A] Plant cells [B] Protozoa
[C] Algae [D] Bacteria cells

26. **A fungus cell wall consists of which of the following components?**
[A] Cellulose [B] Hemicellulose
[C] Chitin [D] Pectin

27. **What is S-layer?**
[A] Solid layer [B] Surface layer
[C] Secondary layer [D] Soluble layer

28. **The cell wall of diatoms is composed of**
[A] Biogenic cadmium [B] Biogenic silica
[C] Biogenic ammonia [D] Biogenic Lead

29. **The cell wall of Mycobacterium tuberculosis consists primarily of what kind of compounds?**
[A] Mycolic acid [B] Nitric acid
[C] Acetone [D] Humic acid

30. **Do any of the following antibiotics have the property of inhibiting the synthesis of cell walls as their mode of action?**
[A] Beta lactams [B] Nalidixic acid
[C] Ciprofloxancn [D] Norfloxancin

31. **Lysozyme is effectively defended by which of the following components of the cell wall?**
[A] Peptidoglycan [B] Pseudomurein
[C] Biogenic opal [D] Chitin

32. **Transverse diffusion (flip-flop) is the movement of**
[A] Amino acid [B] Phospholipid
[C] Protein [D] Cholesterol molecule

33. **It is possible to determine whether integral proteins are mobile by looking at their physical status of**
[A] Amino acids [B] External phospholipids
[C] Membrane phospholipids [D] Membrane appendages

34. **In order for the fusion of two plasma membranes of two different cells to occur, which of the following procedures cannot be used?**
[A] Emulsifier [B] Inactivated viruses
[C] Electric Shock [D] Polyethylene glycol

35. **Living cells can be analyzed using FRAP as a technique. This statement is**
[A] True [B] False
[C] Not known [D] None of the above

36. Which of the following genetic diseases is caused by the mutation of a membrane protein?

[A] Anemia [B] Hemolytic anaemia

[C] Parkinson's; s disease [D] Alzheimer;s disease

37. Glycophorin plays a very important role in the diagnosis of which of the following diseases?

[A] Viral fever [B] Common cold

[C] Malaria [D] Asthma

38. A structure that is responsible for the transfer of macromolecules between the nucleus and the cytoplasm of cells

[A] Microtubules [B] Nuclear pores

[C] Cilia [D] Centrioles

39. The name of the signal that is responsible for assisting a protein to move from the nucleus into the cytoplasm

[A] Notch signal [B] Chemical signals

[C] Nuclear localization signal [D] Paracrine signal

40. Is there a region of the chromatin that is transcriptionally silent?

[A] Nucleoid [B] Centrosome

[C] Euchromatin [D] Heterochromatin

41. Cis-Golgi network (CGN) is closer to the endoplasmic reticulum than the trans-Golgi network (TGN). This statement is

[A] True [B] False

[C] Not Known [D] None of the above

42. What is the part of the Golgi complex responsible for directing the protein to its final destination within the cell?

[A] Cis-Golgi network (CGN)

[B] trans-Golgi network

[C] Tubules

[D] Medial cisternae

43. Is there a family of proteins that does not support the Golgi complex mechanically?

[A] Keratin [B] Actin

[C] Spectrin [D] Ankyrin

44. The trans end of the Golgi stack is represented by which of the following structures?

[A] Ascorbic acid [B] Methanoic acid

[C] Dehydrogenase [D] Sialyl-transferase

45. Which of the following is not fully synthesized by the Golgi complex in a cell?

[A] Pectins [B] Hemicellulose

[C] N-linked oligosaccharide [D] O-linked oligosaccharides

46. In the Golgi complex, how many models exist showing the movement of materials within it?

[A] 1 [B] 2
[C] 3 [D] 4

47. In the Golgi complex, which part of the cis and trans cisternae is responsible for the transport of the materials from one to the other?

[A] Passive diffusion [B] Transport vesicles
[C] Active diffusion [D] Translocon

48. Which is a molecule that is not able to leave the lumen of a cisternal lumen while moving from a cis cisternae to a trans cisternae?

[A] Membrane proteins [B] Procollagen
[C] Oligosaccharides [D] Lysosomal proteins

49. Which is an intracellular compartment that is comprised of a single membrane and bound to at least one other membrane?

[A] Endoplasmic reticulum [B] Golgi apparatus
[C] Ribosome [D] Nucleus

50. There is a modification that takes place in the ER before the protein reaches its final destination. This statement is

[A] True [B] False
[C] Not known [D] None of the above

51. What a microsome is?

[A] Compartment of Golgi
[B] Smaller ribosomes
[C] Small ER compartments
[D] Small vesicles of fragmented ER

52. Identify the location where xenobiotic compounds are detoxified from their toxins.

[A] Cytosol [B] RER
[C] SER [D] Ribosome

53. Name the sequence that allows the resident protein to remain in the lumen of the ER.

[A] KDEL [B] KKXX
[C] KLDE [D] KXXX

54. Name a coated vesicle in the ER that transports proteins from the ER to the Golgi?

[A] Clathrin [B] COP II
[C] COPI [D] COP III

55. There is a coated vesicle that is used to transport proteins from the plasma membrane to the endosomes. What is it called?

[A] Clathrin [B] COP I
[C] COP II [D] COP III

56. **Which of the following enzymes is not a hydrolytic enzyme of a lysosome?**

[A] Lipases [B] Sulfatase
[C] Aldolase [D] Phosphatase

57. **What is amphisme?**

[A] Early endosome
[B] Fusion of endosome and autophagosome
[C] Vacuole
[D] The bigger size of the lysosome

58. **In order for melanocytes to release their melanosomes from their melanocytes, what process is involved in this process?**

[A] Autophagy [B] Endocytosis
[C] Pinocytosis [D] Exocytosis

59. **In a mitochondrial membrane, how the small molecules can pass through?**

[A] ATP pump [B] Carrier protein
[C] Porins [D] Channels

60. **What is the similarity between the following division technique used by mitochondria and bacteria?**

[A] Binary fission [B] Budding
[C] Binary fuson [D] Meiosis

61. **Name the organelle which is used for aerobic respiration and ATP synthesis in Entamoeba histolytica.**

[A] Mitosome [B] Mitochondria
[C] Peroxisome [D] Noen of the above

62. **What would be the part of a living cell that should be alive if the organelles are separated from each other?**

[A] Ribosome [B] Cell wall
[C] Chloroplast [D] Endoplasmic reticulum

63. **Which one of the followings is involved in lipid metabolism?**

[A] RER [B] SER
[C] Golgi apparatus [D] Peroxisomes

64. **Organelle- Glyoxysome is involved in**

[A] Conversion of fatty acids into lipids
[B] Conversion of amino acids into carbohydrates
[C] Conversion of fatty acids into carbohydrates
[D] Conversion of Amino acids into proteins

65. **What is the total number of filamentous structures that make up the cytoskeleton?**

[A] 1 [B] 2
[C] 3 [D] 4

66. Cytoskeleton filaments are polymers of

[A] Proteins [B] Ribonucleic acid

[C] Deoxyribonucleic acids [D] Carbohydrates

67. Mitotic spindles are formed by the microtubules of dividing cells. This statement is

[A] True [B] False

[C] Not know [D] None of the above

68. Cilia and Flagella are

[A] Microtubules [B] Intermediate filaments

[C] Microfilaments [D] Chloroplast

69. The following are some of the types of molecules found in animal cells:

[A] Intermediate filaments [B] Microtubules

[C] Microfilaments [D] Chloroplast

70. Ggap junctions are absent in

[A] Sperm cells [B] Brain cells

[C] Cardiac cells [D] None of the above

71. What organelle of the cell covers the Plasmodesmata of the cell?

[A] Ribosome [B] Nucleus

[C] RER [D] SER

72. Which of these does not belong to the ribosomal protein family?

[A] L21 [B] S34

[C] S21 [D] L4

73. Which ribosome recognizes the transcription initiation site of the messenger RNA?

[A] 23 S [B] 5 S

[C] 18 S [D] 5.8 S

74. 50 S ribosomal subunit is folded in .... major domains.

[A] 20 [B] 4

[C] 10 [D] 6

75. Among the following, which is a toxic by-product found in the vacuoles of plant cells that can be used in clinical trials?

[A] Oxytocin [B] Sialic acid

[C] Digitalis [D] None of the above

76. How many types of cell signallings are there?

[A] 1 [B] 2

[C] 3 [D] 4

77. Paracrine messenger molecules are usually

[A] Larger [B] Stable

[C] Unstable [D] Amphoteric

78. Protein kinases and phosphatase act by altering .... the signalling proteins

[A] Basicity [B] Conformation

[C] Size [D] Acidity

79. What are the residues of the protein kinases that do not have phosphate groups added to them?

[A] Serine [B] Cytosine

[C] Threonine [D] Tyrosine

80. G-protein coupled receptors contain .... transmembrane alpha helicases.

[A] 2 [B] 5

[C] 7 [D] 9

A ANSWERS

Chapter 1— THE BIOLOGICAL CLOCK: RHYTHMS OF LIFE

1. [C] 5. [D] 9. [D] 13. [C] 17. [A] 21. [A] 25. [C] 29. [C]
2. [B] 6. [D] 10. [D] 14. [C] 18. [A] 22. [A] 26. [B] 30. [A]
3. [A] 7. [B] 11. [B] 15. [B] 19. [A] 23. [A] 27. [A]
4. [D] 8. [C] 12. [B] 16. [C] 20. [D] 24. [C] 28. [D]

Chapter 2— PHOTOMORPHOGENESIS

1. [A] 9. [D] 17. [C] 25. [A] 33. [B] 41. [C] 49. [A] 57. [B] 65. [B]
2. [C] 10. [C] 18. [B] 26. [B] 34. [A] 42. [B] 50. [A] 58. [A] 66. [C]
3. [C] 11. [B] 19. [C] 27. [C] 35. [A] 43. [C] 51. [A] 59. [A] 67. [C]
4. [C] 12. [A] 20. [B] 28. [A] 36. [A] 44. [D] 52. [A] 60. [A] 68. [C]
5. [D] 13. [A] 21. [A] 29. [C] 37. [B] 45. [B] 53. [A] 61. [A]
6. [D] 14. [B] 22. [B] 30. [A] 38. [B] 46. [A] 54. [A] 62. [A]
7. [D] 15. [C] 23. [A] 31. [B] 39. [B] 47. [B] 55. [B] 63. [A]
8. [A] 16. [B] 24. [A] 32. [B] 40. [D] 48. [A] 56. [A] 64. [A]

Chapter 3— THE POWER OF MOVEMENT IN PLANTS

1. [B] 9. [B] 17. [B] 25. [A] 33. [B] 41. [A] 49. [A] 57. [A] 65. [A]
2. [D] 10. [A] 18. [A] 26. [A] 34. [A] 42. [A] 50. [A] 58. [A] 66. [A]
3. [C] 11. [C] 19. [D] 27. [A] 35. [A] 43. [A] 51. [D] 59. [A]
4. [D] 12. [A] 20. [A] 28. [A] 36. [A] 44. [C] 52. [A] 60. [A]
5. [D] 13. [A] 21. [A] 29. [A] 37. [A] 45. [D] 53. [A] 61. [A]
6. [B] 14. [C] 22. [C] 30. [A] 38. [C] 46. [A] 54. [B] 62. [B]
7. [B] 15. [B] 23. [C] 31. [A] 39. [A] 47. [B] 55. [A] 63. [C]
8. [C] 16. [A] 24. [A] 32. [A] 40. [C] 48. [B] 56. [C] 64. [C]

Chapter 4— PLANT GROWTH REGULATORS

1. [B]	16. [A]	31. [D]	46. [A]	61. [A]	76. [A]	91. [A]	106. [A]	121. [B]
2. [D]	17. [D]	32. [A]	47. [A]	62. [A]	77. [A]	92. [A]	107. [B]	122. [A]
3. [A]	18. [A]	33. [A]	48. [A]	63. [A]	78. [F]	93. [A]	108. [B]	123. [D]
4. [B]	19. [C]	34. [A]	49. [E]	64. [A]	79. [C]	94. [A]	109. [A]	124. [A]
5. [B]	20. [A]	35. [D]	50. [D]	65. [D]	80. [B]	95. [A]	110. [A]	125. [D]
6. [C]	21. [A]	36. [C]	51. [C]	66. [C]	81. [A]	96. [B]	111. [A]	126. [A]
7. [A]	22. [D]	37. [B]	52. [A]	67. [B]	82. [A]	97. [A]	112. [D]	127. [A]
8. [B]	23. [B]	38. [A]	53. [A]	68. [C]	83. [B]	98. [B]	113. [A]	128. [A]
9. [A]	24. [E]	39. [C]	54. [A]	69. [A]	84. [A]	99. [C]	114. [B]	129. [A]
10. [A]	25. [A]	40. [D]	55. [A]	70. [B]	85. [C]	100. [D]	115. [C]	130. [A]
11. [A]	26. [B]	41. [D]	56. [A]	71. [A]	86. [C]	101. [B]	116. [D]	131. [A]
12. [A]	27. [B]	42. [B]	57. [A]	72. [D]	87. [B]	102. [A]	117. [B]	132. [A]
13. [B]	28. [A]	43. [A]	58. [A]	73. [A]	88. [C]	103. [A]	118. [A]	133. [A]
14. [B]	29. [D]	44. [A]	59. [B]	74. [A]	89. [A]	104. [A]	119. [A]	134. [A]
15. [A]	30. [A]	45. [A]	60. [A]	75. [A]	90. [B]	105. [A]	120. [A]	135. [D]
136. [A]	153. [A]	170. [A]	187. [A]	204. [C]	221. [A]	238. [A]	255. [B]	272. [A]
137. [A]	154. [A]	171. [A]	188. [A]	205. [A]	222. [A]	239. [A]	256. [B]	273. [B]
138. [A]	155. [C]	172. [B]	189. [A]	206. [A]	223. [C]	240. [A]	257. [D]	274. [C]
139. [A]	156. [A]	173. [A]	190. [A]	207. [A]	224. [B]	241. [B]	258. [A]	275. [A]
140. [A]	157. [A]	174. [B]	191. [C]	208. [A]	225. [A]	242. [A]	259. [C]	276. [A]
141. [B]	158. [B]	175. [A]	192. [A]	209. [A]	226. [A]	243. [A]	260. [A]	277. [A]
142. [A]	159. [C]	176. [B]	193. [A]	210. [B]	227. [D]	244. [A]	261. [D]	278. [C]
143. [A]	160. [A]	177. [A]	194. [A]	211. [D]	228. [A]	245. [A]	262. [B]	279. [A]
144. [B]	161. [B]	178. [A]	195. [A]	212. [A]	229. [B]	246. [A]	263. [A]	
145. [A]	162. [C]	179. [A]	196. [A]	213. [A]	230. [A]	247. [A]	264. [B]	
146. [B]	163. [A]	180. [D]	197. [A]	214. [C]	231. [D]	248. [B]	265. [A]	
147. [C]	164. [C]	181. [D]	198. [A]	215. [A]	232. [D]	249. [A]	266. [A]	
148. [A]	165. [D]	182. [B]	199. [A]	216. [A]	233. [C]	250. [B]	267. [A]	
149. [C]	166. [D]	183. [A]	200. [A]	217. [A]	234. [A]	251. [A]	268. [A]	
150. [B]	167. [F]	184. [A]	201. [B]	218. [B]	235. [A]	252. [A]	269. [A]	
151. [A]	168. [E]	185. [B]	202. [C]	219. [B]	236. [A]	253. [A]	270. [A]	
152. [A]	169. [A]	186. [A]	203. [B]	220. [A]	237. [A]	254. [A]	271. [A]	

Chapter 5— PHOTOPERIODISM

1.	[A]	6.	[B]	11.	[A]	16.	[A]	21.	[B]	26.	[A]	31.	[C]	36.	[A]	41.	[A]
2.	[A]	7.	[A]	12.	[A]	17.	[D]	22.	[B]	27.	[A]	32.	[A]	37.	[C]	42.	[A]
3.	[C]	8.	[A]	13.	[A]	18.	[B]	23.	[B]	28.	[A]	33.	[B]	38.	[A]	43.	[A]
4.	[A]	9.	[B]	14.	[A]	19.	[D]	24.	[B]	29.	[A]	34.	[A]	39.	[C]		
5.	[A]	10.	[B]	15.	[B]	20.	[B]	25.	[A]	30.	[A]	35.	[E]	40.	[C]		

Chapter 6— TEMPERATURE: PLANT DEVELOPMENT AND DISTRIBUTION

1.	[D]	4.	[A]	7.	[B]	10.	[A]	13.	[A]	16.	[A]	19.	[B]
2.	[A]	5.	[A]	8.	[C]	11.	[A]	14.	[A]	17.	[A]	20.	[C]
3.	[A]	6.	[A]	9.	[C]	12.	[A]	15.	[A]	18.	[A]		

Chapter 7— SECONDARY METABOLITES

1.	[B]	13.	[A]	25.	[A]	37.	[C]	49.	[C]	61.	[B]	73.	[C]	85.	[A]	97. [A]
2.	[A]	14.	[B]	26.	[A]	38.	[A]	50.	[D]	62.	[C]	74.	[B]	86.	[A]	98. [A]
3.	[A]	15.	[A]	27.	[B]	39.	[B]	51.	[E]	63.	[C]	75.	[C]	87.	[B]	99. [B]
4.	[B]	16.	[A]	28.	[A]	40.	[C]	52.	[A]	64.	[A]	76.	[A]	88.	[A]	100. [B]
5.	[A]	17.	[B]	29.	[C]	41.	[A]	53.	[A]	65.	[A]	77.	[A]	89.	[A]	101. [D]
6.	[A]	18.	[D]	30.	[A]	42.	[A]	54.	[A]	66.	[A]	78.	[A]	90.	[A]	102. [C]
7.	[A]	19.	[A]	31.	[A]	43.	[C]	55.	[A]	67.	[D]	79.	[B]	91.	[A]	
8.	[A]	20.	[D]	32.	[B]	44.	[A]	56.	[B]	68.	[B]	80.	[C]	92.	[D]	
9.	[B]	21.	[A]	33.	[C]	45.	[B]	57.	[A]	69.	[C]	81.	[A]	93.	[A]	
10.	[A]	22.	[A]	34.	[A]	46.	[C]	58.	[B]	70.	[C]	82.	[A]	94.	[B]	
11.	[A]	23.	[B]	35.	[C]	47.	[A]	59.	[C]	71.	[A]	83.	[C]	95.	[C]	
12.	[C]	24.	[B]	36.	[A]	48.	[B]	60.	[A]	72.	[C]	84.	[A]	96.	[A]	

Chapter 8— RESPONSES OF PLANTS TO ENVIRONMENTAL STRESS

1.	[A]	8.	[A]	15.	[B]	22.	[D]	29.	[A]	36.	[A]	43.	[C]	50.	[D]	57.	[B]
2.	[B]	9.	[A]	16.	[A]	23.	[A]	30.	[A]	37.	[A]	44.	[A]	51.	[B]	58.	[D]
3.	[A]	10.	[A]	17.	[A]	24.	[A]	31.	[A]	38.	[C]	45.	[A]	52.	[D]	59.	[A]
4.	[B]	11.	[A]	18.	[A]	25.	[A]	32.	[D]	39.	[A]	46.	[B]	53.	[A]	60.	[A]
5.	[A]	12.	[E]	19.	[A]	26.	[C]	33.	[A]	40.	[B]	47.	[C]	54.	[A]	61.	[A]
6.	[A]	13.	[A]	20.	[B]	27.	[A]	34.	[C]	41.	[A]	48.	[C]	55.	[A]	62.	[A]
7.	[B]	14.	[A]	21.	[C]	28.	[A]	35.	[A]	42.	[B]	49.	[A]	56.	[E]	63.	[A]

64. [A] 67. [A] 70. [D] 73. [A] 76. [D] 79. [D] 82. [D] 85. [C] 88. [C]
65. [C] 68. [A] 71. [C] 74. [A] 77. [A] 80. [A] 83. [C] 86. [A] 89. [D]
66. [C] 69. [D] 72. [C] 75. [A] 78. [B] 81. [D] 84. [E] 87. [B]

Chapter 9— GERMINATION AND DORMANCY

1. [C] 4. [B] 7. [D] 10. [G] 13. [B] 16. [A] 19. [A] 22. [A] 25. [E] 28. [D]
2. [A] 5. [C] 8. [E] 11. [B] 14. [B] 17. [A] 20. [B] 23. [D] 26. [B]
3. [A] 6. [D] 9. [C] 12. [A] 15. [B] 18. [A] 21. [C] 24. [E] 27. [C]

Chapter 10— VERNALIZATION

1. [B] 3. [A] 5. [A] 7. [A] 9. [D] 11. [A] 13. [B]
2. [C] 4. [A] 6. [B] 8. [B] 10. [B] 12. [C]

Chapter 11— TRANSLOCATION OF ORGANIC SOLUTES

1. [A] 5. [D] 9. [A] 13. [A] 17. [D] 21. [B] 25. [A] 29. [E] 33. [A]
2. [A] 6. [A] 10. [A] 14. [D] 18. [C] 22. [A] 26. [B] 30. [E] 34. [A]
3. [B] 7. [B] 11. [B] 15. [A] 19. [A] 23. [B] 27. [D] 31. [E] 35. [A]
4. [E] 8. [B] 12. [C] 16. [B] 20. [A] 24. [C] 28. [C] 32. [A] 36. [A]

Chapter 12— ABSORPTION OF MINERAL SALTS

1. [C] 6. [A] 11. [A] 16. [C] 21. [B] 26. [A] 31. [A] 36. [A] 41. [A]
2. [B] 7. [B] 12. [D] 17. [A] 22. [A] 27. [B] 32. [B] 37. [C] 42. [A]
3. [A] 8. [A] 13. [B] 18. [D] 23. [A] 28. [C] 33. [B] 38. [A] 43. [D]
4. [A] 9. [A] 14. [A] 19. [B] 24. [A] 29. [A] 34. [A] 39. [A] 44. [D]
5. [A] 10. [B] 15. [D] 20. [D] 25. [A] 30. [D] 35. [A] 40. [A]

Chapter 13— MINERAL NUTRITION OF PLANTS

1. [B] 8. [A] 15. [B] 22. [C] 29. [B] 36. [A] 43. [A] 50. [A] 57. [A]
2. [D] 9. [D] 16. [A] 23. [D] 30. [A] 37. [E] 44. [A] 51. [A] 58. [A]
3. [A] 10. [A] 17. [A] 24. [A] 31. [D] 38. [A] 45. [A] 52. [A] 59. [E]
4. [A] 11. [A] 18. [D] 25. [C] 32. [C] 39. [C] 46. [N] 53. [B] 60. [A]
5. [D] 12. [A] 19. [C] 26. [D] 33. [D] 40. [A] 47. [B] 54. [B] 61. [A]
6. [B] 13. [B] 20. [A] 27. [C] 34. [C] 41. [A] 48. [A] 55. [D] 62. [A]
7. [C] 14. [B] 21. [C] 28. [C] 35. [A] 42. [A] 49. [A] 56. [C] 63. [B]

64. [A]	72. [A]	80. [B]	88. [E]	96. [F]	106. [C]	114. [B]	122. [A]
65. [A]	73. [C]	81. [B]	89. [F]	97. [E]	107. [B]	115. [A]	123. [C]
66. [A]	74. [B]	82. [A]	90. [C]	98. [A]	108. [A]	116. [A]	124. [B]
67. [B]	75. [D]	83. [A]	91. [A]	99. [D]	109. [B]	117. [E]	125. [C]
68. [A]	76. [A]	84. [B]	92. [B]	100. [C]	110. [B]	118. [A]	126. [C]
69. [C]	77. [A]	85. [C]	93. [C]	101. [C]	111. [C]	119. [A]	127. [D]
70. [A]	78. [C]	86. [D]	94. [D]	102. [A]	112. [A]	120. [A]	128. [C]
71. [A]	79. [B]	87. [C]	95. [E]	103. [C]	113. [A]	121. [A]	129. [A]

Chapter 14— TRANSPIRATION AND GUTTATION

1. [A]	11. [D]	21. [A]	31. [A]	41. [A]	51. [A]	61. [A]	71. [D]	81. [C]
2. [B]	12. [A]	22. [A]	32. [B]	42. [A]	52. [A]	62. [A]	72. [B]	82. [B]
3. [A]	13. [A]	23. [B]	33. [C]	43. [B]	53. [D]	63. [A]	73. [B]	83. [A]
4. [B]	14. [A]	24. [A]	34. [D]	44. [A]	54. [D]	64. [B]	74. [B]	84. [A]
5. [A]	15. [B]	25. [A]	35. [E]	45. [B]	55. [A]	65. [B]	75. [B]	85. [A]
6. [A]	16. [C]	26. [C]	36. [D]	46. [C]	56. [E]	66. [B]	76. [B]	86. [A]
7. [A]	17. [D]	27. [C]	37. [C]	47. [A]	57. [D]	67. [D]	77. [A]	
8. [E]	18. [A]	28. [A]	38. [A]	48. [C]	58. [B]	68. [C]	78. [C]	
9. [A]	19. [B]	29. [A]	39. [A]	49. [B]	59. [D]	69. [B]	79. [A]	
10. [A]	20. [C]	30. [A]	40. [D]	50. [A]	60. [C]	70. [B]	80. [C]	

Chapter 15— PLANT TISSUE CULTURE

1. [A]	14. [E]	27. [D]	40. [A]	53. [C]	66. [D]	79. [A]	92. [A]	105. [E]
2. [A]	15. [D]	28. [D]	41. [A]	54. [E]	67. [A]	80. [B]	93. [A]	106. [D]
3. [A]	16. [A]	29. [A]	42. [A]	55. [B]	68. [A]	81. [C]	94. [A]	107. [F]
4. [B]	17. [B]	30. [D]	43. [A]	56. [C]	69. [B]	82. [D]	95. [E]	108. [D]
5. [C]	18. [A]	31. [E]	44. [D]	57. [A]	70. [A]	83. [A]	96. [A]	109. [E]
6. [A]	19. [D]	32. [E]	45. [B]	58. [A]	71. [F]	84. [A]	97. [A]	110. [A]
7. [B]	20. [A]	33. [E]	46. [C]	59. [A]	72. [F]	85. [C]	98. [E]	
8. [A]	21. [A]	34. [C]	47. [D]	60. [A]	73. [A]	86. [A]	99. [A]	
9. [G]	22. [A]	35. [C]	48. [B]	61. [A]	74. [C]	87. [B]	100. [A]	
10. [D]	23. [C]	36. [A]	49. [E]	62. [A]	75. [D]	88. [A]	101. [B]	
11. [A]	24. [D]	37. [E]	50. [A]	63. [B]	76. [E]	89. [B]	102. [G]	
12. [A]	25. [A]	38. [D]	51. [B]	64. [A]	77. [A]	90. [C]	103. [B]	
13. [A]	26. [A]	39. [E]	52. [A]	65. [A]	78. [B]	91. [D]	104. [C]	

Chapter 16— ECO-PHYSIOLOGY AND ENVIRONMENTAL BIOLOGY

1.	[H]	6.	[A]	11.	[C]	16.	[A]	21.	[C]	26.	[D]	31.	[C]	36.	[D]	41.	[A]
2.	[A]	7.	[B]	12.	[C]	17.	[C]	22.	[A]	27.	[A]	32.	[D]	37.	[A]	42.	[A]
3.	[B]	8.	[A]	13.	[B]	18.	[B]	23.	[B]	28.	[C]	33.	[A]	38.	[D]	43.	[A]
4.	[A]	9.	[A]	14.	[B]	19.	[B]	24.	[D]	29.	[A]	34.	[B]	39.	[B]	44.	[A]
5.	[C]	10.	[C]	15.	[A]	20.	[A]	25.	[D]	30.	[A]	35.	[A]	40.	[E]		

Chapter 17— PHOTOSYNTHESIS

1.	[A]	29.	[A]	57.	[C]	85.	[F]	113.	[A]	141.	[A]	169.	[D]	197.	[A]	225.	[A]
2.	[A]	30.	[G]	58.	[C]	86.	[B]	114.	[B]	142.	[A]	170.	[C]	198.	[A]	226.	[A]
3.	[A]	31.	[B]	59.	[A]	87.	[B]	115.	[E]	143.	[C]	171.	[D]	199.	[B]	227.	[F, E]
4.	[B]	32.	[D]	60.	[A]	88.	[C]	116.	[A]	144.	[A]	172.	[C]	200.	[A]	228.	[M]
5.	[B]	33.	[E]	61.	[A]	89.	[A]	117.	[B]	145.	[A]	173.	[B]	201.	[B]	229.	[G]
6.	[C]	34.	[F]	62.	[D]	90.	[A]	118.	[A]	146.	[B]	174.	[A]	202.	[B]	230.	[M]
7.	[D]	35.	[A]	63.	[A]	91.	[B]	119.	[A]	147.	[C]	175.	[C]	203.	[A]	231.	[A]
8.	[A]	36.	[B]	64.	[D]	92.	[A]	120.	[A]	148.	[A]	176.	[A]	204.	[A]	232.	[A]
9.	[C]	37.	[A]	65.	[B]	93.	[C]	121.	[D]	149.	[B]	177.	[B]	205.	[A]	233.	[A]
10.	[D]	38.	[A]	66.	[E]	94.	[B]	122.	[D]	150.	[B]	178.	[C]	206.	[B]	234.	[B]
11.	[B]	39.	[A]	67.	[E]	95.	[B]	123.	[B]	151.	[B]	179.	[A]	207.	[C]	235.	[A]
12.	[G]	40.	[D]	68.	[A]	96.	[A]	124.	[B]	152.	[D]	180.	[B]	208.	[A]	236.	[A]
13.	[B]	41.	[A]	69.	[A]	97.	[C]	125.	[C]	153.	[C]	181.	[C]	209.	[A]	237.	[A]
14.	[A]	42.	[C]	70.	[E]	98.	[A]	126.	[C]	154.	[D]	182.	[A]	210.	[A]	238.	[A]
15.	[B]	43.	[B]	71.	[A]	99.	[A]	127.	[D]	155.	[A]	183.	[C]	211.	[B]		
16.	[B]	44.	[B]	72.	[E]	100.	[D]	128.	[A]	156.	[A]	184.	[B]	212.	[C]		
17.	[C]	45.	[B]	73.	[A]	101.	[B]	129.	[B]	157.	[A]	185.	[A]	213.	[A]		
18.	[A]	46.	[C]	74.	[B]	102.	[A]	130.	[A]	158.	[B]	186.	[B]	214.	[A]		
19.	[B]	47.	[A]	75.	[C]	103.	[B]	131.	[C]	159.	[A]	187.	[C]	215.	[B]		
20.	[A]	48.	[D]	76.	[A]	104.	[A]	132.	[B]	160.	[A]	188.	[A]	216.	[C]		
21.	[A]	49.	[A]	77.	[B]	105.	[A]	133.	[D]	161.	[B]	189.	[B]	217.	[A]		
22.	[A]	50.	[B]	78.	[C]	106.	[A]	134.	[D]	162.	[B]	190.	[C]	218.	[B]		
23.	[A]	51.	[A]	79.	[A]	107.	[C]	135.	[C]	163.	[D]	191.	[A]	219.	[B]		
24.	[C]	52.	[A]	80.	[B]	108.	[C]	136.	[D]	164.	[B]	192.	[B]	220.	[B]		
25.	[E]	53.	[A]	81.	[A]	109.	[A]	137.	[A]	165.	[A]	193.	[C]	221.	[C]		
26.	[C]	54.	[A]	82.	[C]	110.	[D]	138.	[A]	166.	[B]	194.	[A]	222.	[D]		
27.	[B]	55.	[C]	83.	[C]	111.	[A]	139.	[A]	167.	[D]	195.	[B]	223.	[B]		
28.	[G]	56.	[D]	84.	[B]	112.	[A]	140.	[B]	168.	[B]	196.	[A]	224.	[A]		

Chapter 18— ENZYMES AND VITAMINS

1. [A] 13. [C] 25. [A] 37. [A] 49. [A] 61. [B] 73. [A] 85. [J] 97. [A]
2. [A] 14. [A] 26. [A] 38. [D] 50. [F] 62. [A] 74. [A] 86. [E] 98. [B]
3. [B] 15. [C] 27. [A] 39. [A] 51. [A] 63. [D] 75. [B] 87. [A] 99. [A]
4. [C] 16. [D] 28. [D] 40. [B] 52. [A] 64. [C] 76. [A] 88. [B] 100. [B]
5. [D] 17. [D] 29. [C] 41. [A] 53. [B] 65. [A] 77. [C] 89. [B] 101. [C]
6. [C] 18. [A] 30. [B] 42. [A] 54. [C] 66. [B] 78. [D] 90. [A] 102. [A]
7. [A] 19. [D] 31. [A] 43. [A] 55. [A] 67. [A] 79. [B] 91. [C] 103. [A]
8. [C] 20. [B] 32. [C] 44. [D] 56. [C] 68. [A] 80. [D] 92. [A] 104. [A]
9. [C] 21. [B] 33. [A] 45. [E] 57. [B] 69. [D] 81. [B] 93. [A] 105. [A]
10. [B] 22. [A] 34. [B] 46. [A] 58. [D] 70. [A] 82. [C] 94. [C] 106. [B]
11. [A] 23. [C] 35. [D] 47. [B] 59. [B] 71. [A] 83. [B] 95. [A] 107. [A]
12. [C] 24. [C] 36. [B] 48. [B] 60. [C] 72. [B] 84. [B] 96. [A] 108. [A]

Chapter 19— INSTRUMENTATION IN PLANT PHYSIOLOGY

1. [A] 7. [C] 13. [A] 19. [D] 25. [A] 31. [A] 37. [A] 43. [D] 49. [A]
2. [C] 8. [D] 14. [A] 20. [A] 26. [A] 32. [F] 38. [E] 44. [E] 50. [A]
3. [A] 9. [A] 15. [A] 21. [A] 27. [D] 33. [A] 39. [A] 45. [A]
4. [B] 10. [A] 16. [A] 22. [A] 28. [A] 34. [H] 40. [A] 46. [A]
5. [E] 11. [D] 17. [A] 23. [B] 29. [A] 35. [A] 41. [A] 47. [A]
6. [A] 12. [A] 18. [D] 24. [A] 30. [A] 36. [B] 42. [A] 48. [C]

Chapter 20— THE PLANT CELL

1. [B] 13. [D] 25. [A] 37. [C] 49. [B] 61. [C] 73. [A] 85. [C] 97. [B]
2. [A] 14. [C] 26. [B] 38. [A] 50. [A] 62. [C] 74. [B] 86. [C] 98. [B]
3. [B] 15. [A] 27. [A] 39. [A] 51. [B] 63. [D] 75. [E] 87. [C] 99. [A]
4. [D] 16. [D] 28. [C] 40. [A] 52. [A] 64. [A] 76. [A] 88. [D] 100. [A]
5. [A] 17. [C] 29. [C] 41. [A] 53. [A] 65. [A] 77. [A] 89. [B] 101. [D]
6. [B] 18. [A] 30. [B] 42. [E] 54. [B] 66. [A] 78. [B] 90. [A] 102. [B]
7. [B] 19. [A] 31. [D] 43. [A] 55. [A] 67. [B] 79. [C] 91. [B] 103. [A]
8. [B] 20. [A] 32. [E] 44. [A] 56. [B] 68. [C] 80. [A] 92. [C] 104. [A]
9. [B] 21. [B] 33. [A] 45. [C] 57. [A] 69. [A] 81. [C] 93. [A] 105. [B]
10. [C] 22. [A] 34. [A] 46. [D] 58. [A] 70. [C] 82. [B] 94. [A] 106. [A]
11. [D] 23. [B] 35. [A] 47. [A] 59. [A] 71. [C] 83. [A] 95. [C] 107. [A]
12. [A] 24. [A] 36. [B] 48. [A] 60. [A] 72. [A] 84. [D] 96. [A] 108. [B]

109. [A] 123. [A] 137. [A] 151. [A] 165. [B] 179. [B] 193. [A] 207. [A] 221. [D]

110. [A] 124. [A] 138. [A] 152. [B] 166. [A] 180. [A] 194. [C] 208. [D] 222. [A]

111. [D] 125. [A] 139. [A] 153. [B] 167. [A] 181. [B] 195. [A] 209. [C] 223. [A]

112. [A] 126. [A] 140. [D] 154. [A] 168. [B] 182. [A] 196. [A] 210. [A] 224. [A]

113. [B] 127. [A] 141. [A] 155. [C] 169. [C] 183. [B] 197. [A] 211. [C] 225. [A]

114. [D] 128. [C] 142. [A] 156. [C] 170. [A] 184. [A] 198. [A] 212. [A] 226. [C]

115. [B] 129. [C] 143. [C] 157. [D] 171. [A] 185. [D] 199. [A] 213. [A] 227. [C]

116. [A] 130. [D] 144. [E] 158. [A] 172. [A] 186. [A] 200. [B] 214. [A] 228. [A]

117. [A] 131. [A] 145. [B] 159. [E] 173. [A] 187. [C] 201. [C] 215. [A]

118. [E] 132. [H] 146. [B] 160. [D] 174. [C] 188. [E] 202. [A] 216. [A]

119. [F] 133. [B] 147. [A] 161. [A] 175. [C] 189. [A] 203. [A] 217. [A]

120. [A] 134. [D] 148. [A] 162. [C] 176. [G] 190. [A] 204. [A] 218. [A]

121. [C] 135. [A] 149. [A] 163. [A] 177. [C] 191. [C] 205. [C] 219. [A]

122. [C] 136. [A] 150. [C] 164. [B] 178. [A] 192. [C] 206. [A] 220. [A]

Chapter 21— PLANT MOLECULAR BIOLOGY

1. [A] 18. [B] 35. [D] 51. [B] 68. [B] 85. [A] 102. [A] 119. [A] 136. [A]

2. [A] 19. [A] 36. [C] 52. [A] 69. [A] 86. [D] 103. [A] 120. [A] 137. [B]

3. [A] 20. [A] 37. [C] 53. [A] 70. [A] 87. [B] 104. [A] 121. [A] 138. [C]

4. [B] 21. [A] 38. [A] 54. [A] 71. [C] 88. [A] 105. [C] 122. [A] 139. [C]

5. [D] 22. [B] 39. [D] 55. [C] 72. [B] 89. [A] 106. [B] 123. [A] 140. [A]

6. [A] 23. [A] 40. [E] 56. [D] 73. [C] 90. [A] 107. [A] 124. [C] 141. [A]

7. [B] 24. [A] 41. [C] 57. [B] 74. [D] 91. [H] 108. [C] 125. [A] 142. [D]

8. [C] 25. [C] 42. [A] 58. [C] 75. [C] 92. [A] 109. [A] 126. [E] 143. [A]

9. [A] 26. [A] 43. [A] 59. [A] 76. [B] 93. [A] 110. [A] 127. [A] 144. [B]

10. [D] 27. [C] 44. [A] 60. [C] 77. [C] 94. [B] 111. [B] 128. [B] 145. [A]

11. [B] 28. [D] 45. [B] 61. [B] 78. [E] 95. [A] 112. [A] 129. [D] 146. [B]

12. [A] 29. [C] 46. [A] 62. [B] 79. [C] 96. [A] 113. [A] 130. [A] 147. [A]

13. [A] 30. [D] 43. [A] 63. [B] 80. [A] 97. [A] 114. [A] 131. [A] 148. [B]

14. [A] 31. [B] 47. [C] 64. [D] 81. [A] 98. [C] 115. [A] 132. [B]

15. [A] 32. [C] 48. [B] 65. [B] 82. [A] 99. [C] 116. [A] 133. [C]

16. [A] 33. [D] 49. [D] 66. [B] 83. [A] 100. [A] 117. [A] 134. [A]

17. [B] 34. [C] 50. [D] 67. [D] 84. [A] 101. [A] 118. [B] 135. [H]

Chapter 22— PHYSIOLOGY OF CROP PLANTS

1. [A] 9. [A] 17. [A] 25. [A] 33. [C] 41. [A] 49. [B] 57. [A]
2. [A] 10. [A] 18. [A] 26. [A] 34. [I] 42. [A] 50. [A] 58. [B]
3. [D] 11. [B] 19. [B] 27. [A] 35. [A] 43. [H] 51. [A] 59. [C]
4. [A] 12. [C] 20. [A] 28. [A] 36. [A] 44. [A] 52. [D] 60. [A]
5. [A] 13. [C] 22. [A] 29. [A] 37. [A] 45. [A] 53. [A] 61. [A]
6. [A] 14. [A] 21. [B] 30. [D] 38. [B] 46. [A] 54. [D] 62. [A]
7. [A] 15. [A] 23. [D] 31. [A] 39. [B] 47. [A] 55. [A] 63. [A]
8. [A] 16. [A] 24. [A] 32. [A] 40. [D] 48. [A] 56. [A] 64. [A]

Chapter 23— WATER RELATIONS

1. [B] 7. [A] 13. [D] 19. [A] 25. [E] 31. [A] 37. [C] 43. [A] 49. [A]
2. [C] 8. [B] 14. [D] 20. [A] 26. [C] 32. [A] 38. [D] 44. [A] 50. [A]
3. [A] 9. [C] 15. [D] 21. [A] 27. [A] 33. [A] 39. [C] 45. [B]
4. [H] 10. [C] 16. [E] 22. [C] 28. [A] 34. [A] 40. [A] 46. [B]
5. [A] 11. [A] 17. [E] 23. [E] 29. [B] 35. [D] 41. [B] 47. [A]
6. [A] 12. [C] 18. [A] 24. [E] 30. [E] 36. [C] 42. [B] 48. [E]

Chapter 24— NITROGEN METABOLISM AND BIOLOGICAL NITROGEN FIXATION

1. [A] 9. [A] 17. [A] 25. [C] 33. [B] 41. [D] 49. [A] 57. [A] 65. [C]
2. [D] 10. [B] 18. [A] 26. [D] 34. [C] 42. [A] 50. [A] 58. [C] 66. [A]
3. [A] 11. [B] 19. [A] 27. [C] 35. [A] 43. [C] 51. [A] 59. [A]
4. [A] 12. [A] 20. [C] 28. [C] 36. [B] 44. [D] 52. [A] 60. [C]
5. [A] 13. [A] 21. [A] 29. [A] 37. [C] 45. [A] 53. [A] 61. [C]
6. [A] 14. [D] 22. [D] 30. [A] 38. [A] 46. [A] 54. [A] 62. [D]
7. [A] 15. [A] 23. [D] 31. [B] 39. [A] 47. [B] 55. [A] 63. [A]
8. [A] 16. [A] 24. [A] 32. [C] 40. [A] 48. [A] 56. [A] 64. [A]

Chapter 25— SULFUR ASSIMILATION AND METABOLISM IN PLANTS

1. [A] 5. [D] 9. [C] 13. [A] 17. [A] 21. [B] 25. [A] 29. [D]
2. [A] 6. [A] 10. [C] 14. [A] 18. [F] 22. [A] 26. [D] 30. [A]
3. [A] 7. [D] 11. [B] 15. [A] 19. [A] 23. [A] 27. [A] 31. [A]
4. [D] 8. [B] 12. [A] 16. [D] 20. [A] 24. [A] 28. [A]

Chapter 26— CARBOHYDRATE METABOLISM IN PLANTS

1. [D] 6. [D] 11. [D] 16. [A] 21. [A] 26. [A] 31. [B] 36. [A] 41. [A]
2. [C] 7. [E] 12. [C] 17. [D] 22. [A] 27. [A] 32. [A] 37. [C]
3. [C] 8. [A] 13. [A] 18. [D] 23. [C] 28. [C] 33. [A] 38. [A]
4. [D] 9. [B] 14. [D] 19. [G] 24. [C] 29. [D] 34. [C] 39. [D]
5. [A] 10. [D] 15. [D] 20. [A] 25. [A] 30. [D] 35. [B] 40. [A]

Chapter 27— SPECIAL MODES OF NUTRITION

1. [A] 4. [B] 7. [D] 10. [C] 13. [A] 16. [A] 19. [B] 22. [A] 25. [A]
2. [E] 5. [A] 8. [A] 11. [A] 14. [A] 17. [A] 20. [C] 23. [A]
3. [A] 6. [C] 9. [D] 12. [A] 15. [A] 18. [E] 21. [A] 24. [A]

Chapter 28— RESPIRATION AND PHOTORESPIRATION

1. [D] 7. [C] 13. [D] 19. [A] 25. [C] 31. [C] 37. [A] 43. [A] 49. [A]
2. [B] 8. [B] 14. [E] 20. [D] 26. [B] 32. [C] 38. [B] 44. [D] 50. [B]
3. [D] 9. [A] 15. [C] 21. [A] 27. [C] 33. [A] 39. [A] 45. [A] 51. [A]
4. [D] 10. [E] 16. [A] 22. [C] 28. [A] 34. [A] 40. [C] 46. [A]
5. [A] 11. [A] 17. [C] 23. [D] 29. [C] 35. [C] 41. [A] 47. [B]
6. [A] 12. [A] 18. [C] 24. [D] 30. [D] 36. [B] 42. [D] 48. [A]

Chapter 29— MATCHING TYPE QUESTIONS

1. [B] 15. [D] 29. [A] 43. [C] 57. [B] 71. [B] 85. [B] 99. [B] 113. [A]
2. [D] 16. [B] 30. [A] 44. [B] 58. [B] 72. [C] 86. [A] 100. [A] 114. [B]
3. [A] 17. [A] 31. [D] 45. [B] 59. [A] 73. [D] 87. [A] 101. [A] 115. [B]
4. [C] 18. [B] 32. [B] 46. [B] 60. [B] 74. [C] 88. [A] 102. [A] 116. [B]
5. [A] 19. [B] 33. [B] 47. [B] 61. [C] 75. [C] 89. [C] 103. [A] 117. [C]
6. [A] 20. [B] 34. [B] 48. [A] 62. [D] 76. [C] 90. [B] 104. [B] 118. [C]
7. [B] 21. [D] 35. [A] 49. [B] 63. [D] 77. [B] 91. [B] 105. [B] 119. [B]
8. [D] 22. [C] 36. [C] 50. [B] 64. [C] 78. [B] 92. [C] 106. [A] 120. [B]
9. [D] 23. [A] 37. [A] 51. [B] 65. [D] 79. [B] 93. [C] 107. [B] 121. [B]
10. [C] 24. [B] 38. [B] 52. [A] 66. [B] 80. [A] 94. [C] 108. [A] 122. [C]
11. [B] 25. [B] 39. [C] 53. [B] 67. [A] 81. [A] 95. [C] 109. [B] 123. [C]
12. [A] 26. [B] 40. [C] 54. [B] 68. [A] 82. [A] 96. [B] 110. [B] 124. [B]
13. [B] 27. [D] 41. [B] 55. [C] 69. [A] 83. [B] 97. [B] 111. [B] 125. [C]
14. [C] 28. [C] 42. [A] 56. [D] 70. [B] 84. [B] 98. [B] 112. [B] 126. [B]

127. [B] 163. [C] 199. [B] 235. [B] 271. [D] 307. [A] 353. [C] 389. [B] 425. [B]

128. [B] 164. [B] 200. [A] 236. [A] 272. [D] 308. [A] 354. [C] 390. [C] 426. [B]

129. [B] 165. [B] 201. [B] 237. [B] 273. [D] 309. [A] 355. [B] 391. [C] 427. [B]

130. [B] 166. [C] 202. [C] 238. [C] 274. [B] 310. [A] 356. [A] 392. [C] 428. [A]

131. [C] 167. [C] 203. [D] 239. [D] 275. [D] 311. [A] 357. [A] 393. [C] 429. [B]

132. [C] 168. [C] 204. [D] 240. [A] 276. [D] 312. [A] 358. [B] 394. [C] 430. [A]

133. [C] 169. [B] 205. [C] 241. [B] 277. [B] 313. [A] 359. [B] 395. [C] 431. [B]

134. [B] 170. [B] 206. [C] 242. [B] 278. [D] 314. [A] 360. [A] 396. [C] 432. [A]

135. [B] 171. [D] 207. [C] 243. [C] 279. [D] 315. [A] 361. [B] 397. [C] 433. [A]

136. [B] 172. [B] 208. [C] 244. [D] 280. [B] 316. [A] 362. [A] 398. [C] 434. [B]

137. [B] 173. [D] 209. [C] 245. [A] 281. [D] 317. [A] 363. [B] 399. [B] 435. [A]

138. [B] 174. [D] 210. [B] 246. [B] 282. [B] 318. [A] 364. [A] 400. [B] 436. [B]

139. [D] 175. [B] 211. [C] 247. [C] 283. [D] 319. [A] 365. [A] 401. [C] 437. [D]

140. [A] 176. [C] 212. [C] 248. [D] 284. [B] 320. [A] 366. [B] 402. [C] 438. [B]

141. [D] 177. [D] 213. [B] 249. [A] 285. [D] 321. [A] 367. [A] 403. [C] 439. [D]

142. [C] 178. [A] 214. [B] 250. [B] 286. [D] 322. [A] 368. [A] 404. [C] 440. [B]

143. [B] 179. [B] 215. [C] 251. [C] 287. [D] 323. [A] 369. [A] 405. [C] 441. [D]

144. [B] 180. [C] 216. [C] 252. [D] 288. [D] 334. [C] 370. [B] 406. [B] 442. [B]

145. [B] 181. [D] 217. [C] 253. [B] 289. [D] 335. [C] 371. [B] 407. [C] 443. [D]

146. [A] 182. [C] 218. [C] 254. [A] 290. [D] 336. [A] 372. [A] 408. [C] 444. [B]

147. [A] 183. [B] 219. [B] 255. [B] 291. [C] 337. [A] 373. [B] 409. [C] 445. [B]

148. [B] 184. [A] 220. [B] 256. [C] 292. [C] 338. [A] 374. [A] 410. [C] 446. [D]

149. [B] 185. [B] 221. [C] 257. [D] 293. [B] 339. [A] 375. [B] 411. [C] 447. [D]

150. [A] 186. [C] 222. [C] 258. [C] 294. [B] 340. [A] 376. [B] 412. [C] 448. [D]

151. [B] 187. [B] 223. [C] 259. [B] 295. [B] 341. [A] 377. [B] 413. [C] 449. [D]

152. [B] 188. [B] 224. [B] 260. [A] 296. [C] 342. [C] 378. [B] 414. [C]

153. [A] 189. [B] 225. [B] 261. [B] 297. [B] 343. [C] 379. [B] 415. [B]

154. [B] 190. [B] 226. [B] 262. [C] 298. [C] 344. [C] 380. [B] 416. [B]

155. [A] 191. [A] 227. [B] 263. [B] 299. [C] 345. [C] 381. [B] 417. [B]

156. [A] 192. [B] 228. [B] 264. [D] 300. [C] 346. [C] 382. [B] 418. [B]

157. [A] 193. [A] 229. [B] 265. [B] 301. [C] 347. [C] 383. [B] 419. [B]

158. [B] 194. [B] 230. [B] 266. [B] 302. [C] 348. [C] 384. [B] 420. [B]

159. [C] 195. [B] 231. [B] 267. [D] 303. [C] 349. [C] 385. [B] 421. [B]

160. [B] 196. [A] 232. [D] 268. [B] 304. [B] 350. [C] 386. [B] 422. [B]

161. [C] 197. [A] 233. [C] 269. [B] 305. [C] 351. [C] 387. [B] 423. [B]

162. [C] 198. [B] 234. [B] 270. [C] 306. [A] 352. [C] 388. [B] 424. [B]

Chapter 30— SEED PHYSIOLOGY OF CROPS

1. [D] 17. [C] 33. [A] 49. [A] 65. [A] 81. [A] 97. [D] 113. [C] 129. [D]
2. [C] 18. [B] 34. [A] 50. [C] 66. [B] 82. [C] 98. [C] 114. [C] 130. [C]
3. [D] 19. [A] 35. [D] 51. [A] 67. [D] 83. [D] 99. [A] 115. [A] 131. [I]
4. [D] 20. [B] 36. [D] 52. [A] 68. [D] 84. [D] 100. [E] 116. [A] 132. [A]
5. [A] 21. [A] 37. [D] 53. [C] 69. [A] 85. [E] 101. [A] 117. [C] 133. [A]
6. [D] 22. [A] 38. [A] 54. [A] 70. [D] 86. [A] 102. [C] 118. [A] 134. [A]
7. [D] 23. [D] 39. [B] 55. [B] 71. [C] 87. [E] 103. [A] 119. [D] 135. [A]
8. [A] 24. [A] 40. [B] 56. [C] 72. [D] 88. [D] 104. [D] 120. [A] 136. [A]
9. [D] 25. [B] 41. [C] 57. [A] 73. [C] 89. [D] 105. [A] 121. [D]
10. [A] 26. [A] 42. [D] 58. [A] 74. [A] 90. [A] 106. [B] 122. [A]
11. [B] 27. [A] 43. [D] 59. [A] 75. [E] 91. [A] 107. [A] 123. [A]
12. [C] 28. [B] 44. [D] 60. [A] 76. [E] 92. [A] 108. [D] 124. [B]
13. [D] 29. [A] 45. [D] 61. [B] 77. [D] 93. [A] 109. [B] 125. [A]
14. [D] 30. [A] 46. [D] 62. [A] 78. [A] 94. [A] 110. [A] 126. [B]
15. [A] 31. [A] 47. [D] 63. [A] 79. [A] 95. [A] 111. [B] 127. [B]
16. [A] 32. [A] 48. [E] 64. [B] 80. [F] 96. [A] 112. [A] 128. [A]

Chapter 31— CELL BIOLOGY

1. [A] 10. [A] 19. [B] 28. [B] 37. [C] 46. [B] 55. [A] 64. [C] 73. [C]
2. [C] 11. [D] 20. [C] 29. [A] 38. [B] 47. [B] 56. [C] 65. [C] 74. [D]
3. [D] 12. [A] 21. [C] 30. [A] 39. [C] 48. [B] 57. [B] 66. [A] 75. [C]
4. [D] 13. [B] 22. [B] 31. [B] 40. [D] 49. [A] 58. [D] 67. [A] 76. [C]
5. [A] 14. [D] 23. [A] 32. [B] 41. [A] 50. [A] 59. [C] 68. [B] 77. [C]
6. [C] 15. [C] 24. [A] 33. [C] 42. [B] 51. [D] 60. [A] 69. [A] 78. [B]
7. [D] 16. [A] 25. [B] 34. [A] 43. [A] 52. [C] 61. [A] 70. [A] 79. [B]
8. [B] 17. [B] 26. [C] 35. [A] 44. [D] 53. [A] 62. [C] 71. [D] 80. [C]
9. [C] 18. [D] 27. [B] 36. [D] 45. [C] 54. [B] 63. [B] 72. [B]

◈ ◈ ◈

Related Books

• Basic Concepts of Plant Biotechnology (With MCQs)	Vijay Prakash
• Basic Concepts of Plant Science	S.K. Bangarwa
• CSIR NET: Part A (Hindi)	Christy Varghese
• CSIR-NET General Aptitude: A New Outlook	Christy Varghese
• Fundamentals of Agriculture	R.L. Arya
• Glimpse on General Agriculture	P. Laxman Rao
• Indiras Objective Agricultural Biotechnology *2nd Ed*	R.L. Arya
• Indiras Objective Agricultural Extension	R.L. Arya
• Indiras Objective Agriculture	R.L. Arya
• Indiras Objective Agronomy *2nd Ed*	R.L. Arya
• Key to Success in Agriculture: Objective MCQ's for JRF, SRF, NET & Other Competitive Exams	B.N. Kalsariya
• कृषि जगत : JET एवं कृषि पर्यवेक्षक परीक्षा उपयोगी	M.L. Jat
• MCQs for Food Technology	P.M. Kotecha
• MCQs in Plant Breeding Biotechnology and Seed Science	K. Vanangamudi
• Objective Agribusiness Management *2nd Ed*	S.R. Panigrahy
• Objective Environmental Studies and Disaster Management	M. Prasanthrajan
• Objective Food Science & Technology *3rd Ed*	Deepak Mudgil
• Objective Forestry: For All Competitive Examination *2nd Ed*	K.T. Parthiban
• Objective Home Science at a Glance	Shanti Balda
• Objective Life Science MCQs for Life Science Examination *3rd Ed*	Kailash Choudhary
• Objective Plant Pathology *2nd Ed*	Susanta Banik
• Objective Seed Science and Technology	K. Vanangamudi
• Question Bank: Seed Science and Technology	K. Vanangamudi
• Questions and Answers in Environmental Science Practical	R.K. Kamble
• SET Life Science: Solved Exam Questions	Kailash Choudhary
• Vedic Maths and Mental Arithmetic: A New Outlook	Christy Varghese
• Vegetable Crops at a Glance	S.K. Tyagi